Handbook on

Ultrasonic and Dielectric Characterization Techniques for Suspended Particulates

Handbook on

Ultrasonic and Dielectric Characterization Techniques for Suspended Particulates

Edited by

Vincent A. Hackley
National Institute of Standards and Technology

John Texter
Strider Research Corporation

Published by
The American Ceramic Society ■ 735 Ceramic Place ■ Westerville, Ohio 43081

Proceedings of the International Workshop on Ultrasonic and Dielectric Characterization Techniques for Suspended Particulates, held at the National Institute of Standards and Technology in Gaithersburg, Maryland, August 4–6, 1997.

Library of Congress Cataloging-in-Publication Data

A CIP record for this title is available from the Library of Congress

Cover micrograph of silica particles was taken by James Kelly at the National Institute of Standards and Technology. Line drawing of overlapping scattered thermal and shear fields was provided by Malcolm Povey of the University of Leeds, United Kingdom.

For information on ordering titles published by the American Ceramic Society, or to request a publications catalog, please call 614-794-5890 or visit us at www.acers.org.

Printed in the United States of America.

1 2 3 4–01 00 99 98

ISBN 1-57498-034-3

Contents

Applications in Inorganic Particulate Suspensions

Applications in Emulsions and Organic Particulate Suspensions

Preface

Finely dispersed systems such as emulsions and colloidal suspensions are recognized as playing a critical role in many important industrial processes. Many of these systems are quite concentrated, and significant needs exist to develop measurement techniques for probing the physical and chemical properties of these systems and processes. Meeting such needs has set forth numerous technological challenges. These challenges have been underscored by the increasing demand for nondestructive evaluation and on-line monitoring capabilities in intelligent processing applications. Conventional methods, such as laser diffraction and single particle counting, generally fall short in practical applications because of their limitations to dilute systems.

Recent advances in the use of high-frequency electric and acoustic fields have led to the development of commercial instrumentation for characterizing nondilute dispersed systems. However, these emerging techniques are often unfamiliar to researchers and process engineers. Furthermore, there are no comprehensive texts currently available which specifically address the theory and application of these measurements. The objective of this volume is to present a series of reviews and expositions that give a thorough and well-rounded exposure to these techniques and the attendant measurement fundamentals. These contributions have also been selected so as to illustrate the application of these techniques in the characterization of a wide variety of material and process applications.

The papers herein were contributed by participants at an International Workshop on Ultrasonic and Dielectric Characterization Techniques for Suspended Particulates, held at the National Institute of Standards and Technology in Gaithersburg, Maryland, on August 4–6, 1997. This workshop began with the simple notion of bringing together at one time participants from academia, government, and industry, including instrument manufacturers, theoreticians, applications development specialists, and current and potential users of this technology, to stimulate communication and interaction in this rapidly emerging, but highly fractured, area. This volume is organized by topic with collections of papers emphasizing each of the following areas: fundamental aspects, measurements in concentrated systems, sensor development and process control, inorganic systems applications, and organic systems applications. Measurement techniques include electrokinetic sonic amplitude, colloid vibration potential, ultrasonic attenuation, ultrasonic velocity, and dielectric spectroscopy.

This workshop and these proceedings required the involvement and efforts of a number of individuals and organizations. We would like to express our gratitude to the National Institute of Standards and Technology, The American Ceramic

Society, and the Division of Colloid and Surface Chemistry of the American Chemical Society for co-sponsoring the workshop. We also gratefully acknowledge the donors of the Petroleum Research Fund, Eastman Kodak Company, Dispersion Technology Inc., and Malvern Instruments Inc., for financial support of the workshop. This support made it possible for the participation at the workshop to be truly global. Additional thanks are extended to Kathy Kilmer, Tammie Grice, and the Public and Business Affairs Division at NIST, for their tireless efforts behind the scenes, which allowed this conference to run smoothly. Most importantly, the success of this endeavor rests largely upon the shoulders of the workshop participants, particularly the speakers, who brought with them the ideas, insight, experience, and critical perspective necessary for such an undertaking. We especially thank the authors for the high quality of their contributed manuscripts and for their active participation in the peer review process, which made our job much easier. Finally we express our appreciation to Sarah Godby, Mary Cassells, and other individuals in the publications department at The American Ceramic Society for their help, advice, and patience in producing this volume.

Vincent A. Hackley
John Texter

Contributors

Felix Alba
Felix ALBA Consultants Inc., Murray, Utah, USA

Koji Asami
Institute for Chemical Research, Kyoto University, Kyoto, Japan

Lennart Bergström
Institute for Surface Chemistry, Stockholm, Sweden

Timothy C. Bliss
Foxboro Company, Foxboro, Massachusetts, USA

Y.L. Chen
Bell Labs, Lucent Technologies, Murray Hill, New Jersey, USA

Andrei S. Dukhin
Dispersion Technology Inc., Mount Kisco, New York, USA

Leonard V. Dullea
OSRAM SYLVANIA Development Inc., Danvers, Massachusetts, USA

Arunava Dutta
OSRAM SYLVANIA Development Inc., Danvers, Massachusetts, USA

Yuri Feldman
Department of Applied Physics, The Hebrew University of Jerusalem, Jerusalem, Israel

Philip J. Goetz
Dispersion Technology Inc., Mount Kisco, New York, USA

Richard Greenwood
Birchall Centre, Department of Chemistry, Keele University, Staffordshire, United Kingdom

Vincent A. Hackley
Powder Characterization and Processing Group, National Institute of Standards and Technology, Gaithersburg, Maryland, USA

Wei Han
Electronic Materials and Devices Group, Los Alamos National Laboratory, Los Alamos, New Mexico, USA

David Higgs
Malvern Instruments Ltd., Malvern, Worcestershire, United Kingdom

Frank Hinze
Institut für Verfahrenstechnik und Umwelttechnik, Technische Universität Dresden, Dresden, Germany

Robert J. Hunter
Colloidal Dynamics Pty Ltd., National Innovation Centre, Eveleigh, New South Wales, Australia

Robert Jack
Malvern Instruments Ltd., Malvern, Worcestershire, United Kingdom

Paul Kippax
Malvern Instruments Ltd., Malvern, Worcestershire, United Kingdom

Udo Kräuter
Bayer AG, Leverkusen, Germany

Eric Laarz
Institute for Surface Chemistry, Stockholm, Sweden

David Julian McClements
Department of Food Science, University of Massachusetts, Amherst, Massachusetts, USA

Tõnis Oja
Malvern Instruments Inc., Southborough, Massachusetts, USA

Ungyu Paik
Department of Ceramic Engineering, Chang Won National University, Chang Won, South Korea

Sanjay S. Patel
Bell Labs, Lucent Technologies, Murray Hill, New Jersey, USA

Hemant P. Pendse
Industrial Process Control Sensor Systems (IPCoSS) Program, Chemical Engineering Department, University of Maine, Orono, Maine, USA

Bjørnar Pettersen
Department of Chemistry, University of Bergen, Bergen, Norway

Malcolm Povey
Procter Department of Food Science, University of Leeds, Leeds, United Kingdom

Ulrich Riebel
Lehrstuhl Mechanische Verfahrenstechnik, Brandenburgische Technische Universität Cottbus, Cottbus, Germany

Siegfried Ripperger
Institut für Verfahrenstechnik und Umwelttechnik, Technische Universität Dresden, Dresden, Germany

David M. Scott
Particle Science & Technology (PARSAT), Experimental Station, DuPont Central Research & Development, Wilmington, Delaware, USA

Arvind Sharma
Earth and Environmental Sciences Division, Los Alamos National Laboratory, Los Alamos, New Mexico, USA

Johan Sjöblom
Department of Chemistry, University of Bergen, Bergen, Norway

Tore Skodvin
Department of Chemistry, University of Bergen, Bergen, Norway

Michael Stintz
Institut für Verfahrenstechnik und Umwelttechnik, Technische Universität Dresden, Dresden, Germany

Shin-ichi Takeda
Department of Applied Chemistry, Okayama University, Okayama, Japan

Jorge L. Valdes
Bell Labs, Lucent Technologies, Murray Hill, New Jersey, USA

Fundamental Aspects of Ultrasonic and Dielectric Characterization Techniques

The Application of Acoustics to the Characterisation of Particulate Suspensions

Malcolm J. W. Povey

Procter Department of Food Science, The University of Leeds, Leeds, LS2 9JT, UK

1. Introduction

In order to understand propagation of ultrasound through particulate suspensions it is necessary to throw away common sense and instead rely on a truly scientific approach[1]. The reasons for this are to do with the complex interactions between the pressure wave that comprises ultrasound and the particles, the interactions between the particles themselves and the interactions between particles and the surrounding continuum. In systems of interest to colloid scientists, the situation is usually complicated further by the heterogeneous nature of the particles and the surrounding material.

Colloidal particles will be in constant motion due to Brownian diffusion. As a result particles may collide and even enter a permanent or temporary association. In Figure 1, this situation is illustrated graphically, for the example of oil particles suspended in water, in the center of the picture a floc of particles is beginning to form. A floc may be characterized as a group of particles held together with forces that are of the order of kT in magnitude. Also present in Figure 1 is a solid oil particle initiating crystallization in a liquid drop. Crystallization often occurs in colloidal systems and when it does it also has an impact on ultrasonic propagation[2].

Ultrasound characterisation of particulate suspensions must also cope with issues that apply to any particle characterisation technique. For example, particles are very often non-spherical. Particle size distributions can be anything but log-normal and hence impossible to measure with model dependent[3] methods, which can only deal with log-normal and bi-disperse log-normal distributions.

A further difficulty that any characterization technique must overcome is the issue of particle sizes 'out-of-bounds'. Light microscopy is a good example of how this dilemma can give very misleading results. Since it is difficult to see objects less than 1 μm in size, the sight of a few emulsion droplets can give the impression of a low particle density, whereas the entire volume of the dispersed phase could easily be invisible. A further complication arises when particles network, forming a particle gel or other type of space filling network. In Figure 2, a particle network formed from the fusing of needle crystals that have grown from oil particles is shown.

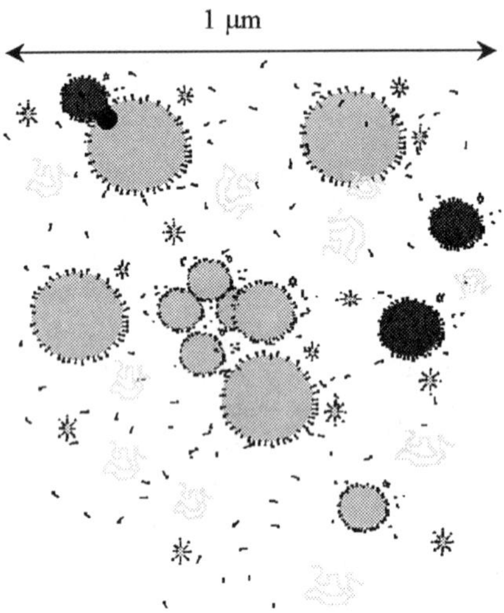

Figure 1. A fluid containing colloidal particles. The gray spheres represent liquid oil particles and the black spheres solid oil particles. The gray strings represent large protein molecules and polymers; the black 'tadpoles' are amphiphilic surfactant molecules.

It might be suggested that the situation in Figure 1 and Figure 2 is a hopeless one from the point of view of obtaining any information at all from ultrasound propagation but this is not at all the case. It is the purpose of this paper to demonstrate that, on the contrary, a great deal may be learnt from studying complex colloidal systems with ultrasound.

2. Ultrasound Measurement Techniques

2.1 Ultrasound Scanning

There are a variety of ultrasound measurement techniques available for particle characterization, a number of which are now in commercial production.

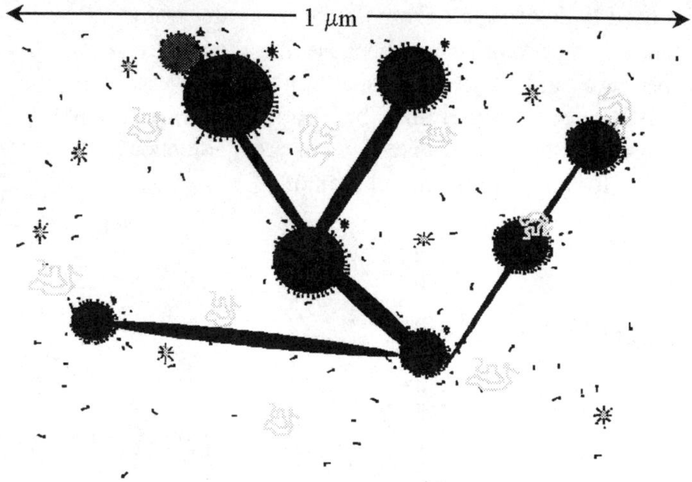

Figure 2. A colloidal particle network. In addition to the conventions adopted in Figure 1, the black lines represent needle crystals crossing the water phase and penetrating nearby oil drops, nucleating crystallization.

In Figure 3 is shown the ultrasound profiling apparatus that we have used extensively to characterise particulate suspensions. A pitch and catch technique is used in which the velocity of sound through the emulsion may be determined. The transducers are scanned from the bottom to the top of the glass cell containing the emulsion, building up a picture of the time and spatial evolution of the dispersed phase in the emulsion under examination. The volume fraction of the suspended phase may then be determined from a relationship that we have called the modified Urick equation. A detailed description of the apparatus and the results of the technique are given in Povey[1].

The modified Urick equation can be derived from scattering theory[1] and takes the following form:

$$\frac{1}{v^2} = \frac{1}{v_1^2}\left(1 + \alpha\phi + \beta\phi^2 + O(\phi^3)\right) \qquad (1)$$

where v is the ultrasound velocity in the dispersion; v_1 is the ultrasound velocity in the continuous phase and φ is the volume fraction of the dispersed phase. It is therefore possible to determine the coefficients α and β with the scanner, because the volume fraction ϕ of the emulsion is initially known. Subsequently, provided the particle size distribution does not change drastically, the known values of α

and β can be used to determine ϕ from the known values of v and v_1.

This is an important result, because the quantities α and β are directly related to the zero and first order scattering coefficients, A_0 and A_1. These scattering coefficients may therefore be determined experimentally[4]. This is a great advantage because otherwise 14 different physical constants must in principle be used for an ab-initio calculation of ϕ.

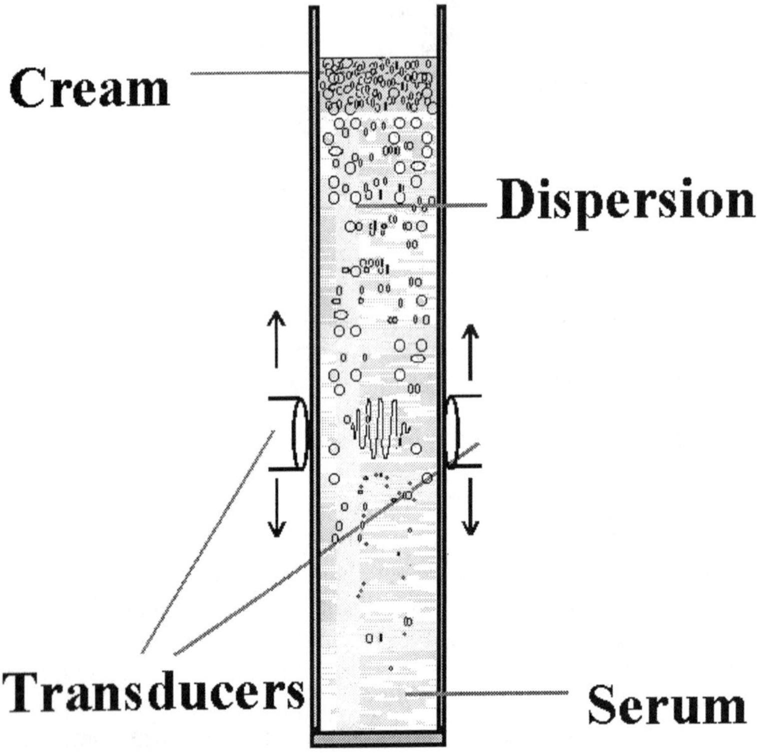

Figure 3. Schematic illustration of Acoustiscan ultrasonic scanner.

A further advantage of an experimental determination of A_0 and A_1 is that particle shape effects are automatically accounted for. Once A_0 and A_1 have been determined a measurement of the frequency dependent velocity and attenuation may be used to obtain the particle size distribution in the emulsion. The particle size distribution may also be determined, in the unhindered creaming case, by comparing the time dependent volume fraction of the dispersed phase with a computer model of particle diffusion and creaming[5].

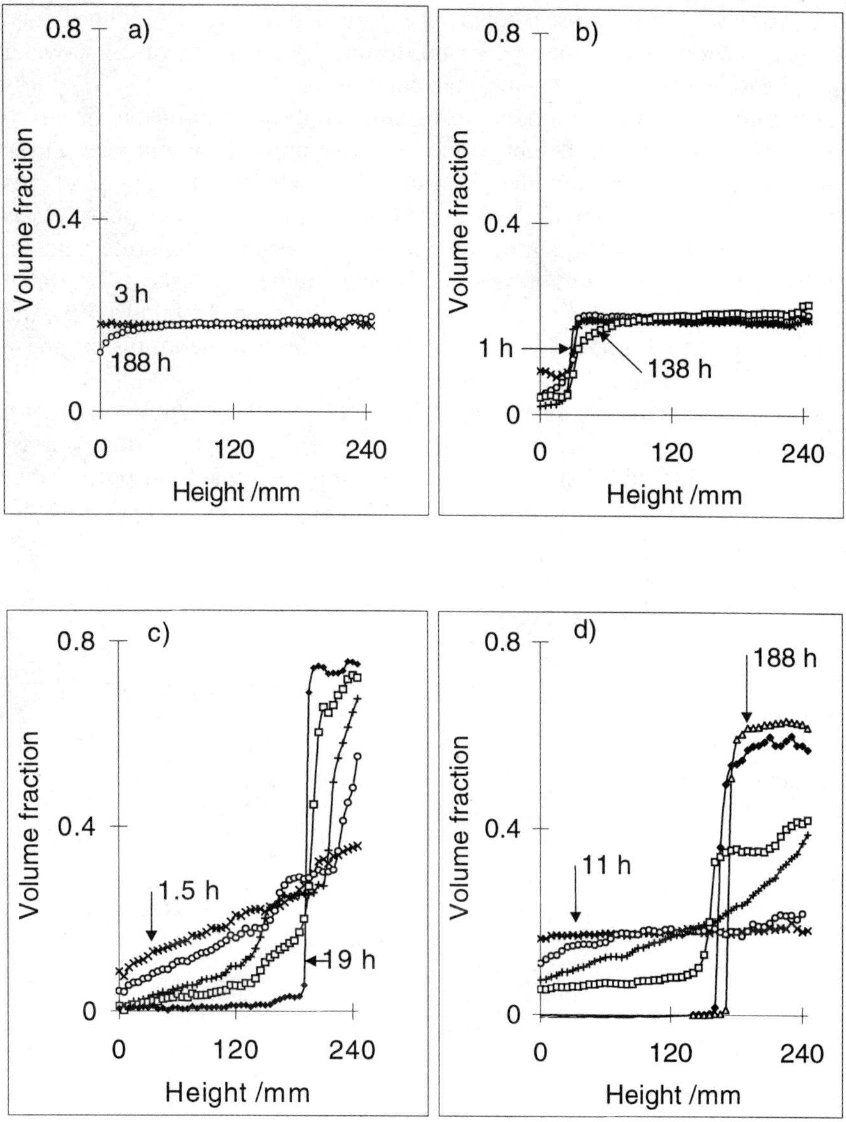

Figure 4. Differing creaming profiles in sunflower oil-in-water emulsions: (a) 18% v/v sunflower oil-in-water with 2 wt % Tween 20 at 30 °C; (b) as (a) but with 0.017 wt % xanthan added; (c) as (a) but with 0.089 wt % xanthan added; (d) as (a) but with 0.173 wt % xanthan added.[6]

Creaming profiles such as those shown in Figure 4 illustrate the richness and diversity of behaviour which may be found during the creaming of oil-in-water emulsions due to relatively small changes in composition.

In Figure 4, the vertical axis represents the concentration of the dispersed phase, the horizontal axis the height in the vertical sample tube and each curve represents a different time at which a scan was taken. So the set of curves illustrates the time evolution of the distribution of the dispersed phase. It is interesting to note how small quantities of the microbial polysaccharide xanthan dramatically alter the stability of the emulsion[6]. Figure 4(a) shows a stable emulsion, although a small degree of instability can be seen. In Figure 4(b), the addition of a very small amount of xanthan has destabilised the emulsion and a serum has developed, although no cream can be observed.

Depletion flocculation destabilises the emulsion. In Figure 4(c), rapid, unhindered cream takes place as the depletion effect reaches its full effectiveness. Figure 4(d) shows that additional xanthan is slowing the creaming, presumably due to the increased viscosity of the continuous phase and compression of the cream may also be observed.

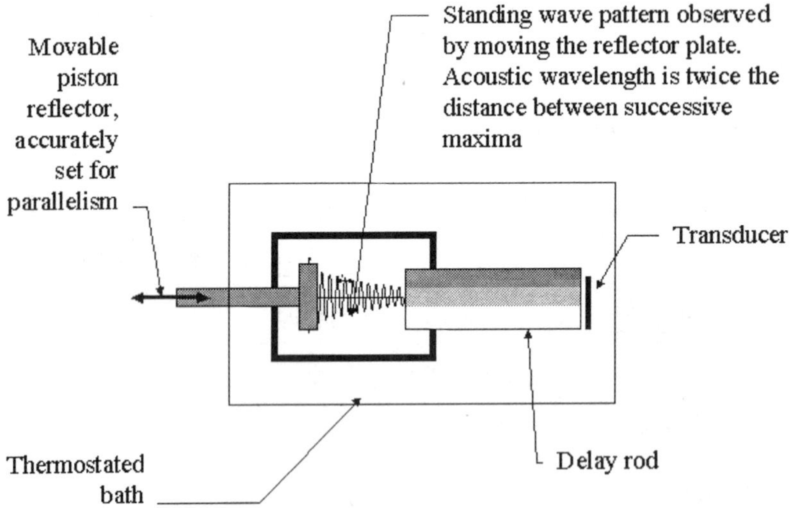

Figure 5. Schematic diagram of piston interferometer. (Adapted from Povey[1])

2.2 Interferometers

There are a very wide variety of interferometric techniques that are described in a recent paper by Eggers and Kaatze[7]. One instance of the technique is illustrated in Figure 5. Interferometric techniques require a reference wave and a detected signal. A standing wave pattern can be formed by interference between the outgoing and reflected portions of a continuously emitted ultrasound signal (Fig. 5) or in the case of pulsed signals a beat pattern appears when different parts of the wave are combined together. The most accurate measurements of both velocity and attenuation may be made using these methods, although they are generally expensive to implement and do not readily lend themselves to in-line application.

2.3 Ultrasound Spectroscopy

Ultrasound spectroscopy involves the measurement of the properties of a material over a range of frequencies and need not necessarily involve interferometry. Pulse techniques offer the prospect of lower cost implementations of ultrasound spectroscopy and the frequency scanning pulse echo reflectometer (Fig. 6) designed by McClements and Fairley[8] at Leeds is a good example. Here, the amplitude and phase of a signal are obtained by FFT techniques.

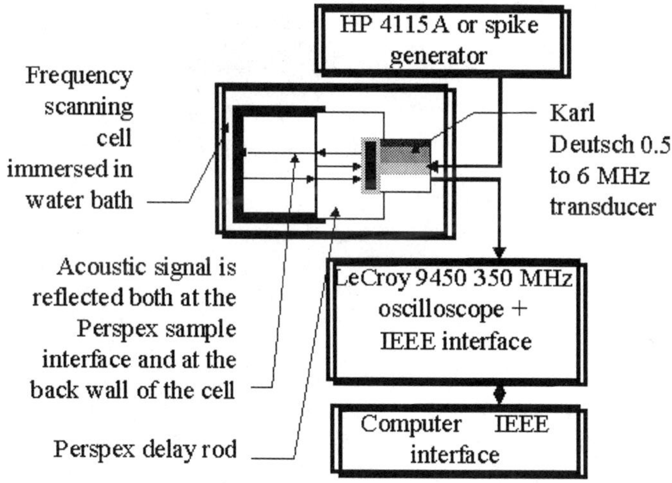

Figure 6. Frequency Scanning Pulse Echo Reflectometer. (Adapted from Povey[1])

One signal is reflected at the interface between a buffer rod and the liquid sample and the other comes from a backwall reflection, having traveled through the sample. Comparison of the frequency dependent amplitude and phase with scattering theory models of ultrasound propagation[9] permits the determination of particle size in complicated systems such as mayonnaise[10]. This technique has the advantage that the ultrasound cell is relatively small, robust and cheap to manufacture and has the potential to be used in-line. Its restricted frequency range (0.5 to 6 MHz) is a disadvantage when attempting to size particle suspensions with a large particle size range.

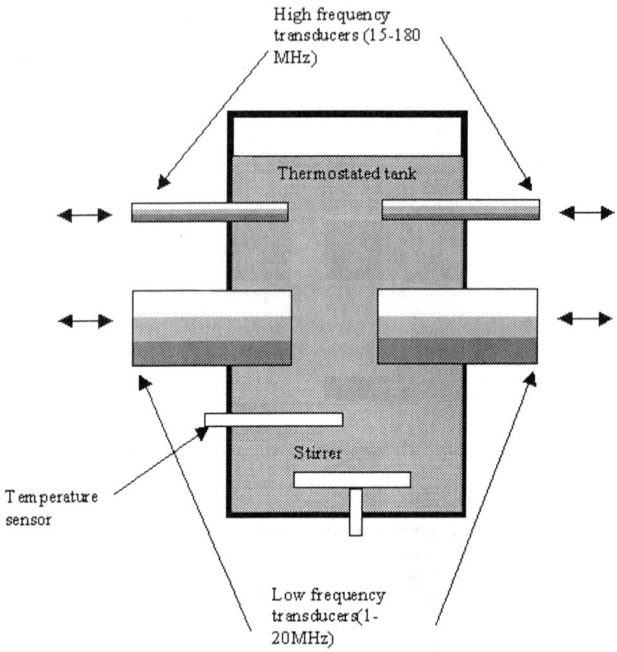

Figure 7. Schematic diagram of Malvern Ultrasizer.

A commercial interferometer, which can be used, for particle sizing particulate suspensions in the size range 10nm to 1mm is illustrated in Figure 7. Operating at frequencies between 1 MHz and 200 MHz, the Malvern Ultrasizer requires around 3 litre of sample liquid. Good temperature control, measurement or compensation is vital to the success of ultrasound particle sizing and stirring of the tank and the use of a heater achieves this. Very great care in transducer alignment and calibration is necessary to achieve high accuracy in attenuation measurement and this is obtained with an automated optical bench system. The proprietary software[11] employs the Epstein-Carhart/Allegra-Hawley scattering theory model[12 13] of acoustic propagation in particulate suspensions. Extremely good results may be obtained with this system. For example, Povey, Golding, Higgs and Wang[14] have sized casein submicelles and micelles and obtain extremely good agreement with photon correlation spectroscopy results in these systems (Fig. 8).

We have also studied the aggregation behaviour of these micelles[14] (Fig. 9). Figure 9 illustrates the sensitivity of casein aggregation to temperature. The casein submicelles had been stored at 4 °C prior to measurement at room temperature whereas the aggregated submicelles were dispersed at 30 °C and stored at room temperature prior to measurement. In Figure 10 the effects of varying concentrations of calcium ions on micelle aggregation may be seen.

3. Theoretical Analysis of Ultrasound Propagation

3.1 Scattering and Losses

When making measurements over such wide frequency ranges a number of issues must be borne in mind. Examination of Figure 2 and of Figure 3 indicates that a variety of particles that may scatter and absorb ultrasound are present in most particulate suspensions. The scattering contribution is obtained by measuring the frequency dependent attenuation in the continuous and dispersed phases and combining them to obtain a figure for the contribution to the total attenuation in the absence of scattering, viz:

$$\alpha_m = \alpha_1(1-\phi)+\alpha_2\phi \qquad (2)$$

Here α is the attenuation and ϕ the volume fraction occupied by particles, subscripts $_1$ and $_2$ refer to the continuous and dispersed phase respectively and $_m$ refers to the mean attenuation. The scattering contribution α_s is then obtained from the total attenuation α_{tot} as follows:

$$\alpha_s = \alpha_{tot} - \alpha_m \qquad (3)$$

One problem with this is that the mean attenuation given by Equation (2) is not correct once the particles are mixed with the continuous phase. Additional boundary effects, not accounted for by the scattering formulation appear. A good example can be seen in Figure 8. Close examination of the figure indicates that the scattering data for the submicelles departs from the model prediction for 21nm particles over a region between approximately 80 and 120MHz. This leads to a poor fit to the model but this may be explained as relaxation between the submicelle and the surrounding solvent. Once the relaxation has been subtracted it might be presumed that the model fit would improve considerably.

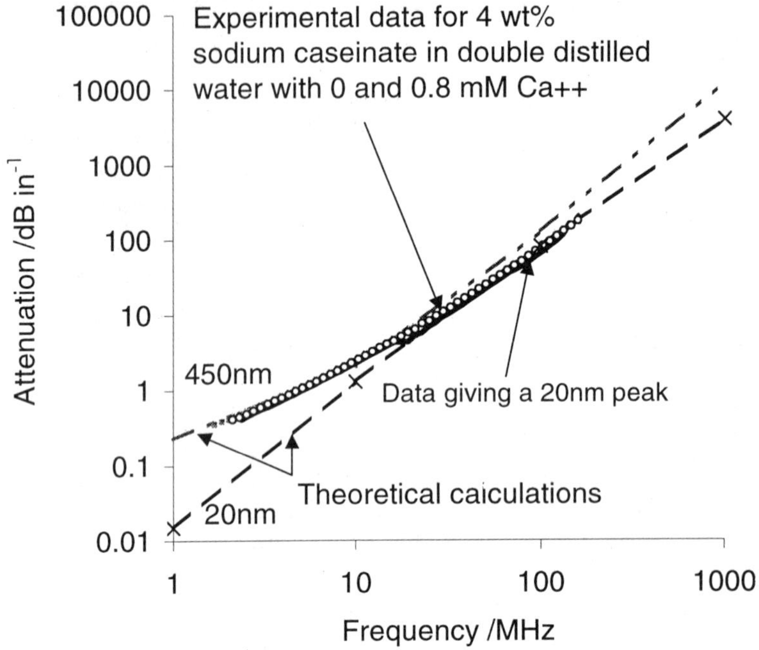

Figure 8. Ultrasound spectra for casein solutions.

Principally Frindi and co-workers have used ultrasound spectroscopy for many years to study surfactant exchange[15]. For small surfactant molecules Frindi et al[16] have found relaxation frequencies of between 0.5 MHz and 3 MHz which is much lower than that observed in Figure 8. It may be that here we are seeing exchange of water between the solvation layer of the protein and the surrounding water. This is because we would expect the relaxation frequency of the relatively large casein molecules to be much lower than that observed for the small molecule surfactants such as C_6E_3 studied by Frindi et al[15].

Equation 2 also suffers from the fact that multiple scattering theories such as that of Waterman and Truell[17] and Lloyd and Berry[18] compensate for the presence of other particles. This is accounted for by using another form of Equation 2 in which we put:

$$\alpha_s = \alpha_{tot} - \alpha_1 \tag{4}$$

However, this equation does not satisfactorily account for the 'non-scattering' contributions of the two phases either. For these and other reasons referred to later, the 'loss' and 'scattering' contributions to the attenuation should not be treated separately. The primary contribution to the mean attenuation are relaxation processes such as viscosity, surfactant and solvent exchange, and phase transitions like crystallization These are frequency dependent and may be contributed by either or both the dispersed and continuous phase. The dispersed phase may diffract the acoustic field if it is ordered in some way and most authors have so far ignored this potential contribution from the acoustic structure factor. Interactions between the entities comprising the dispersed phase may also make very significant contributions to the mean attenuation and can introduce the possibility of new modes of propagation, such as the second sound mode described by Biot[19 20].

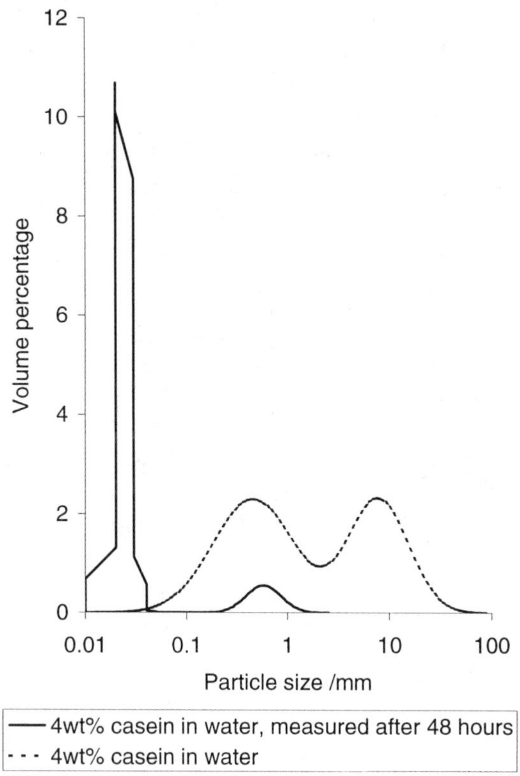

Figure 9. Aggregation of casein.

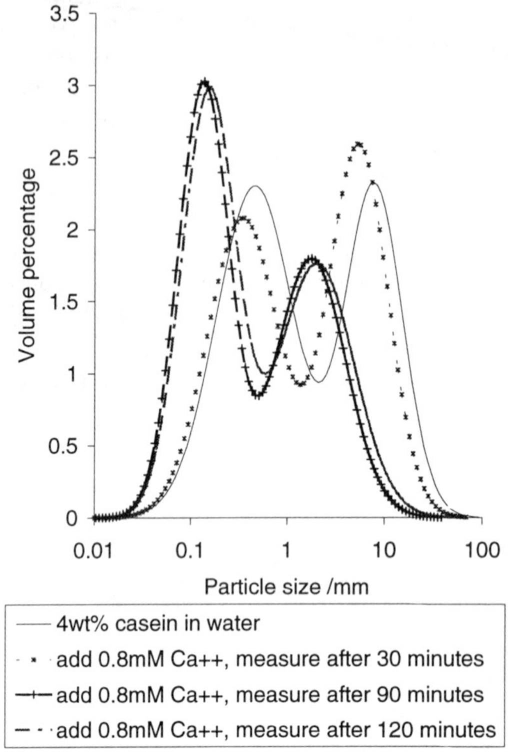

Figure 10. Aggregation of casein in the presence of calcium ions.

Some authors have shown that the multiple scattering theory model described later in this paper generally overestimates the attenuation in the sample, and that this overestimate increases at higher concentrations (McClements[21], Evans and Attenborough[22]). The reasons for this are a matter for argument. The proponents of coupled phase theories (Harker and Temple[23,24], Evans and Attenborough[22]) suggest that it is due to the lack of self-consistency in the scattering treatment, which has been alluded to above. However, most comparisons between scattering theory and experimental results have been carried out using plane wave expansions that are truncated at the first or second order and the inclusion of higher order terms may improve the accuracy.

Another possible reason for the overestimate is the contribution of the overlapping thermal waves as particles approach each other (Fig. 11 and Fig. 12). This overlap occurs, even at low concentrations since particle diffusion means that a proportion of the particles is always in close proximity to each other.

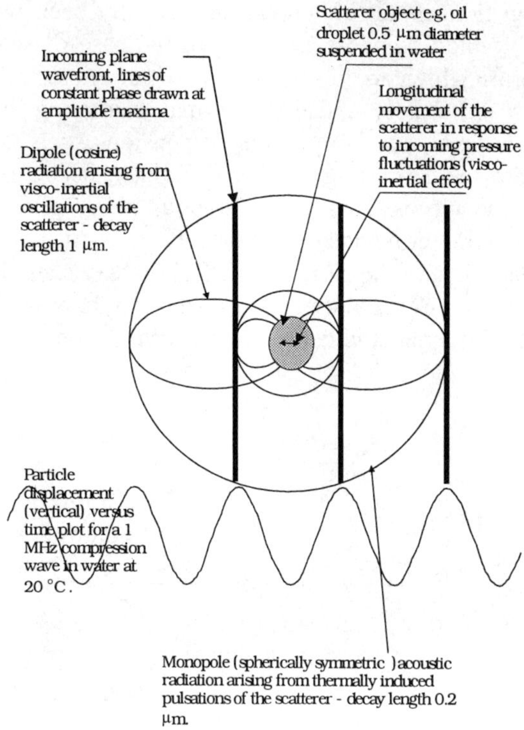

Figure 11. Single particle scattering a plane acoustic field.

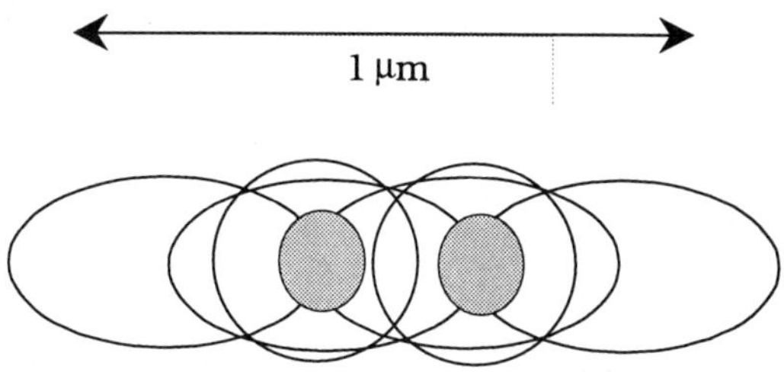

Figure 12. Overlapping scattered thermal and shear fields.

A mean field theory treatment has recently been published (Herrmann[25], Hemar et al[26]) which demonstrates that, in the case of scattering dominated by thermal scattering which applies to many cases of interest, the particle size can be predicted very accurately indeed. Thermal scattering can be the dominant scattering mechanism, even in particle suspensions such as those often found in ceramics, where there is a big density difference between the two phases. The reason is that the viscosity of the continuous phase plays an important role in scattering, hence the term 'visco-inertial scattering'. At low viscosity, the reaction between the particle and the surrounding fluid is weak and hence the scattering is small and at high viscosity the relative motion between the particle and the surrounding fluid is small and hence the scattering is once more weak.

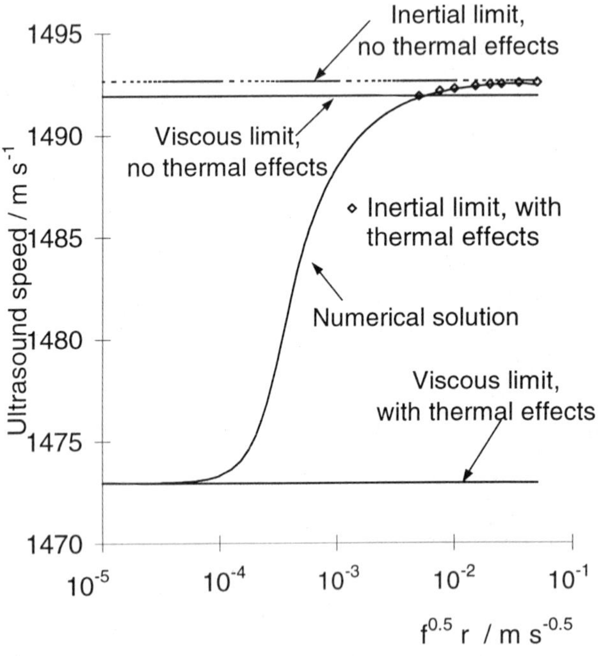

Figure 13. Ultrasound velocity plotted against the scaling parameter $f^{0.5}r$. This figure illustrates the relative contributions of thermal and viscous effects for a 18 % v/v sunflower oil-in-water emulsion at 30 °C.

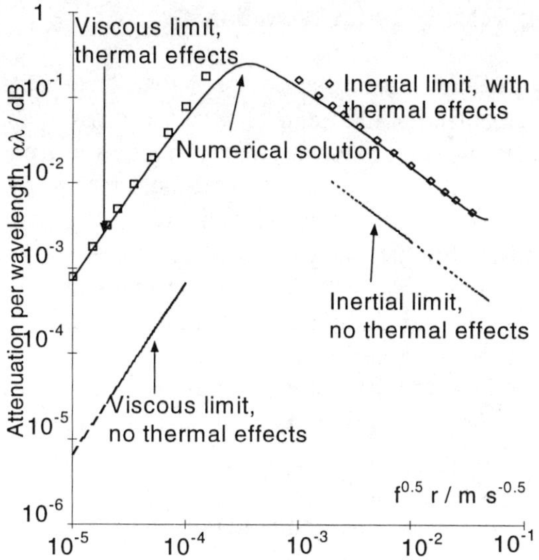

Figure 14. Ultrasound attenuation plotted against the scaling parameter $f^{0.5}r$. This figure illustrates the relative effects of viscous and thermal effects on the attenuation. Conditions are as for Figure 13.

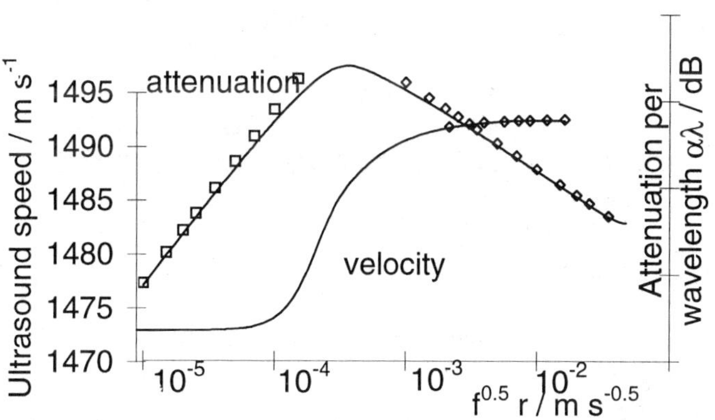

Figure 15. $f^{0.5}r$ dependence of velocity and attenuation compared. Conditions are as for Figure 13.

4. The Relation Between Velocity and Attenuation

One aspect of ultrasound propagation data that has yet to be fully exploited is the fact that both amplitude and phase of the wave are measured. Commercial interferometers measure attenuation as a function of frequency. If an infinite bandwidth is available it is only necessary to measure the amplitude and the phase can be calculated via the Kramers-Kronig relations, and vice versa. However, an infinite bandwidth is never available. In this case measurements of both amplitude and phase can greatly improve the accuracy of the fit between the ultrasonic data and the particle size distribution. If the frequency dependence of the velocity and the attenuation are compared (Fig.s 13-15) it is possible to see that the shape of the curves is quite different. For example, it is easier to distinguish monodisperse and bi-disperse distributions if both amplitude and phase data is available.

4.1 Frequency Independent Velocity Measurement

A great deal of information can be obtained from ultrasonic velocity measurements at a single frequency. The compressibility of particles can be obtained from measurement of the concentration dependence of the velocity of sound and the modified Urick equation (Equation 1) has already been alluded to. The interesting fact here is that, whilst the attenuation in solutions is unaffected by thermal scattering (the limit of zero particle radius in Figure 13), the velocity may be considerably affected. Thus thermal scattering must be accounted for when determining the compressibility of particles (Pinfield and Povey[30]). However, once this is realised, then the modified Urick equation may be used to obtain extremely accurate fits between particle concentration and velocity and hence accurate measurements of compressibility may be made without the need to resort to high accuracy and specialist interferometric equipment for velocity measurement. The additional advantage that the scattering coefficients may be obtained in this way was pointed out earlier.

4.2 Comparisons with Light Scattering

The propagation of ultrasound through systems of particles is best understood as scattering phenomena. Rayleigh first developed the mathematical apparatus of scattering theory[27] in the last century and his discovery of the reason for the blueness of the sky came about through these acoustical investigations. The mathematical formalism of acoustics and optics bears very many similarities, so much so that Jones[28] has written a book on Electromagnetic and Acoustic

Propagation. The reason for this similarity lies in the wave-like part of the nature of both phenomena. Acoustical and optical propagation are compared in Table I, since many people have some familiarity with light scattering and light scattering metaphors may illuminate acoustic phenomena.

Table I. Comparison of Electromagnetic and Acoustic Propagation. The symbols used for electromagnetic propagation are used for this purpose only in this table and have a different meaning elsewhere in this work..

Ultrasound	Light
Transducers are phase sensitive	Transducers are phase insensitive
Wavelength between cm and μm	Wavelength between 0.5 and 1 μm
Frequency between 0.1 and 10^{13} Hz	Frequency between 3×10^{16} and 6×10^{16} Hz
Coherence between pulses	No coherence between pulses
Responds to elastic, thermophysical, and density properties	Responds to dielectric and permeability properties
Particle motion parallel to the direction of propagation; no polarization	Field displacement perpendicular to direction of propagation; polarization is therefore possible
Propagates through optically opaque materials	Sample dilution is normally required
Mathematical Formalism	
The velocity of propagation, v, is $v^2 = \frac{1}{k\rho}$,	The velocity of propagation, c, is $c^2 = \frac{1}{\varepsilon\mu}$,
where k is complex compressibility and ρ is density.	where ε is permittivity and μ is permeability.
Assume $\rho = constant;$	Assume $\mu = constant;$
then $k = \kappa' + i\kappa''$.	then $\varepsilon = \varepsilon' + i\varepsilon''$.
Define the wave vector, $\mathbf{k} = k' + i\alpha$, $v^2 = \frac{\omega^2}{\mathbf{k}^2}$.	Define the wave vector, $\mathbf{k} = k' + i\alpha$, $c^2 = \frac{\omega^2}{\mathbf{k}^2}$.
Comparing real and imaginary parts gives $\kappa' = \frac{1}{\rho\omega^2}\left[k'^2 - \alpha^2\right]$	Comparing real and imaginary parts gives $\varepsilon' = \frac{1}{\mu\omega^2}\left[k'^2 - \alpha^2\right]$
And $k'' = \frac{2k'\alpha}{\rho\omega^2}$	And $\varepsilon'' = \frac{2k'\alpha}{\mu\omega^2}$

The analogy between light scattering and acoustical scattering should not be taken too far. Since ultrasound transducers are phase sensitive, it is the amplitude of the signal which is detected, not its intensity which is the case in light detectors. So the attenuation of a light signal should not be confused with that of an acoustic signal. The insensitivity of light detectors to phase means that light is detected even when multiple scattering mixes up the phase. In the case of ultrasound, multiple scattered signals that have undergone significant phase shifts may be undetectable, even though the energy is still present as an ultrasound signal.

Despite the preceding discussion, developments in light scattering theory are likely to be of great benefit to the development of acoustical techniques. For example, Rayleigh-Gans[29] theory and Mie theory[29] can help the treatment of ultrasonic data in the region where particle size is of the order of the acoustic wavelength.

5. Scattering Theory in Brief

The reader is directed to Povey[1] for a comprehensive discussion of scattering theory in the context of the ultrasound measurement of particulate suspensions. Scattering may be defined as the redirection of one mode or direction of propagation into another. Defined in this way, scattering includes both dissipative and non-dissipative processes. In the absence of dissipation, scattering is called *elastic*. If dissipation is significant, scattering is called *inelastic*. A mode such as the acoustic compressional wave is called *propagational* because it conveys energy over distances of very many wavelengths. On the other hand, modes such as the thermal and shear modes convey energy over distances of a fraction of a wavelength and are called non-propagational.

Non-propagational modes need not dissipate significant amounts of energy despite having a profound effect on the propagational mode and for this reason it is misleading to separate the contributions to acoustic propagation into a scattering contribution and various loss contributions. The loss mechanisms also act to couple the various propagational and non-propagational modes together. So the viscous drag on a particle couples the particle motion to the sound field in the fluid, creating the resultant dipolar field pattering due to visco-inertial scattering (Fig. 11). It was pointed out earlier that if the viscosity is small or zero there is no reaction between the particle and fluid motion and consequently no scattering, regardless of the inertial term. The same is true in the case of very high viscosity.

The thermal wave associated with the acoustic pressure wave is coupled to particle pulsation by the differing volume expansivity and heat capacity of the dispersed and continuous phases, resulting in a predominantly monopolar scattered field pattern (Fig. 11). Thermal dissipation itself is the result of the scattering of low frequency acoustic waves into the high frequency phonons, which constitute heat. Thermal scattering in water-continuous systems may be the major scattering mechanism, even in some high-density contrast systems in which visco-inertial scattering may otherwise be expected to dominate (Fig. 13 and Figure 14). This is a result of the anomalous volume expansivity of water that is negative below 4 °C and positive above 4 °C, attaining very high values.

The pressure wave that constitutes ultrasound causes a temperature difference to come into existence at the boundary between the particle and its surroundings. This temperature difference is related to the volume expansivity and the specific heat differences between the particle and its surroundings. The boundary layer is heated, causing it to expand and contract and hence the particle becomes a source of sound. The result is that the temperature ratio R between the particle and its surroundings in water can be larger than 7[30] and is highly temperature dependent (Equation 5).

$$R = \left[\frac{\frac{\beta_2}{\rho_2 C_{p2}} - \frac{\beta_1}{\rho_1 C_{p1}}}{\frac{\beta_1}{\rho_1 C_{p1}}} \right] \tag{5}$$

Where β is the volume expansivity, ρ is density and C_p is the specific heat. Quantities subscripted $_1$ apply to the continuous phase and quantities subscripted $_2$ apply to the particles. The complex propagation constant (wave-vector) for the propagational compressional acoustic mode, K is given by:

$$K = k_1 \left[1 + \frac{\phi}{2}(a_0 + 3a_1) - \frac{\phi^2}{8}\left(a_0^2 - 6a_0a_1 - 15a_1^2\right) \right] \tag{6}$$

where K is the wave vector for acoustic propagation in the suspension (Povey[1]), k_1 is the wave-vector in the pure continuous phase, a_0 and a_1 are the zero and second order scattering coefficients. This equation includes both the propagational part of the wave-vector k' and the dissipative part, k''. So the velocity and attenuation of the wave can be obtained from Equation 6, using the equations for acoustic propagation given in Table I. In this way, the contribution to the attenuation from scattering mechanisms can be shown to be:

$$\alpha_s = \frac{\omega}{v_1} \left[\frac{\phi}{2} \mathrm{Im}(a_0 + 3a_1) - \frac{\phi^2}{8} \mathrm{Im}\left(a_0^2 - 6a_0a_1 - 15a_1^2\right) \right]$$

$$+ \alpha_1 \left[1 + \frac{\phi}{2} \mathrm{Re}(a_0 + 3a_1) - \frac{\phi^2}{8} \mathrm{Re}\left(a_0^2 - 6a_0a_1 - 15a_1^2\right) \right] \tag{7}$$

Where ω is the radial frequency and v_l is the velocity of sound in the continuous phase. This equation incorporates both the real (propagational) and imaginary (loss) components into a single equation derived from scattering theory. This approach still requires that the various possible loss mechanisms be enumerated and correctly incorporated into the theory. Finally, it should be emphasised that these equations apply in the long wavelength limit and that as wavelength approaches the particle diameter, numeric solution of the scattering equations[1] incorporating terms up to a very high order may be necessary for a satisfactory solution.

6. Conclusion

Ultrasound spectroscopy is unique amongst particle sizing techniques in a number of respects. Firstly it gives results over a remarkable range of particle sizes and concentration. Secondly it has the potential to make a robust, accurate, in-line, particle size measurement system. Thirdly it can detect particle aggregation and fourthly particle hydration and aggregation may be measured. Finally, the properties of micelles and molecules can also be studied.

References

1. M.J.W. Povey, *Ultrasonic Techniques for Fluids Characterization*, Academic Press, San Diego, 1997

2. D.J. McClements, M.J.W. Povey and E. Dickinson, "Absorption and Velocity Dispersion due to Crystallization and Melting of Droplets," *Ultrasonics*, **31**, 433-437 (1993).

3. Y.D. Yan, and J.H.R Clarke "In-Situ Determination of Particle Size Distributions in Colloids," *Adv. Colloid Interface Sci.*, **29**, 277-318 (1989).

4. V.J. Pinfield, M.J.W. Povey, and E. Dickinson "Interpretation of Ultrasound Velocity Creaming Profiles," *Ultrasonics,* **34**, 695-69 (1996).

5. D.J Hibberd, A.M Howe, A.R. Mackie, P.W. Purdy, and M.M. Robins, "Measurement of Creaming Profiles in Oil-in-Water Emulsions" in *Food Emulsions and Foams*, Edited by E. Dickinson, RSC, Cambridge, UK (1987).

6. E. Dickinson, J. Ma and M.J.W. Povey, "Creaming of Concentrated Oil-in-Water Emulsions Containing Xanthan," *Food Hydrocolloids*, **8**, 481-497 (1994).

7. F.G. Eggers and U Kaatze, "Broad-Band Ultrasonic Measurement Techniques for Liquids," *Measur. Sci. Technol.*, **7**, 1-19 (1996).

8. D.J. McClements, and P. Fairley, "Ultrasonic Pulse Echo Reflectometer," *Ultrasonics* **29**, 58-62 (1991).

9. D.J. McClements, and M.J.W. Povey, "Scattering of Ultrasound by Emulsions," *J. Phys. D: Appl. Phys.*, **22**, 38-47 (1989).

10. D.J. McClements, M.J.W. Povey, M. Jury, and E. Betsanis, "Ultrasonic Characterization of a Food Emulsion," *Ultrasonics*, **28**, 266-272 (1990).

11. E. Alba, "*Method and Apparatus for Determining Particle Size Distribution and Concentration in a Suspension Using Ultrasonics*", U.S. Patent No. 5121629 (1992).

12. P.S. Epstein, and R.R. Carhart, "The Absorption of Sound in Suspensions and Emulsions. I. Water Fog in Air," *J. Acoust. Soc. Am.,* **25**, 553-565 (1953).

13. J.R. Allegra, and S.A. Hawley, "Attenuation of Sound in Suspensions and Emulsions: Theory and Experiments," *J. Acoust. Soc. Am.*, **51**, 1545-1564 (1972).

14. M.J.W. Povey, M. Golding, D. Higgs, and Y. Wang, "Ultrasonic Spectroscopy Studies of Casein in Water," in *Casein and Caseinates: Structures, Interactions, Networks*. Edited by D. Horne. RSC, Cambridge, in press.

15. M. Frindi, B. Michels, and R. Zana, "Ultrasonic Absorption Studies of Surfactant Exchange between Micelles and Bulk Phase in Aqueous Micellar Solutions of Nonionic Surfactants with Short Alkyl Chains. 1. 1,2-Hexanediol and 1,2,3-Octanetriol," *J. Phys. Chem.*, **95**, 4832-4837 (1991).

16. M. Frindi, B. Michels, and R. Zana, "Ultrasonic-Absorption Studies of Surfactant Exchange Between Micelles and the Bulk Phase in Aqueous Micellar Solutions of Amphoteric Surfactants," *J. Phys. Chem.*, **98**, 6607-6611 (1994).

17. P.C. Waterman, and R. Truell, "Multiple Scattering of Waves," *J. Math. Phys.*, **2**, 512-537 (1961).

18. P. Lloyd, and M.V. Berry, "Wave Propagation through an Assembly of Spheres IV. Relations Between Different Multiple Scattering Theories," *Proc. Phys. Soc.*, **91**, 678-688 (1967).

19. M.A. Biot, "The Theory of Propagation of Elastic Waves in a Fluid-Saturated Porous Solid. I. Low Frequency Range," *J. Acoust. Soc. Am.*, **2**, 168-178 (1956).

20. M.A. Biot, "The Theory of Propagation of Elastic Waves in a Fluid-Saturated Porous Solid. II. Higher Frequency Range," *J. Acoust. Soc. Am.*, **2**, 179-191 (1956).

21. D.J. McClements, "Ultrasonic Determination of Depletion Flocculation in Oil-in-Water Emulsions Containing a Non-Ionic Surfactant," *Colloids Surf. A*, **90**, 25-35 (1994).

22. J. M. Evans, and K. Attenborough, "Coupled Phase Theory for Sound Propagation in Emulsions," *J. Acoust. Soc. Am.*, **102**, 278-282 (1997).

23. A.H. Harker, and J.A.G. Temple, "Propagation and Attenuation of Ultrasound in Suspensions," in *Developments in Acoustics and Ultrasonics,* Edited by M.J.W. Povey and D.J. McClements, IOPP, Bristol, UK, 1992.

24. A.H. Harker, and J.A.G. Temple, "Velocity and Attenuation of Ultrasound in Suspensions of Particles in Fluids," *J. Phys. D: Appl. Phys.*, **21**, 1576-1588 (1988).

25. N. Herrmann, "Application of Ultrasonic Techniques to the Study of Dispersions" PhD. Thesis, Louis Pasteur University, Strasbourg, France, 1997.

26. Y. Hemar, N. Herrmann, P. Lemarechal, R. Hocquart, and F. Leqeux, "Effective Medium Model for Ultrasonic Attenuation due to the Thermo-elastic Effect in Concentrated Emulsions," *J. Phys. II (France)*, **7**, 637-647 (1997).

27. J.W. Strutt, "Investigation of the Disturbance Produced by a Spherical Obstacle on the Waves of Sound," *Proc. London Math. Soc.*, **4**, 253-283 (1872).

28. D.S. Jones, *Acoustic and Electromagnetic Waves,* OUP, Oxford, U.K. 1986.

29. Y.D. Yan, and J.H.R. Clarke, "In-Situ Determination of Particle Size Distributions in Colloids," *Adv. Colloid Interface Sci.*, **29**, 277-318 (1989).

30. V. Pinfield and M.J.W. Povey, "Thermal Scattering Must be Accounted for in the Determination of Adiabatic Compressibility," *J. Phys. Chem. B*, **101**, 1110-1112 (1997).

The Electroacoustic Characterization of Colloidal Suspensions

Robert J. Hunter
Colloidal Dynamics Pty Ltd., National Innovation Centre, Australian Technology Park, Eveleigh, NSW 1430, Australia

1. Introduction

Electroacoustics in this context refers to two sorts of processes: (i) when a sound wave passes through an electrolyte solution or a colloidal suspension or emulsion, it creates a macroscopic potential difference (called the ultrasonic vibration potential or UVP effect or, for colloids, the CVP) and (ii) the obverse effect when an alternating electric field is applied to an electrolyte or a colloidal suspension or emulsion and this generates a sound wave (called the electrokinetic sonic amplitude or ESA effect). In both cases the applied field and the resulting response occur at the same frequency.

The early work on the theory and applications of the UVP has been adequately reviewed by Zana and Yeager[1] who also describe the theory of the CVP and its use in the study of polyelectrolytes[2] and proteins. Some more recent material on the CVP is reviewed by Marlow *et al*[3] who describe the basis of a commercial CVP instrument design (the Pen Kem, System 7000 Acoustophoretic Titrator) and draw attention to the similarities between the CVP effect and the sedimentation potential (or Dorn effect). In both cases the external field produces a dipole due to the distortion of the ionic atmosphere around the particle surface and although one is static and the other dynamic, the results show many correspondences. Marlow *et al.* also describe the extension of the CVP effect to more concentrated suspensions and show that a cell model correction can give a satisfactory description of the behaviour up to concentrations of order 30% by volume. We will return to that point when we discuss the ESA effect. There has been little in the way of theoretical development of the CVP since that time so here we will concentrate on reviewing the development of the ESA effect since its discovery in the early 1980s[4].

The ESA technique has, in some respects, transformed the characterisation of colloidal dispersions, at least for the more concentrated systems (i.e. for particle concentrations in excess of about 1% by volume) because it allows a simultaneous measurement of both the zeta potential and the particle size, from the same measurement. Much of that work, however, has appeared only in conference proceedings or the manufacturer's technical notes; the only reviews of developments have been limited in scope[5] or focus[6].

The ESA technique was introduced by Matec Applied Sciences in the mid 1980s using a device called the ESA-8000. That instrument is still in commercial production and is widely used, especially for the measurement of the properties of non-aqueous suspensions, even though it operates at only one frequency and estimates only the zeta potential of the suspended material (and can do that only when provided with the particle size).

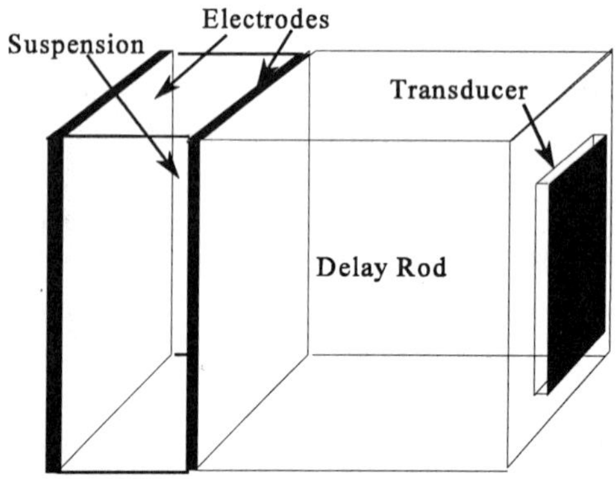

Figure 1. Schematic electroacoustic cell.

Figure 1 shows a typical, if minimal, schematic arrangement required for either a CVP or an ESA measurement. For CVP, an alternating voltage is applied to the transducer and the resulting sound wave travels down the delay rod and passes into the suspension. The resulting potential difference is measured between the electrodes. For ESA measurements, an electric field is applied between the electrodes, a sound wave is generated, and this travels out of the suspension and down the delay rod where it is detected by the transducer.

In the ESA effect, the applied electric field causes the (charged) particles to oscillate backwards and forwards at the same frequency as the field, and this generates tiny *acoustic dipoles* associated with each particle, provided that there is a density difference between the particles and the medium. These dipoles cancel one another throughout the body of the suspension but, in the neighbourhood of the electrodes, no such cancellation occurs, and the acoustic dipoles there generate a sound wave which can emerge from the suspension and move down the delay rod where it is detected by the transducer.

2. Theory of Electroacoustic Effects

2.1 The Relation between CVP and ESA Effects

The first theoretical treatment of the ESA effect was provided by O'Brien[7] who showed that the CVP and ESA effects were linked through a reciprocal relationship, analogous to that established by Mazur and Overbeek[8] between the electrophoresis and electroosmotic effects. An experimental verification of this reciprocal effect, for electrolytes and polyelectrolytes was provided by O'Brien *et al.*[9] That result shows that the CVP and ESA effects are intrinsically capable of yielding the same information but to date the measurements of CVP have been confined to a single frequency (usually around 200 kHz) whereas the ESA effect is now commonly measured over a range of frequencies from about 0.3 to 11 MHz. The quantity in which we are most interested is the dynamic mobility, μ_d and it transpires that the ESA effect gives μ_d immediately, whereas one needs not only the CVP but also the complex conductivity to obtain μ_d by that route.

2.2 Relation between ESA and dynamic mobility

To obtain the complete relation between the ESA and the dynamic mobility for any particular experimental set-up requires the solution of the differential equations given by O'Brien[7] for the macroscopic pressure, velocity, and electric field in the suspension. In the instrument which was designed to take advantage of these new developments (the AcoustoSizer, Matec Applied Sciences, Hopkinton, MA, USA) the solution is simplified by the fact that the electrode dimensions are large compared to the wavelength of sound (of order mm) at the frequencies employed. O'Brien *et al.*[10] give the following expression for μ_d:

$$\mathrm{ESA} = A(\omega)\ \phi\ (\Delta\rho/\rho)\ \mu_d.[Z_sZ_r/(Z_s+Z_r)] \tag{1}$$

where $A(\omega)$ is an instrument constant. The Z function, in the square brackets, is a measure of the acoustic impedance match between the suspension and the delay rods through which the signal passes to be measured.

The acoustic impedance function in Equation (1) is made up of the impedance of the suspension, Z_s, and that of the rod into which the sound must travel, Z_r. In dilute suspensions, Z_s is the same as that for the suspension medium so for such systems the impedance function can be incorporated into the general calibration function A. That is what is done in the Matec ESA-8000 device. For more concentrated systems, the acoustic impedance depends upon the volume fraction and density of the particles and for such systems it is necessary to monitor Z_s directly. That is done in the AcoustoSizer, and so it becomes possible to determine accurately the dynamic mobility at any particle concentration up to values in excess of 60% by volume, which is near to close pack.

As noted above, the dynamic mobility is a complex quantity, having both a magnitude and a phase angle and it measures the velocity of the particle motion per unit field strength. It is the analogue of the d.c. mobility, μ_E, as measured by electrophoresis and, indeed, for colloidal particles less than one micron in diameter, it has a similar magnitude even at frequencies approaching 1MHz. The phase angle measures the time lag between the applied field and the subsequent motion of the particle and this is zero at low frequencies and rises with increase in frequency and increase in particle size. The dynamic mobility is itself an important property of the particle and it can, in principle, be used to characterise a particulate dispersion.

2.3 Relation Between Dynamic Mobility and Particle Properties

O'Brien[7] showed that, for a dilute suspension of spherical particles (less than about 4% by volume, say) the dynamic mobility could be related to the particle properties as follows:

$$\mu_d = \frac{2\epsilon\zeta}{3\eta} G\left(\frac{\omega a^2}{\nu}\right)[1 + f(\lambda,\omega')] \tag{2}$$

where f and G are complex function of the particle properties.

Here a is the particle radius, η is the viscosity and ϵ the permittivity of the liquid, ν ($=\eta/\rho$) is the kinematic viscosity and K^∞ is the electrical conductivity of the liquid. The quantity λ is the (dimensionless) surface conduction parameter and is the ratio of the surface conductivity to the bulk conductivity. Provided that electrical conduction occurs only in the diffuse part of the double layer (which is the usual situation) λ can be calculated from the zeta potential.

The G factor represents the effect of inertia forces on the dynamic mobility. At low frequencies, that is when $\omega a^2/\nu$ is much less than 1, $G \approx 1$ and the inertia forces are unimportant. It is not possible to assess particle size from the dynamic mobility at such low frequencies but zeta can still be obtained. As $\omega a^2/\nu$ approaches unity the particle inertia becomes significant and as a result the magnitude of μ_d drops, and the particle motion begins to lag behind the applied field. Both the magnitude and phase lag depend on particle size at these frequencies and so measurements of μ_d can be used for getting size as well as ζ. If the frequency is increased further the magnitude of μ_d diminishes until eventually the ESA signal is too small to measure. In this limit the phase lag approaches 45°. The AcoustoSizer has a frequency range from 0.3 to 11 MHz which allows sizing of particles in the range 0.1 to 10 microns diameter. For smaller particles ζ can be determined, but the inertia forces are too small to allow their size to be determined.

For many colloids the formula [2] can be simplified by putting $f = 0.5$. The dynamic mobility is then given by :

$$\mu_d = (\epsilon\zeta/\eta)\, G\, (\omega a^2/\nu) \tag{3}$$

In this case the frequency dependence and phase lag of the dynamic mobility are entirely determined by the inertia factor G and ζ is calculated from a modified Smoluchowski formula to take account of the inertia effect for the larger particles, especially at the higher frequencies. The determination of size and charge from the dynamic mobility is particularly simple in this case, for the argument of μ_d is then independent of ζ, and so, for a monodisperse suspension, the size can be obtained from the argument of μ_d, and ζ can then be obtained from the magnitude of μ_d.

2.4 Effect of Particle Shape

Loewenberg and O'Brien[11] have calculated the mobility of spheroidal particles from which one can infer the behaviour of discs (oblate spheroids) and rods (prolate spheroids). In general they behave like spheres but with a size intermediate between their small and large diameters. Loewenberg and O'Brien[11] state that for a disk of axial ratio 0.1, the estimated radius is approximately $1/\sqrt{10}$ of the true disk radius.

For prolate particles the dynamic mobility becomes very insensitive to the length of the particle, so rods of axial ratio >5 all behave like spheres with a radius of 1.5 times the true cross-sectional radius. This result has recently been confirmed by Ohshima[12] who has derived approximate formulae for cylindrical particles with zero permittivity and low zeta potentials and for all κa values.

3. Instrumentation

3.1 The Matec ESA-8000

The ESA-8000 instrument has been described briefly by Babchin *et al.*[5] and by Klingbiel *et al.*[13] and its main design features have been more fully described by

Cannon.[14] A general block diagram is also given by Sayer.[15] The device can operate either in CVP or in ESA mode, depending on the way the terminals are connected. In the ESA arrangement, the field is applied across the electrode gap and the resulting sound wave travels along the delay line and excites the transducer on the end of the probe. The field is applied in the form of a pulse for a few microseconds and one then waits for the response. The delay line is introduced to eliminate the signal which is generated directly by the applied field in the measuring tranducer. The signal processing scheme takes the delayed electroacoustic signal and estimates its amplitude and phase angle using quadrature detection with a signal processing scheme which is described briefly by Cannon.[14] When the ESA-8000 was first introduced, the theory did not exist for relating the ESA to fundamental particle properties like the dynamic mobility. With the introduction of O'Brien's theory[7], it became possible to calculate the dynamic mobility directly from the ESA signal, and the procedure is described by Cannon.[14] For the dilute suspensions used in the ESA-8000, the acoustic impedance is equal to the impedance of water and only the terms ϕ and $\Delta\rho$ are required in order to estimate the dynamic mobility of the sample. It is, however, necessary to input the particle radius in order to calculate the true zeta potential. Details of the procedure are given by James[16] who has shown how consistent calibration standards can be produced with suspensions of polystyrene, poly (methyl methacrylate) latices and mineral oxide systems.

The ESA-8000 provides a rapid method of determining the 1MHz dynamic mobility of colloidal particles in either aqueous or non-aqueous media, and we will review some of the data acquired below. The technique is accurate to about ±2% for suspension concentrations up to about 5% by volume. For higher concentrations the acoustic impedance of the suspension becomes dependent on the particle concentration and must be measured. That facility is not available on the ESA-8000. Nevertheless, the estimates of dynamic mobility in such concentrated systems can still give very useful qualitative information and have been used for that purpose in some industrial situations. When the particle size is significantly larger than 1 μm, the 1 MHz mobility becomes a strong function of particle size and one must take care to use the appropriate distribution function to represent that size.[17] A procedure for introducing any arbitrary or measured particle size distribution and calculating the corrected dynamic mobility and, hence, the zeta potential has been developed as a spreadsheet in Lotus-123.[18]

Although the ESA-8000 is capable of making very accurate estimates of zeta potential of particles in both water and non-aqueous environments, it suffers from a number of limitations. To begin with, it is limited to measurements on dilute suspensions. Since it measures at only one frequency (1 MHz) it does not collect enough information to measure both particle size and charge. With two sufficiently different frequencies it would be possible, in favourable cases, to estimate the size but only if the phase angle could be measured very accurately and that is not possible with the ESA device. It was designed to detect the 180^0 phase change which accompanies a change in the sign of charge but to determine the size with the necessary degree of accuracy requires a precision in the phase angle measurement of order 1^0. That raises a number of problems which were addressed in the design of the AcoustoSizer.

3.2 The AcoustoSizer

As noted above the AcoustoSizer measures both the magnitude and the phase angle of the dynamic mobility of suspensions of essentially any particle concentration. The frequency range of the instrument (0.3 to 11.2 MHz) is chosen to give a particle size range of 0.1 to 10 μm. For the smallest measurable particles (0.1 μm) the inertia effect only becomes significant at the highest frequencies, whilst for the largest particles (10 μm) the amplitude of the particle motion at the lowest frequency is still sufficiently large to be detected. The amplitude of the particle motion is extremely small. At 10 MHz, the amplitude of a 1μm particle in a typical field (10 V/cm) is only a few picometres, which is much less that the size of an atom. The frequency range chosen corresponds to values of the function α (= $\omega a^2/\nu$, the argument of G) of order unity for the quoted size range.

The special design features of the AcoustoSizer, which distinguish it from the ESA-8000, are described by Cannon[14] and its calibration and use, and the validation of the underlying theory are treated fully by O'Brien *et al.*[10] To measure over a wide frequency range, the acoustic resonance technique must be abandoned, so the measuring system needs to be more sensitive. To allow measurements at arbitrary concentrations it is necessary to incorporate facilities for measuring the acoustic impedance, Z_s, of the suspension in the cell. The most important design problem was to measure the phase angle precisely.

3.3 Behaviour of Polydisperse Systems

In a polydisperse system, the ESA signal is related to the volume average dynamic mobility of the particles, $<\mu_d>$ which is defined as :

$$<\mu_d(\omega)> = \int_0^\infty \mu(\omega,a)p(a)\mathrm{d}a \qquad (4)$$

where $p(a)\mathrm{d}a$ is the mass fraction of particles with radius between a + da/2 and a –

da/2. For a dilute system, Equation (2) can be used for $\mu_d(\omega)$ for each value of a provided only that the double layer is thin. The only unknown terms in equation [4] are then ζ and the function $p(a)$ which must be adjusted until the best fit is found to the dynamic mobility spectrum.

The AcoustoSizer software assumes that the size follows a log-normal distribution and adjusts the median and spread of the distribution, along with the ζ-potential to give the best fit to the mobility spectrum, by minimising the relative rms error. O'Brien *et al.*[10] show what excellent agreement can be achieved between the ESA log-normal size distribution and the 'true' distribution for a pair of ground quartz standards supplied by the Bureau of Common Reference of the EEC. In the same paper they also describe the results on a variety of industrial samples of ceramics, paper coatings and pigments, indicating good agreement between sizes obtained by the ESA method and by an alternative sedimentation technique (the Horiba Capa-700).

4. Studies with the ESA-8000

Although the earliest version of the ESA-8000 device was limited in its ability to measure absolute mobilities and zeta potentials, it proved of great value in the study of moderately concentrated (up to, say, 5%v/v) suspensions of mineral particles in aqueous media[15] and the mechanisms of charging of dispersions in non-aqueous media. The link between the UVP and the more conventional electrokinetic methods was well established and Marlowe *et al.*[3] showed a comparison of electrophoretic measurements on a titania sol which agreed reasonably well with UVP measurements, certainly over the acid pH range. By the late 1980s the connection between the ESA and other electrokinetic effects had also been established[7] and Hunter[19] described the first quantitative comparison of mobility of polymer latex particles estimated from the high frequency behaviour (after correction for the inertia effect) and d.c. electrophoresis results. Shortly afterwards, Babchin *et al*[5] were able to demonstrate near quantitative agreement between ESA measurements and electrophoresis of a laboratory silver iodide preparation.

4.1 Studies in Non-Aqueous Media

One of the earliest publications referring to the use of the ESA-8000 apparatus in non-aqueous media was that of Isaacs *et al.*[20] who successfully monitored the coalescence of water-in-oil emulsions. The studies of Larson, Pearlstine, El-Sayed and their colleagues on liquid electrostatic toners were particularly interesting[21,22] though only a fraction of that material has passed into the more general colloids literature. Felder *et al.*[23] compared the results obtained from

introducing the concept of an *anomalous surface conduction* process by which ions in the Stern layer are able to contribute to the electrical conductivity, even though they are in a region where there is no fluid flow. This has an important effect on the estimation of the parameter λ, which can no longer be calculated from the zeta potential. We will take this matter up again in considering the experimental work using the AcoustoSizer since it is now possible to obtain a much more definitive solution of the problem than was possible by considering only the 1 MHz results from the ESA-8000.

5. AcoustoSizer Studies at Modest Concentration

Some of the earliest studies with the AcoustoSizer were recorded in the proceedings of the special meeting of the National Institute of Standards and Technology[45] and further work on mineral systems was described in the ceramics literature.[46] The latter paper describes the typical study in which a sample is titrated over a certain pH range and the zeta potential and size recorded at each pH, to determine the isoelectric point (*iep*). This is an important defining characteristic of many colloidal systems, particularly those for which the hydrogen and hydroxide ion are potential determining. Studies of this sort have also been undertaken by Hackley and Malghan[47] who obtained excellent agreement between size measurements using the AcoustoSizer, the Sedigraph and the Horiba LA900 laser diffraction analyser for a well dispersed system. They also showed that the zeta potential determined by the AcoustoSizer as a function of pH, on a silica sample, was in excellent agreement with the results of conventional d.c. electrophoresis measurements.

Very stable, insoluble oxides, like titania, typically show a quite reversible titration curve with respect to zeta potential. The size of the particles tends to a minimum value at the extremes of the pH range where the particles are strongly charged, either positively or negatively. In the region of the *iep* the size appears larger, but this is to some extent an artefact. As one passes through the *iep* and the system coagulates, the size of the 'particle' is essentially undefined. The size of the aggregates formed will depend on time and stirring regime. One might characterise these aggregates as very large porous particles and O'Brien[48] has shown that such particles have some very interesting properties. The porous aggregate, made up of many small particles all with thin double layers, moves in phase with the field at high frequencies, and the magnitude of the mobility only falls below 50% of the Smoluchowski value at extremely high frequencies (well above the range of the AcoustoSizer). A solid particle of the same size would have practically no mobility and would have a phase lag at the maximum value of around 45°. The formula given by O'Brien could be used to monitor the behaviour of coagulating systems but that has so far not been attempted.

O'Brien et al.[46] also describe the measurement of zeta potential and size of bentonite clay as a function of pH and also the effect of calcium ion on the bentonite surface. Bentonite is a very important clay mineral used, for example, in oil drilling muds. Even the dispersion of fine cement can be studied as a function of added dispersing agent; the point at which dispersion is maximised can be determined quite precisely, despite the highly reactive nature of the material and the high pH (over 12) involved.[46]

Another mineral system of great interest is the metal sulfides which have frequently been studied using conventional electrophoresis. Prestidge and Rowlands[49] have compared electrophoretic and electro-acoustic assessments of zeta potential for a sphalerite (zinc sulfide) and a galena (lead sulfide) sample and found them to be in good agreement. These systems are complicated by the possibility of surface oxidation which not only produces time effects but which can make the result depend significantly on the surface area to volume ratio in the suspension. The capacity of the electroacoustic method to make measurements at the normal suspension concentration then becomes of great value.

6. Effect of Concentration of Particles

6.1 The Acoustic Impedance

One immediate effect of increasing the particle concentration in the suspension is that the acoustic impedance, Z_s, can no longer be approximated as equal to that of the suspension medium. Since Equation (1) remains valid at all concentrations commonly encountered, it is important that the correct value of Z_s is used, so that the correct value of the dynamic mobility is obtained from the measured ESA signal. In principle, the value of Z_s for the suspension could be a complex function of the frequency and the properties of the suspension but the exact behaviour is of little consequence for measurements with the AcoustoSizer, since it measures the value at each frequency before calculating μ_d from the ESA signal.

Equation (2) shows that the ESA effect should be proportional to the volume fraction ϕ for dilute systems and the measurements of Klingbiel *et al.*[13] suggest that that holds for at least some spherical particles up to concentrations of order 5%. Texter's non-spherical particles were linear only up to about 2% by volume.[39] Even so this is a great deal more concentrated than the normal concentrations at which d.c. electrophoresis is conducted but it is at the lower end of the range of the ESA method. Traditional methods of determining the electrophoretic mobility in a d.c. electric field[50] have involved particle concentrations with $\phi << 0.001$. This provides the infinite dilution limiting value and the appropriate theoretical analysis is for an isolated particle in an infinite volume of electrolyte.

The tracer and mass transport methods allow the electrophoresis procedure to be used on more concentrated systems and have been reviewed in reference 50 Sections 4.3.3 and 4.3.4. Reed and Morrison[51] have shown that even in highly concentrated systems the hydrodynamic and electrostatic interactions cancel one another when the double layers are thin and the only effect which must be taken into account is the reverse flow of fluid displaced by the moving particles. Zukowski and Saville,[52] using red blood cells mixed with ghosts, have verified that this is so and that the d.c. mobility, μ_c of a concentrated system of volume fraction ϕ is given by the simple relation:

$$\mu_c = \mu_0[1 - g\phi] \tag{5}$$

where μ_0 is the mobility at infinite dilution and g is, within the limits of experimental error, equal to unity.

Marlow and Rowell[53] working with coal/water slurries and using the CVP technique have shown that, at the frequencies of their measurements (200 kHz), the effect of particle concentration can be adequately described by introducing a factor $(1 - g\phi)$ into their equivalent of Equation (4) where again, g was very close to unity. In their review article[3] Marlow *et al.* discuss the way the Levine *et al.* cell model[54] is introduced into the CVP theory and show that, for thin double layers, the result is that the hydrodynamic and electrostatic interactions essentially cancel one another and one is left with only the factor $(1-\phi)$ to take account of the backflow of liquid caused by the particle motion.

Unfortunately, no such simple solution is available for the ESA effect at the high frequencies at which it is currently used. Goetz and El-Aasser[55] attempted to compare the electroacoustic and electro-phoretic behaviour of concentrated mini-emulsion systems of toluene in water, stabilised by cetyl alcohol and sodium lauryl sulfate. They concluded that the simple correction which works well for CVP does not produce a similar reconciliation in the case of the ESA effect. Their conclusions are, however, suspect because of uncertainties arising from the dilution of the emulsion system; that can so easily lead to changes in surface properties, no matter how carefully it is done. Texter's results[39] referred to above are perhaps more definitive in this case. He showed that in the range from 2 to 5% by volume where his particles showed a non-linear dependence of the ESA signal on volume fraction, the Levine and Neale model was not able to account for the non-linearity. His particles were, however, non-spherical and that may at least partially explain the discrepancy.

Nonetheless there are good reasons to believe that the concentration correction for the ESA method is not as simple as Equation (5) would suggest. The Levine approach uses Kuwabara's "zero vorticity" model[56] in which the vorticities of both the hydrodynamic and the electric fields are zero on the defining surface of the cell which encloses each particle (ref 50 p.116). At frequencies in the megahertz range, the vorticity of the flow field stretches out beyond the confines of the cell so that hydrodynamic interactions between the particles are very much more significant. O'Brien *et al.*[45] showed that the Levine cell model drastically underestimates the effect of concentration on both the magnitude and the phase angle of the dynamic mobility in the range from 0.5 to 11 MHz. It should be noted that the reciprocal relationship between CVP and ESA makes it clear that precisely the same limitations would apply to the CVP in the same frequency range.

Rider and O'Brien[57] have extended the dilute solution theory to incorporate the order ϕ correction which allows one to describe the ESA behaviour up to particle concentrations of order 10% by volume. For higher concentrations, different treatments are required for different particle densities. The denser particles show a higher degree of hydrodynamic interaction than the less dense ones. For near neutrally buoyant particles it is only necessary to take account of the near-neighbour interactions. Even so the result is far from simple. Using the Percus-Yevick approximation to estimate the distribution function for the nearest neighbours $(g(r))$ and assuming additivity of the contributions from each particle in the vicinity of the central particle, O'Brien has shown[58] that the ESA signal is given by:

$$\mu_d = \frac{\epsilon\zeta}{\eta} H \left[1 - \frac{2\beta^2}{(2+\phi)(3+3\beta+\beta^2)} - \frac{3\phi}{2+\phi} F \right] \qquad (6)$$

where the factors H and F are functions of the complex variable
$\beta = (1 + I)\sqrt{(\omega a^2/2\nu)}$.

The efficacy of this analysis is demonstrated by the data in Table I which shows a comparison of the estimated values of zeta potential and of size using the dilute solution theory and the more elaborate theory of Equation (6). The system used in this study was a parenteral/intravenous emulsion used by Carasso et al.[59] This is a very stable material and it is supplied as a 20% or 10% (w/v) emulsion which was carefully diluted with the suspending fluid and measurements made at varying particle concentrations. It is clear from the table that using the concentrated suspension theory gave rise to almost identical zeta and size values at all dilutions whereas the dilute theory caused variations of order 15% in zeta potential and up to 30% in size. We have successfully used Equation (6) to size bitumen particles at concentrations up to 60% by volume in water.

For particles with densities significantly different from that of the solvent, it is not possible to use this approach. The interactions between particles are not confined to nearest neighbours and are greater than is predicted by Equation (6). O'Brien has obtained limiting solutions for particles of arbitrary density at high and low frequency and is currently completing calculations for the more general concentration effect. In the meantime, the AcoustoSizer uses a semi-empirical method to correct for concentration effects, based on the behaviour of the common oxide systems.[45]

Table I. Comparison of dilute theory with concentrated theory for a near neutrally buoyant emulsion system.

Concen-tration	Volume Fraction	Concentrated formula				Dilute Formula			
		ζ	d_{50}	d_{15}	d_{85}	ζ	d_{50}	d_{15}	d_{85}
20% w/v	0.217	-40.8	0.40	0.36	0.4	-29.4	0.28	0.2	0.31
20% w/v	0.163	-40.3	0.40	0.36	0.4	-31.9	0.31	0.2	0.35
20%w/v	0.109	-40.2	0.40	0.30	0.5	-34.2	0.35	0.3	0.38
20%w/v	0.054	-40.6	0.40	0.26	0.6	-37.5	0.39	0.2	0.53
10% w/v	0.109	-40.4	0.33	0.23	0.4	-34.4	0.28	0.2	0.38

7. Particle Conductivity

7.1 Semiconductor Behaviour

O'Brien has shown[60] that, in an intrinsic semiconductor, the positive and negative charge carriers are easily able to keep in phase with the field at the frequencies used in electroacoustics. The result is that such particles behave as though they had an enhanced permittivity given by:

$$\epsilon^* = \epsilon \frac{[2 + (\kappa a)^2]\sinh(\kappa a) - 2\kappa a \cosh(\kappa a)}{\kappa a \cosh(\kappa a) - \sinh(\kappa a)} \qquad (7)$$

where κ is in this case the reciprocal of the *internal* Debye length, defined by :

$$\kappa = \sqrt{\frac{2ne^2}{\epsilon k T}} \qquad (8)$$

and n is the number of electrons per unit volume. The effective permittivity is the same as that of the particle at $\kappa a = 0$, and ϵ^* increases monotonically with κa, approaching the limiting form $\epsilon^*/\epsilon \sim \kappa a$ as κa approaches infinity. Thus in these suspensions, the larger particles have larger effective permittivities. For doped semiconductors the analysis is very similar and leads to the same equation (7) but in Equation (8) the factor 2 is replaced by unity. Hunter and O'Brien[61] have applied this theory to analyse the electroacoustic behaviour of a suspension of silicon particles of about 4 μm diameter with excellent results (Figure 2).

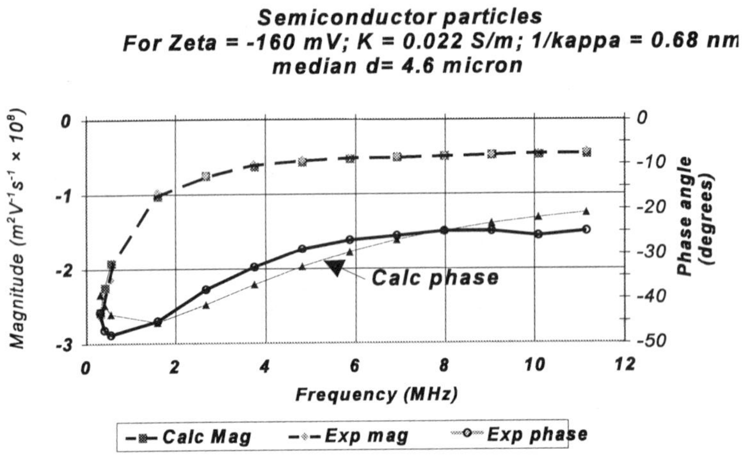

Figure 2. Comparison of theory and experiment for silicon particles.

7.2 Anomalous Surface Conduction in Clay Minerals

We referred above to some work by Shubin *et al.*[44] on the phenomenon of anomalous surface conduction in some polymer latex systems. Recall that the surface conduction parameter λ measures the relative importance of the double layer (compared to the bulk electrolyte) in conducting current past the particle surface. λ may be large, and therefore significant, when the particle size and/or the electrolyte concentration are small. In most systems, the conduction in the double layer is due to the ions in the diffuse part, above the shear plane and λ can then be calculated from the zeta potential. For particles around 1μm in size the resulting λ value would seldom be significant since for $\zeta < 100$ mV, the λ value is less than 0.05 in 10^{-3} M salt. There are, however, some systems for which there is a significant contribution to the conduction from ions in the Stern layer, below the shear plane. We refer to this as *anomalous* surface conduction[44] since those ions are in a part of the fluid which is stationary and one might not expect them to contribute to the conduction process. It seems, however, that they are able to do so, even at the low frequencies used in normal conductance measurements. When λ becomes large (of order unity) it affects the function f in Equation (2) and can lead to significant alterations in the form of the phase angle function for the dynamic mobility. Instead of the phase lag increasing steadily from low values to its limiting value of -45^0, it may go through a maximum at low frequency, and even turn into a phase lead over part of the range.

In the AcoustoSizer software, the value of λ is normally estimated from the zeta potential and clearly this will lead to erroneous sizing results if λ is much larger than the value predicted from the zeta potential. Such a case is that of the clay mineral kaolinite. Fortunately, when this happens it is usually readily apparent from the form of the dynamic mobility spectrum. O'Brien and Rowlands[62] measured the surface conduction of kaolinite directly, using a high frequency bridge and found it to be much higher than would be expected from the usual values for the zeta potential. In a subsequent study of the dynamic mobility, Rowlands and O'Brien[63] showed that an approximate relation derived by Loewenberg[64] for the dynamic mobility of a spheroidal particle with arbitrary λ value was in good agreement with their measured results, using the values for λ derived from the high frequency conduction information. In addition, the shape of the particle turned out to be very critical, so that the measurement provided a value for the axial ratio of the disc-like kaolinite particles. That work has now been extended to the clay mineral bentonite[65] with similar results. The axial ratio (ratio of disc diameter to thickness) was found to be 0.1 for the kaolinite studied and 0.02 for the bentonite and these are consistent with values estimated by other methods.

8. Effect of High Salt Content

The AcoustoSizer is, in principle, capable of making electrokinetic measurements at elevated salt concentrations (above 1 M) but the factory calibration covers only the range from 0.01 to 1 S/m (which is from about 5×10^{-4} to 0.1M salt). We noted earlier that as the conductivity increases, the field applied across the cell by the gated amplifier decreases, since it is unable to supply the necessary current to the very low impedance at high salt concentrations. The high currents also set up magnetic fields within the cell which alter the distribution of the electric field in the cavity and so alter the consequent motion of the particles. To extend the range requires the provision of standards of known ESA signal and higher conductivity. Caesium chloride solutions are suitable for the purpose and Rowlands *et al.*[66] give details on how the calibration range can be extended to about 20 S/m (which is the conductivity of 3M sodium chloride). The calibration[10] involves the assumption that the ionic mobilities are independent of frequency and Rowlands *et al.* give supporting evidence for this assumption, at least for CsCl and some other common salt solutions. To calculate the ESA for CsCl, the ratio of ionic mobilities for the Cs^+ and the Cl^- ion are required and these are not available at such concentrations. They have assumed, therefore, that the infinite dilution value of unity still holds at high concentration.*

The earliest attempt to tackle this region of very high salt concentration was that of Kosmulski and Rosenholm[67] who studied the adsorption of ions on anatase and zirconia from very concentrated electrolytes. Unfortunately, they were unaware of the pitfalls of working at very high salt concentration, so the quantitative results quoted in their investigations cannot be relied upon, although the more qualitative conclusions seem consistent enough to warrant further examination.

Rowlands *et al.*[66] studied the effect of high salt concentrations on the behaviour of a suspension of aluminium hydroxide. It was found that the isoelectric point *(iep)* of this material was at pH 9.1 in dilute salt solution, but it moved up to over pH 11 in 0.5 M sodium chloride, and in 3M sodium chloride there was no *iep* at all; the colloid showed a positive zeta potential at all pHs measured (up to well over pH 12). Thus, what is normally considered to be an 'indifferent' electrolyte shows a distinct tendency to specific adsorption at high enough concentrations, something that has always been considered possible in the current descriptions of the oxide-solution interface[68] but has not previously been observed. The *iep* of calcined alumina (α-alumina) is usually quoted at pH 9.1 and for the slightly more hydrated γ- form a value of 8.5 has been given.[68] It is perhaps surprising then that the even more highly hydrated aluminium hydroxide has an *iep* at pH 9.1 in dilute salt solution. There can, however, be little doubt about the result. In the course of their study, Rowlands *et al.*[66] developed a useful new technique for determining the *iep* using the Argand diagram for the dynamic mobility. When one plots the real against the imaginary part of the dynamic mobility, the points at different frequencies for a given pH fall on a smooth curve and the points at the same frequency can be joined by a straight line passing through the origin. It is a simple matter to interpolate to an *iep* of 9.1 for the alumina in dilute salt solution. The corresponding figure for 0.5 M sodium chloride, after correction for the background salt effect, is not quite so straightforward, presumably because the background electrolyte signal correction† is a little faulty. This has the effect of shifting the intersection point away from the origin, but the *iep* remains in little doubt and it would certainly seem to be above

---

* Departures due to differing frequency dependence of the interactions of the two ions at the higher concentrations are likely to cancel in taking the ratio.

† It should be noted that this procedure enables one to make a final adjustment to the background correction, to bring the intersection point into the origin.

pH 11. A follow-up study of the same material with potassium chloride as the supporting electrolyte gave a slight upward movement in *iep* in 0.5 M salt with no further change at the higher concentration (which was 2M in this case). [2M potassium chloride has approximately the same conductivity as 3M sodium chloride.]

9. Summary

The ESA effect, in which a sound wave is generated in a colloidal system by the application of an alternating electric field, provides us with a powerful method for characterising colloidal suspensions. The key parameter is the particle dynamic mobility which can be determined accurately in suspensions of essentially any concentration above about 1% by weight. This quantity can then be related to such important characteristics as the particle size and the zeta potential on the particle surface. Theoretical analyses have been provided to cover particles of almost any shape. The currently available apparatus can measure particle sizes in the range from about 0.1 to 10 μm and the zeta potential of particles down to nanometer dimensions. Since the electroacoustic measurement provides both the magnitude and the phase angle of the dynamic mobility over a range of frequencies it is, in principle, capable of supplying much more information than is a d.c. electrophoretic mobility measurement.

Acknowledgments

I am indebted to my colleagues Richard O'Brien and James Beattie for helpful discussions when preparing this manuscript.

References

1. R. Zana and E.B. Yeager, "Ultrasonic Vibration Potentials," pp. 1-61 in *Modern Aspects of Electrochemistry*, No. 14. Edited by J.O.M. Bockris, B.E. Conway and R.E. White, 1982.
2. R. Zana and E. Yeager, *J. Phys. Chem.* **71**, 3502 (1967).
3. B.J. Marlow, D. Fairhurst, and H.P. Pendse, *Langmuir*, **4**, 611-626 (1988).
4. T. Oja, D. Cannon, and G.L. Petersen, *US Patent* 4 497 208 (1985).
5. A. Babchin, R.S. Chow, and R.P. Sawatsky, *Adv. Colloid Interface Sci.*, **30**, 111-151 (1989).
6. A.S. Dukhin and P.J. Goetz, *Langmuir*, **12**, 4336-4344 (1996).
7. R.W. O'Brien, *J. Fluid Mech.*, **190**, 71-86 (1988); **212**, 81 (1990).
8. P. Mazur, and J.Th.G. Overbeek, *Rec. Trav. Chim. (Pays-Bas)*, **70**, 83 (1951) .
9. R.W. O'Brien, P. Garside, and R.J. Hunter, *Langmuir*, **10**, 931-935 (1994).

10. R.W. O'Brien, D.W. Cannon, and W.N. Rowlands, *J. Colloid Interface Sci.*, **173**, 406-418 (1995).

11. M. Loewenberg, and R.W. O'Brien, *J.Colloid Interface Sci.*, **150**, 158-168 (1992).

12. H. Ohshima, *J. Colloid Interface Sci.*, **185**, 131-139 (1997).

13. R.T. Klingbiel, H. Coll, R.O. James, and J. Texter, *Colloids Surfaces*, **68**, 103-109 (1992).

14. D.W. Cannon, "New Developments in Electroacoustic Methods and Instrumentation," pp. 40-66 in *Electroacoustics for Characterization of Particulates and Suspensions*. NIST Special Publication 856. Edited by S.G. Malghan. National Institute of Standards and Technology, Gaithersburg, MD, USA, 1993.

15. T.S.B. Sayer, *Colloids Surfaces A: Physicochem. Eng. Aspects*, **77**, 39-47 (1993).

16. R.O. James, "Calibration of Electroacoustic (ESA) Apparatus Using Lattices and Oxide Colloids," pp. 67-91 in *Electroacoustics for Characterization of Particulates and Suspensions*. NIST Special Publication 856. Edited by S.G. Malghan. National Institute of Standards and Technology, Gaithersburg, MD, USA, 1993.

17. M. James, R.J. Hunter, and R.W. O'Brien, *Langmuir*, **8**, 420-423 (1992).

18. R.J. Hunter, Available from the author.

19. R.J. Hunter, *Journal and Proceedings Royal Society of New South Wales*, **121**, 165-178 (1988). (Liversidge Lecture).

20. E.E. Isaacs, H. Huang, A.J. Babchin, and R.S. Chow, *Colloids Surfaces*, **46**, 177-192 (1990).

21. J.R. Larson, G.A. Lane, J.R. Swanson, T.J. Trout, and L. El-Sayed, "Aminoalcohols as additives for liquid electrostatic toners," in Proc. 5th Internat. Cong. Adv. in Non-Impact Printing Tech., San Diego CA, USA November 1989. Society for Imaging Science and Technology, 1990.

22. J.R. Larson, W.A. Houle, K.A. Pearlstine, J.R. Swanson, T.J. Trout, and L. El-Sayed, "Acidic Additives for Positive Liquid Electrostatic Toners" pp. 585-597 in Proc. 6th Internat. Cong. Adv. in Non-Impact Printing Tech., Florida, USA October 1990. Society for Imaging Science and Technology, 1990.

23. T.C. Felder, S.M. Marcus, and K.A. Pearlstine, *Int. Symp. on Surface Charge Characterization*, 21st Annual Meeting Fine Particle Society, 1990.

24. J.R. Larson, G.A. Lane, J.R. Swanson, T.J. Trout, and L.M. El-Sayed, *5th International Congress on Advances in Non-impact Printing Technologies*, Society for Imaging Science and Technology (1989).

25. F.M. Fowkes, H. Jinnai, M.A. Mostafa, F.W. Anderson,. and R.J. Moore, Chapter 15 in *Colloids and Surfaces in Reprographic Technology*. Edited by M. Hair and M.D Croucher. ACS Symposium Series 200, 1982.

26. K.A. Pearlstine, L. Page, and L.M. El-Sayed, *J. Imaging Sci.*, **35**, 55-58 (1991).

27. K.A. Pearlstine, and J.R. Swanson, *J. Colloid Interface Sci.*, **151**, 343-350 (1992).

28. R.E. Kornebrekke, I.D. Morrison, and T. Oja, "Electrokinetic Measurements of Concentrated Suspensions in Apolar Liquids," p.92-110 in *Electroacoustics for Characterization of Particulates and Suspensions*. NIST Special Publication 856. Edited by S.G. Malghan. National Institute of Standards and Technology, Gaithersburg, MD, USA, 1993.

29. R.O. James, J. Texter, and P.J. Scales, *Langmuir*, **7**, 1993 (1991).

30. W.N. Rowlands, and R.J. Hunter, *Clays and Clay Minerals*, **40**, 287-291 (1992).

31. R.J. Hunter, and M. James, *Clays and Clay Minerals*, **40**, 644-649 (1992).

32. R.O. James, and T.W. Healy, Parts I, II, and III. *J.Colloid Interface Sci.*, **40**, 42-52, 53-64, 65-81 (1972).

33. *Electroacoustics for Characterization of Particulates and Suspensions*. NIST Special Publication 856. Edited by S.G. Malghan. National Institute of Standards and Technology, Gaithersburg, MD, USA, 1993.

34. V.A. Hackley, and S.G. Malghan, *J. Mater. Sci.*, **29**[17] 4420-4430 (1994).

35. S.G. Malghan, P.S. Wang, A. Sivakumar, and P. Somasundaran, *Compos. Interfaces*, **1** [3] 193-210 (1993).

36. H.M. Wang, and R.A. McCauley, *Ceram. Trans.*, **62**, 149-156 (1996).

37. V.A. Hackley, R. Premachandran, and S.G. Malghan, *Key Eng. Mater.*, **89-91**, 679-682 (1994).

38. J. Schroder, *Farbe Lack*, **11**, 957 (1991).

39. J. Texter, *Langmuir*, **8**, 291 (1992).

40. N.D. Sanders, and K.J. Roth, *Tappi J.*, **76** [11] 209-214 (1993).

41. N.D. Sanders, and K.J. Roth, *Nord. Pulp Pap. Res.*, **8** [1] 148-152 (1993).

42. J. Gron, and P. Dahlvik, *Ind. Carta*, **33** [8] 395-398, 400-402 (1995). (in Italian)

43. B.R. Midmore, R.J. Hunter, and R.W. O'Brien, *J. Colloid Interface Sci.*, **120**, 210 (1987); **122,** 521 (1988).

44. V.E. Shubin, R.J. Hunter, and R.W. O'Brien, *J. Colloid Interface Sci.*, **159**, 174-183 (1993).

45. R.W. O'Brien, W.N. Rowlands, and R.J. Hunter, "Determining Charge and Size with the AcoustoSizer," pp. 1-22 in *"Electroacoustics for Characterization of Particulates and Suspensions"* NIST Special Publication 856. Edited by S.G. Malghan. National Institute of Standards and Technology, Gaithersburg, MD, USA, 1993.

46. R.W. O'Brien, W.N. Rowlands, and R.J. Hunter, " Characterization of Ceramic Materials using Electroacoustics," in *Science, Technology and Applications of Colloidal Suspensions*. Edited by J.H. Adair, J.A. Casey, C.A. Randall, and S. Venigall. *Ceram. Trans.*, **54**, 53-66, American Ceramic Society, Westerville, Ohio, 1994.

47. V.A. Hackley, and S.G. Malghan, *Ceram. Trans.*, **56**, 283-290 (1995).

48. R.W. O'Brien, *J. Colloid Interface Sci.*, **174**, 495-504 (1995).

49. C.A. Prestidge, and W.N. Rowlands, *Submitted to Minerals Eng.*

50. R.J. Hunter, *Zeta Potential in Colloid Science*, Chap 4, Academic Press, New York & London, 1981.

51. L.D. Reed, and F.A. Morrison, *J. Colloid Interface Sci.*, **54**, 117 (1976).

52. C.F. Zukoski, and D.A. Saville, *J Colloid Interface Sci.*, **115**, 422-436 (1987).

53. B.J. Marlow, and R.L. Rowell., *J. Energy and Fuels*, **2**, 125-131 (1988).

54. S. Levine, and G. Neale, *J. Colloid Interface Sci.*, **47**, 520 (1974); **49**, 332 (1974); S. Levine, G. Neale, and J. Epstein, *J. Colloid Interface Sci.*, **57**, 424 (1976).

55. P.J. Goetz, and M.S. El-Aasser, *J. Colloid Interface Sci.*, **150**, 436-452 (1992).

56. S. Kuwabara, *J.Phys. Soc. Japan.*, **14**, 527 (1959).

57. P. Rider, and R.W. O'Brien, *J. Fluid Mech.*, **257**, 607-636 (1993).

58. R.W. O'Brien, T.A. Wade, M.L. Carasso, R.J. Hunter, W.N. Rowlands, and J.K. Beattie *Electroacoustic Determination of Droplet Size and Zeta potential in Concentrated Emulsions*. Amer. Chem. Soc. Conference Proceedings, Orlando Florida, 1996. Submitted to *J. Colloid Interface Sci.*

59. M.L. Carasso, W.N. Rowlands, and R.A. Kennedy, *J.Colloid Interface Sci.*, **174**, 405-413 (1995).

60. R.W. O'Brien, *J. Colloid Interface Sci.*, **177**, 280-281 (1996).

61. R.J. Hunter, and R.W. O'Brien, "Electroacoustic Characterization of Colloids with Unusual Particle Properties," Presented at the 4th Australia-Japan Colloid and Surface Chemistry Conference, Sorrento, Victoria (April 1996). *Colloids Surfaces A: Physicochem. Eng.* (in press).

62. R.W. O'Brien, and W.N. Rowlands, *J. Colloid Interface Sci.*, **159**, 471-476 (1993).

63. W.N. Rowlands, and R.W. O'Brien, *J. Colloid Interface Sci.*, **175**, 190-200 (1995).

64. M. Loewenberg, *J. Fluid Mech.*, **278**, 149 (1994).

65. M. Rasmusson, W.N. Rowlands, R.W. O'Brien, and R.J. Hunter, *J. Colloid Interface Sci.*, **189**, 92-100 (1997).

66. W.N. Rowlands, R.W. O'Brien, R.J. Hunter, and V. Patrick, *J. Colloid Interface Sci.*, **188**, 325-335 (1997).

67. M. Kosmulski and J.B. Rosenholm, *J. Phys. Chem.*, **100** [28] 11681-11687 (1996).

68. R.O. James, and G. Parks, in *Characterization of Aqueous Colloids by their Electrical Double Layer and Intrinsic Surface Chemical Properties*, Surface and Colloid Science, Vol. 12. Edited by E. Matijevic. Plenum, New York, 1982.

Dielectric Properties of w/o Emulsions and Particulate Suspensions

Johan Sjöblom, Bjørnar Pettersen and Tore Skodvin
Department of Chemistry, University of Bergen, Allegaten 41, N-5007 Bergen, NORWAY

1. Introduction

Lyophobic colloids like w/o emulsions and solid suspensions are kinetically stabilized which means that they, as a function of time, will approach an equilibrium state. For water-in-oil emulsion this means a two-phase system with water in equilibrium with an oil phase containing most of the surfactant. For a solid particulate suspensions a sediment will settle out and the top water phase contains only some suspended fine particles. The processes that lead to the final equilibrium state are several. Flocculation leads to strongly reduced droplet-droplet distances and to the built up of new aggregated states, i.e. flocs. These flocs can under normal gravity settle either at the bottom or at the top of the test tube. This is called sedimentation or creaming and results in a droplet gradient in the emulsion sample. Finally when the droplet-droplet distance is short enough coalescence will bring the droplets together, integrate them and accelerate phase separation (Fig. 1). The time required for the different steps in the breakdown process depends from system to system. However, the flocculation process can be very fast, as a matter of fact primary flocs with a small number of droplets can already exist during the homogenization process. Normally it is the resistance against coalescence that will determine the lifetime of the emulsion because this process is irreversible.[1-3].

In order to counter-act the processes leading to phase separation different stabilization mechanisms exist. These mechanisms vary from oil-in-water to water-in-oil emulsions with regard to components used. For oil-in-water emulsions electrostatic stabilization is a common mechanism and constitutes the background for the use of ionic surfactants in the emulsification process. It is questionable to what extent this mechanism is of importance in oil-continuous systems. In polymeric and nonionic surfactant systems steric stabilization will most likely be the reason for increased stability. It is also well known that addition of small particles will raise the emulsion stability as compared to particle-free systems. Finally, multilayer constellations of surfactants or surfactant mixtures at the o/w interface will create less favorable conditions for coalescence to take place. The multilayers are due to surfactant association patterns.[1-3]

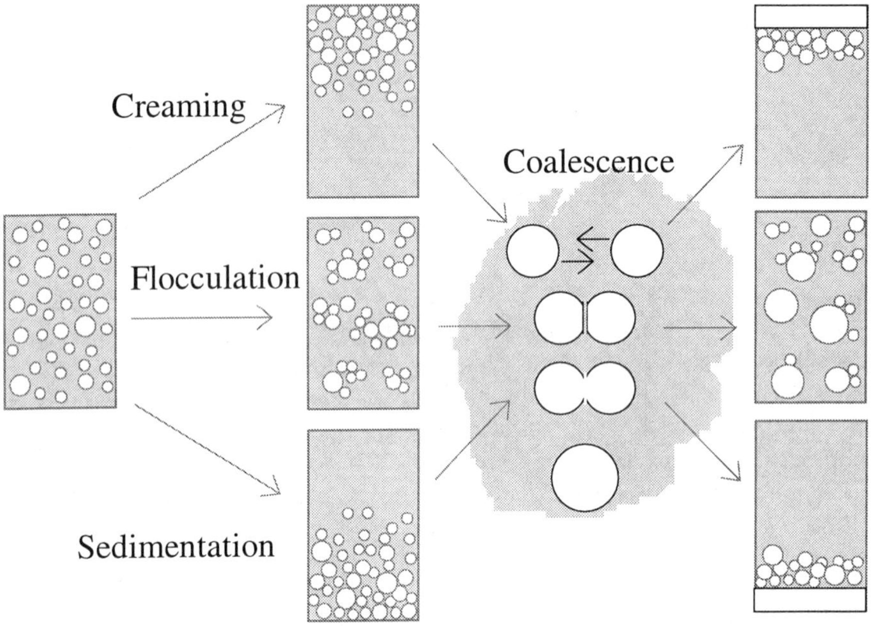

Figure 1. Destabilization processes in emulsions resulting in phase separation. In order to increase the stability of particulate suspensions electrostatically or sterically stabilized systems are normally prepared.

A challenge in process industry is to follow the stability of the dispersed or suspended state. There is a constant chase for improved techniques in order to make better predictions of the colloidal stability. In this paper we present dielectric spectroscopy as a technique to follow the breakdown of water-in-oil emulsions and to follow the sedimentation of particulate suspensions. It seems that dielectric spectroscopy combined with a multivariate design is a good approach to this problem.

2. Dielectric Spectroscopy

The permittivity ε of a substance may be defined as $\varepsilon = 1 + P/\varepsilon_0 E$, where P is an electric field (the polarization field) set up inside the substance as a response to an applied electric field E. ε_0 is the permittivity of free space. If the direction of the applied field alternates with an angular frequency equal to ω, i.e. $E = E_0 \exp(i\omega t)$, a phase lag may appear between the polarization and the field E. Thus the permittivity may be frequency dependent, and this is expressed by means of complex permittivities as

$$\varepsilon^* (\omega) = \varepsilon' (\omega) - i\varepsilon'' (\omega) \tag{1}$$

where the imaginary part ε'' is also called the loss factor.

Normally the polarization response follows an exponential law, i.e. $P(t) = P_o\left(1 - e^{-t/\tau}\right)$, where τ is the relaxation time characteristic of the system. From the definition of permittivity, the time dependence of the polarization and Equation (1) one can obtain the Debye equation

$$\varepsilon^*(\omega) = \varepsilon_\infty + \frac{\varepsilon_S - \varepsilon_\infty}{1 + i\omega\tau} \tag{2}$$

describing the dielectric behavior of a system displaying one relaxation time. Equation (2) can be separated into one real part and one imaginary part:

$$\varepsilon (\omega) = \varepsilon_\infty + \frac{\varepsilon_s - \varepsilon_\infty}{1 + \omega^2\tau^2}, \quad \varepsilon'' (\omega) = \varepsilon_\infty + \frac{(\varepsilon_s - \varepsilon_\infty)\omega\tau}{1 + \omega^2\tau^2} \tag{3}$$

In Equations (1) and (2) ε_S is the static permittivity, ε_∞ the limiting permittivity at high frequencies and τ the dielectric relaxation time. The maximum in the dielectric dispersion occurs at $\omega = 1/\tau$.

In order to extract the dielectric parameters from an experimentally obtained dielectric spectrum a number of different model functions are used. In many cases the simple Debye function (or a sum of several Debye functions) or a Cole-Cole function may be applied.[4-8]

3. Dielectric Time Domain Spectroscopy

Contrary to traditional spot frequency measurements the time domain technique makes it possible to cover a large frequency range in one single measurement. The Dielectric Time Domain Spectroscopy (TDS) is based on the study of the change in the shape of a fast-rising (≈ 25 ps) pulse propagating in a coaxial line when the sample is inserted into the line. From Fourier analysis it is known that the response pulse will contain information on the influence of the dielectric (i.e., its permittivity) over the entire frequency range up to some limiting frequency determined by the rise time of the step pulse.

Figure 2 shows the principle of a TDS spectrometer based on the total reflection method.[9] The pulse generator repetitively generates step pulses with a short rise time, < 25 ps, which propagates in a coaxial line. The pulse is reflected from the sample placed at the end of an open - ended coaxial line. In the sampling oscilloscope the influence of the sample on the shape of the reflected pulse is recorded. Finally a computer calculates the Fourier transforms and solves the reflection coefficient equations. From the reflection coefficient the permittivity as a function of frequency can be calculated and the whole spectrum presented.

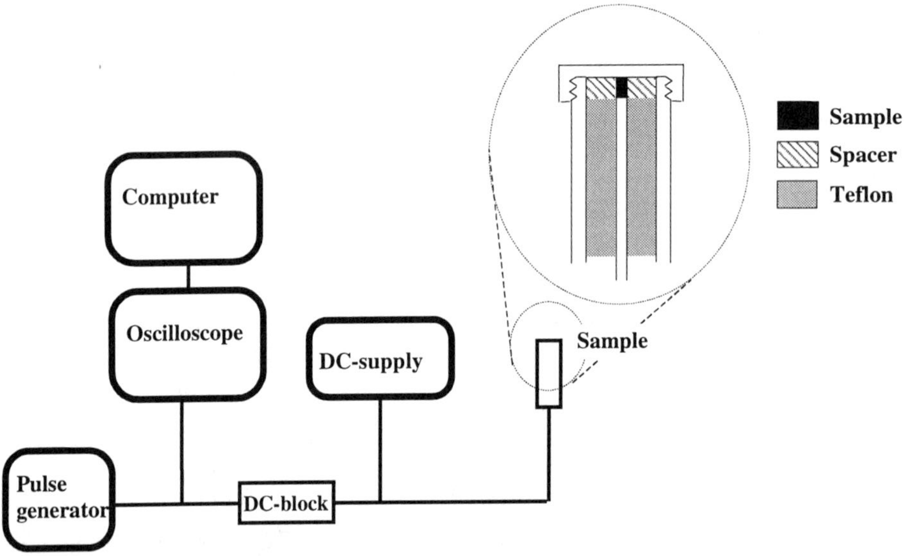

Figure 2. Experimental set-up for dielectric measurements according to the time-domain principle. For emulsion stability measurements (electrocoalescence) the external dc power supply is used.

Instead of utilizing the reflection method also transmission methods can be used. In this case the transmission coefficient equations are solved.

We have recently extended the dielectric set-up to include measurements of systems that are exposed to high external electric fields, in order to monitor the electrocoalescence mechanism of w/o emulsions. In addition to the fast-rising ac step pulse, an external electric dc field is applied through a bias tee (Picoseconds Pulse Labs, 5530A), as shown in Figure 2. The bias tee includes a dc block which allows the application of a voltage over the sample cell while it passes the fast-rising step pulse used to measure the permittivity without appreciably degrading the pulse.

4. Dielectric Models of Heterogeneous Systems

To describe heterogeneous systems like w/o emulsions and particulate suspensions by means of dielectric models has been a classic problem. Already Maxwell discussed this topic during the last century.[10] Most of the models presented are matrix-inclusion models.[11] In this presentation we will focus on emulsions, where the emulsion droplets are the inclusion. Wagner described in 1914 heterogeneous systems with low concentration of the dispersed phase in form of spherical droplets.[12] Sillars and Fricke extended the Wagner approach to involve oriented and randomly oriented ellipsoidal inclusions.[13,14] The Sillars equation for ellipsoidal particles parallel with the K-axis reads:

$$\varepsilon^* = \varepsilon_1^* + \frac{\varepsilon_1^* (1-\Phi)(1-A_K) + \varepsilon_2^* [\Phi + A_K(1-\Phi)]}{\varepsilon_1^* + A_K(1-\Phi)(\varepsilon_2^* - \varepsilon_1^*)} \tag{4}$$

where ε_1^* and ε_2^* are the complex permittivities of the continuous and dispersed phase, respectively. Φ is the volume fraction dispersed phase, while A_K is the depolarization factor along the K-axis. In the case of spherical inclusions A_K = 1/3, otherwise this factor is a function of the axial ratio in the aggregate, normally between the major and minor axis. For oblate spheroids $a/b < 1$ and $A_K < 1/3$, while for needle-shaped spheroids $A_K > 1/3$.

The dielectric dispersion predicted from Equation (4) is the so-called Maxwell-Wagner-Sillars relaxation, or the interfacial polarization. The corresponding relaxation time is given by

$$\tau = \varepsilon_0 \frac{\varepsilon_1 + A_K(1-\Phi)(\varepsilon_2 - \varepsilon_1)}{\sigma_1 + A_K(1-\Phi)(\sigma_2 - \sigma_1)} \tag{5}$$

where ε_1 and ε_2 are the static permittivities of the continuous and dispersed phase, respectively. σ_1 and σ_2 are the corresponding dc conductivities. The mixture formula has also been extended to high volumes of the dispersed phase. Bruggeman introduced an asymmetric integration procedure,[15] which was extended to complex permittivities by Hanai.[16,17] The Bruggeman formula reads:

$$\varepsilon^* = \varepsilon_1^* + \frac{\varepsilon_1^* (\varepsilon_2^* - \varepsilon_1^*)\Phi}{\varepsilon_1^* + A_a(\varepsilon_2^* - \varepsilon_1^*)} \tag{6}$$

Starting from Equation (6), Boyle[18] has derived the formula

$$\left[\frac{\varepsilon^* - \varepsilon_2^*}{\varepsilon_1^* - \varepsilon_2^*}\right]\left[\frac{\varepsilon_2^*}{\varepsilon^*}\right]^{A_a} = 1 - \Phi \tag{7}$$

which is reduced to the classical Hanai equation for spheres, if $A_a = 1/3$. The Boyle equation is originally derived for parallel oriented spheroids. Boned and Peyrelasse[19] have derived an equation for randomly oriented spheroids, i.e.

$$\left[\frac{\varepsilon_2^*}{\varepsilon^*}\right]^{3d}\left[\frac{\varepsilon_2^* - \varepsilon^*}{\varepsilon_2^* - \varepsilon_1^*}\right]\left[\frac{\varepsilon_1^*(1-3A) + \varepsilon_2^*(2-3A)}{\varepsilon^*(1+3A) + \varepsilon_2^*(2-3A)}\right]^{3K} = 1 - \Phi \tag{8}$$

where $A = A_b = A_c$ and $A_a = 1 - 2A$. K and d are both functions of A. The Boned and Peyrelasse equation in the case of spherical aggregates simplifies to the Hanai equation.

5. Dielectric Properties of w/o Emulsions

We will restrict the presentation to involve oil-continuous emulsions stabilized by means of nonionic surfactants. The reason for excluding water-continuous systems is that when these o/w emulsions are stabilized by means of ionic surfactants (or surfactant mixtures) the electric double layer effects are very large. In order to clarify the origin of the dielectric dispersion lower frequencies than accessible for TDS measurements are normally necessary. For the TDS technique, water-in-oil emulsions stabilized by means of nonionic surfactants are very good model systems. Table I gives a simple experiment visualizing the effect of stirring on the dielectric properties of emulsified systems.[20] It is seen that the difference in the static permittivity between the emulsions during stirring and at rest can be as high as 50 %. In this case the volume fraction of water is constant, so the only difference between the measurements is in the state of the

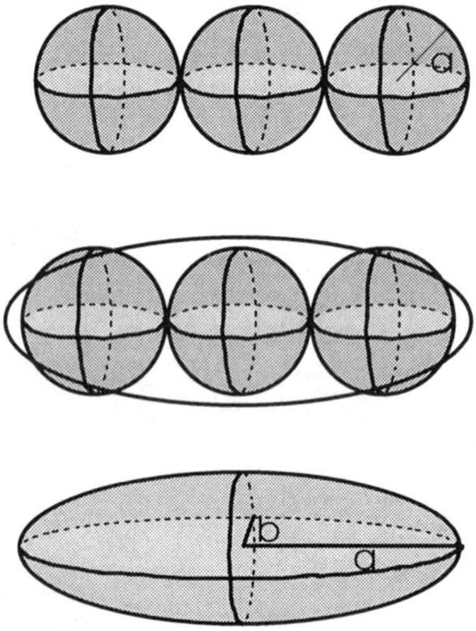

Figure 3. Linear floc treated as a spheroid (schematic).

emulsions. Visually no phase separation could be detected, so the difference must be in the flocculation/sedimentation. In order to explore this we undertook a systematic study of water-in-crude oil emulsions, containing different amounts of water (Table II).[21]

Table I . Dielectric parameters for a w/o emulsion containing 50 vol % water, stabilised by 1% Berol 26. The aqueous phase contains 5 wt % NaCl. The measurements have been performed both during emulsification (using different methods) and at rest.

		ε_S		ε_∞	
Device	Rotation speed	During mixing	At rest	During mixing	At rest
Laboratory Mixer	Max. (≈ 2500 rpm)	15.6	24.2	9.4	11.0
	Min. (≈ 500 rpm)	16.9	20.9	10.4	11.3
Vortex Mixer		15.4	23.9	11.0	12.9

Table II. Dielectric parameters obtained experimentally for different crude oil emulsions, together with the corresponding shape factors found from comparison with theoretical parameters.

Experimental values				Oblate spheroids (0 < A < 1/3)			Prolate spheroids (1/3 < A < 1/2)		
ϕ	ε_s	ε_∞	τ(ns)	$A^{\varepsilon s}$	$A^{\varepsilon\infty}$	A^{τ}	$A^{\varepsilon s}$	$A^{\varepsilon\infty}$	A^{τ}
0.17	17.5	6.5	0.8	0.06	0.06	0.06	0.48	0.49	0.48
0.23	22.4	7.4	1.0	0.08	0.09	0.07	0.48	0.48	0.48
0.28	28.2	9.6	1.0	0.09	0.09	0.08	0.47	0.48	0.47
0.38	35.7	13.2	1.3	0.12	0.12	0.10	0.46	0.47	0.47
0.44	46.0	13.5	1.2	0.14	0.14	0.14	0.46	0.45	0.46
0.50	60.0	14.0	1.3	0.15	0.16	0.13	0.45	0.43	0.46

It is easy to show that dielectric models based upon spherical, noninteracting droplets cannot predict the permittivities found. In order to obtain agreement between measured values and predicted ones a shape factor A must be introduced. The physical meaning of a shape factor indicating a geometrical asymmetry cannot be related to the individual droplets but to the aggregated droplet states, i.e., the flocs. The easiest way to interpret A is to consider a linear flocculation (see Fig. 3) and to define an axial ratio between the major and minor axis. When doing so, the corresponding axial ratios are in the range 1:3 to 1:10.[21] However, a small linear flocculation is not entirely consistent with the complex structures a flocculated system can build up. Other more complex flocculation models have also been developed.[22]

5.1 Electrocoalescence and Dielectric Spectroscopy

The pioneering impact in the field of high electric fields and emulsified systems was done by Cottrell, who also had the first patents as well as the first electrical coalescers in an oil field.[23,24] In recent years electrocoalescence has been combined with a lot of experimental techniques.[25-31]

Dielectric spectroscopy has proved to be an appropriate experimental technique for investigating w/o emulsions. The conducting water droplets dispersed in oil become polarized in an electric field, and the polarization will show a frequency dependence as illustrated in Figure 4. The time domain dielectric spectroscopy technique is fast, and hence suitable for investigations of systems that change with time, such as w/o emulsions. In this way the effect of the electric field on the dielectric response of the emulsion can be monitored. The instrumentation is presented in Figure 2. The emulsions studied include w/o model emulsions stabilized by nonionic surfactants as well as interfacially active crude oil compounds and true water-in-crude oil systems.

In a strong electric field the aqueous droplets in a w/o emulsion become polarized as oppositely charged ions separate due to the field. These polarized droplets tend to form chains oriented parallel to the external field. The local electric field across the film separating the droplets may reach a considerable strength and, provided this local field is sufficiently high, induce a rupture of the film. The study in Ref. (32) showed that at low external field strengths, the applied field would not induce any coalescence. Hence, when the electric field is switched off the system returns to a random distribution of the aqueous droplets, and the system shows a reversibility of the dielectric parameters. However, if the external electric field exceeds a critical value, Ecr, the membrane protecting the aqueous droplets ruptures and a coalescence of droplets takes place.

In a real process where the emulsions are causing problems, like in the exploitation or separation stage in offshore petroleum production, a crucial point is what operational parameters can be adjusted without too large modifications of the whole process. In order to investigate the parameters responsible for the stability/ instability of water-in-crude oil emulsions it is important to understand the chemistry and behavior of the different constituents on crude oils, especially the interfacially active ones, as well as creating reliable models for crude oil emulsions.

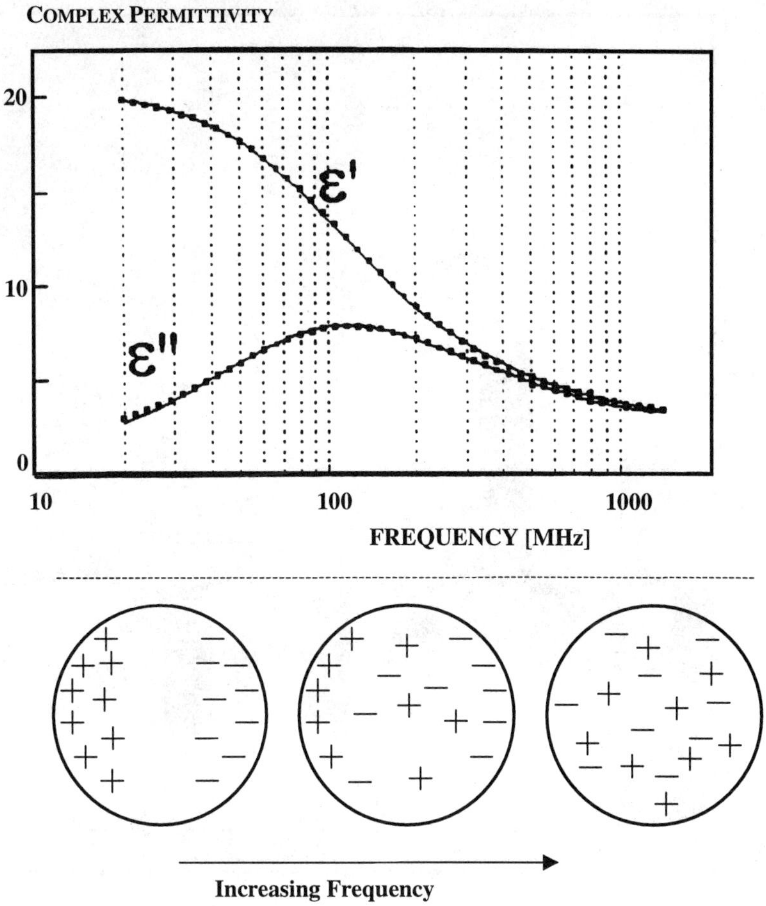

Figure 4. Dielectric spectrum of w/o emulsion stabilized by means of indigenous crude oil components. The inserted aqueous droplets illustrate the relaxation process at different frequencies.

In Bergen we have during the past years developed procedures to separate and isolate interfacially active components from crude oils. The basic idea behind the separation of the interfacially active fractions is to simulate interfacial conditions of relevance for true crude oil systems. In order to undertake more detailed studies of the stabilizing film, we have separated the components responsible for the emulsion stability into asphaltenes and resins[33] and undertaken chemical characterizations of these fractions.[34-39]

In order to visualize the importance of different parameters on the stability of model emulsions and crude oil emulsions, we undertook reduced factorial designs on these systems.[40-42] The two studies show that for a typical model

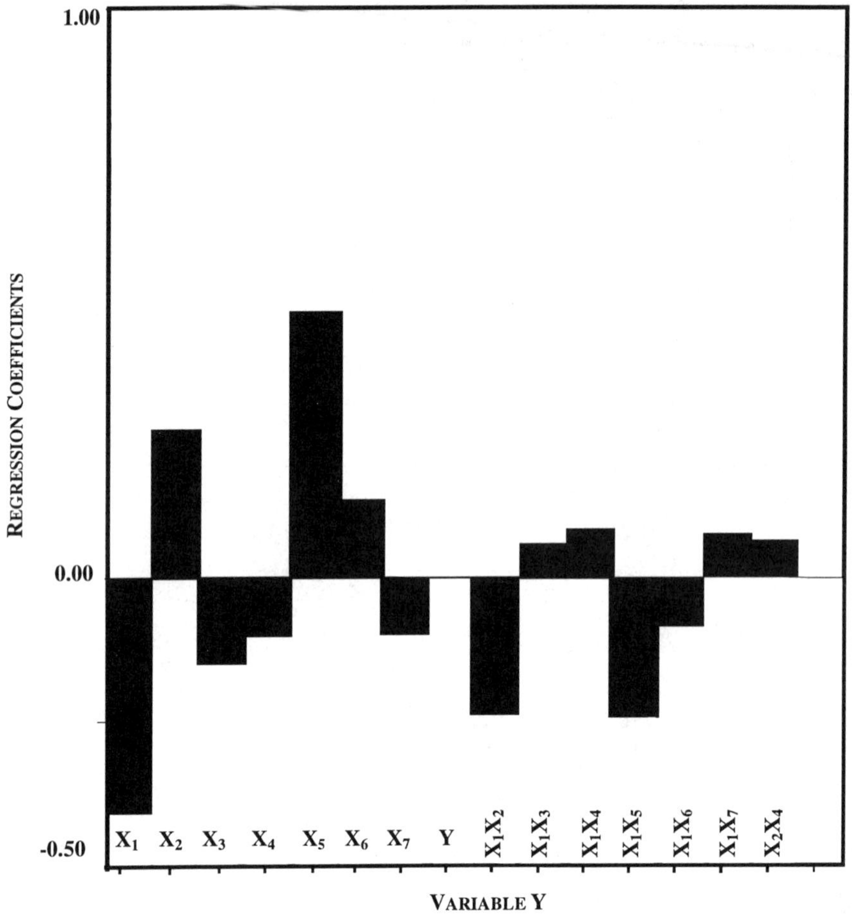

Variables		Levels -	Levels +	Centerpoints
Disperse phase (wt%)	X_1	40	80	60
Salinity (wt% NaCl)	X_2	1	4	2.5
Temperature (°C)	X_3	4	23	13.5
Time (h)	X_4	0.25	24	12
Surfactant (wt%)	X_5	2	4	3
Ratio, surfactant (S:T)	X_6	0.25	0.75	0.5
Laponite (wt% of ϕ_2)	X_7	0.5	2.0	1.25

Note. Ratio of surfactant (S:T) is the ratio between Span 80 and Triton N-42

Figure 5. Definition of variables and their levels together with regression coefficients for electrocoalescence of model nonionic w/o emulsions based on reduced factorial design.

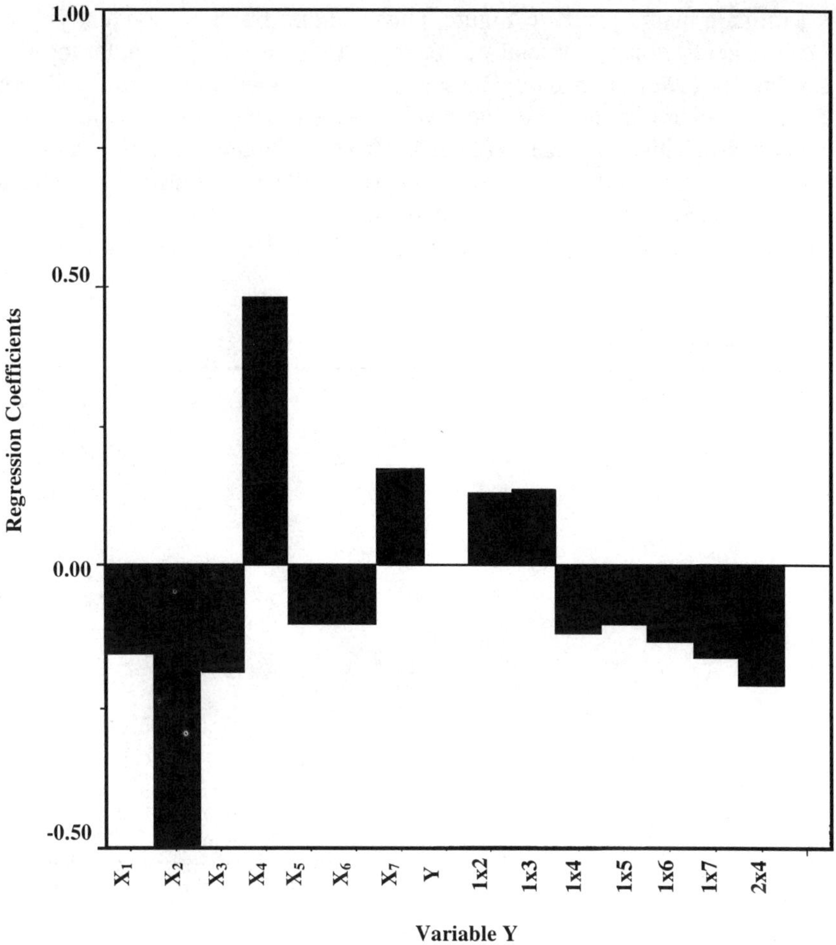

Variables		Levels -	Levels +	Centerpoints
Flocculation solvent	X_1	Pentane	Isooctane	50:50 mixture
Continuous phase	X_2	Decane	Toluene	50:50 mixture
Disperse phase (wt%)	X_3	30	60	45
Asphaltenes (wt%)	X_4	1	5	3
Salinity (wt%)	X_5	1	4	2.5
Temperature (°C)	X_6	4	23	13.5
Emulsification time (min.)	X_7	0.5	3.0	1.75

Figure 6. Definition of variables and their levels together with regression coefficients for electrocoalescence of model crude oil emulsions emulsions based on reduced factorial design.

system with parameters given in Figure 5 the volume fraction, conductivity and surfactant concentration are crucial parameters. Also, cross terms including these variables are essential. For model crude oil systems (Fig. 6) essentially two parameters are of importance, i.e. the level of asphaltenes and the nature of the continuous phase. This is further explored in the next figure where the stability is plotted vs. concentration of toluene in the oil phase. The reason for this behavior is that the asphaltenes exist as small stabilizing particles in paraffinic hydrocarbons and are dissolved in toluene. Hence they loose their stabilizing role (see Fig. 7).

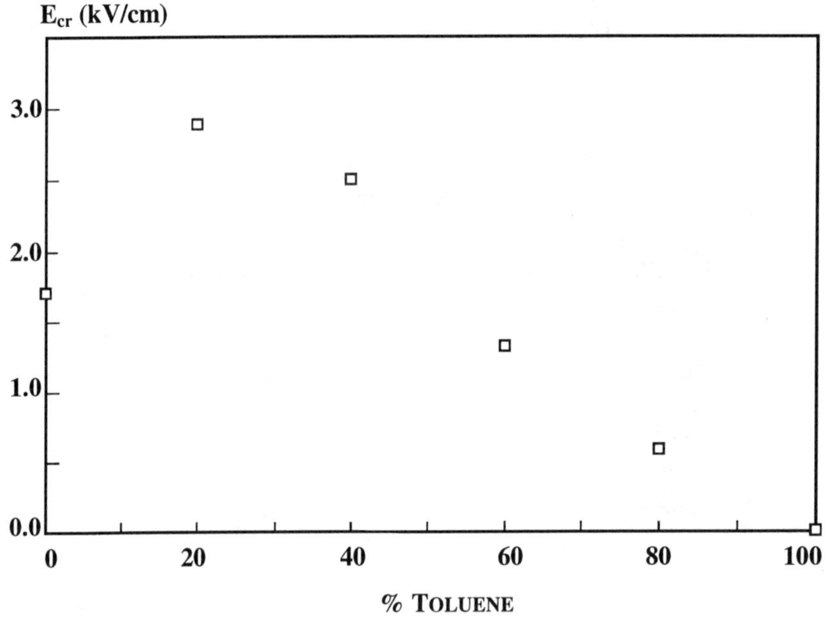

Figure 7. Emulsion stability (expressed as E_{cr}) in model systems consisting of brine and decane/toluene with varying contents of toluene. Extracted asphaltenes (2 %) are used as stabilizers.

5.2 Percolation Phenomena in w/o Emulsions in High Electric Fields

Figure 8 shows the static permittivity of an asphaltene stabilized model emulsion vs. the applied external electric field.[43] The static permittivity increases with the applied electric field. First, rather slowly, then more rapidly as the critical electric field is approached. However, when the critical electric field is exceeded, a steep decline in the static permittivity to a value of 5-7, is observed. This can be viewed upon as a percolation phenomenon according to a static model.[44,45]

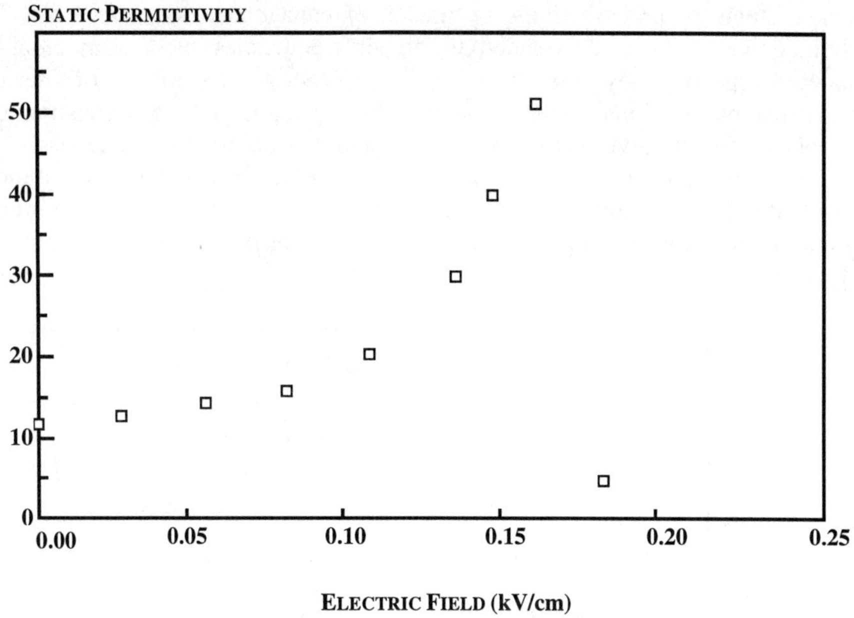

Figure 8. Percolation behaviour in model crude oil emulsions stabilized by extracted asphaltenes as a function of the external electric field.

From studies of the percolation phenomena in w/o microemulsions it has been introduced that permittivity follows a scaling law with regard to temperature.[53] However, in our emulsion system we induce the percolation effect by applying a high electric field, i.e. the following scaling law may apply:

$$\varepsilon_s \sim (E_{cr} - E)^{-s} \tag{9}$$

when E is lower than the electric field where the system collapses, Ecr. From linear regression analysis, the critical exponent s can be estimated from the slope. Our data give a slope of 0.52, which is close to the theoretical value expected from a static percolation model (s=0.6), containing bicontinuous oil and water structures.[44]

5.3 Dielectric Spectroscopy on Technical Emulsions

Literature reports on a variety of applications of dielectric measurements in different types of technical processes. The classical application is to determine water contents in process fluids by means of capacitance measurements. This technique has also been extended to higher frequencies by Wasan et al.[46,47] Dielectric spectroscopy has been used to follow reactions, for instance polymerization reactions, in organic media. The group in Bergen has also applied the technique to involve margarine systems and models for food emulsions.[48] In the following we present a technically very important problem that combines a controlled reaction inside a w/o emulsion and dielectric spectroscopy as a process on-line instrumentation. This problem concerns the formation and transport of gas hydrates in pipelines.

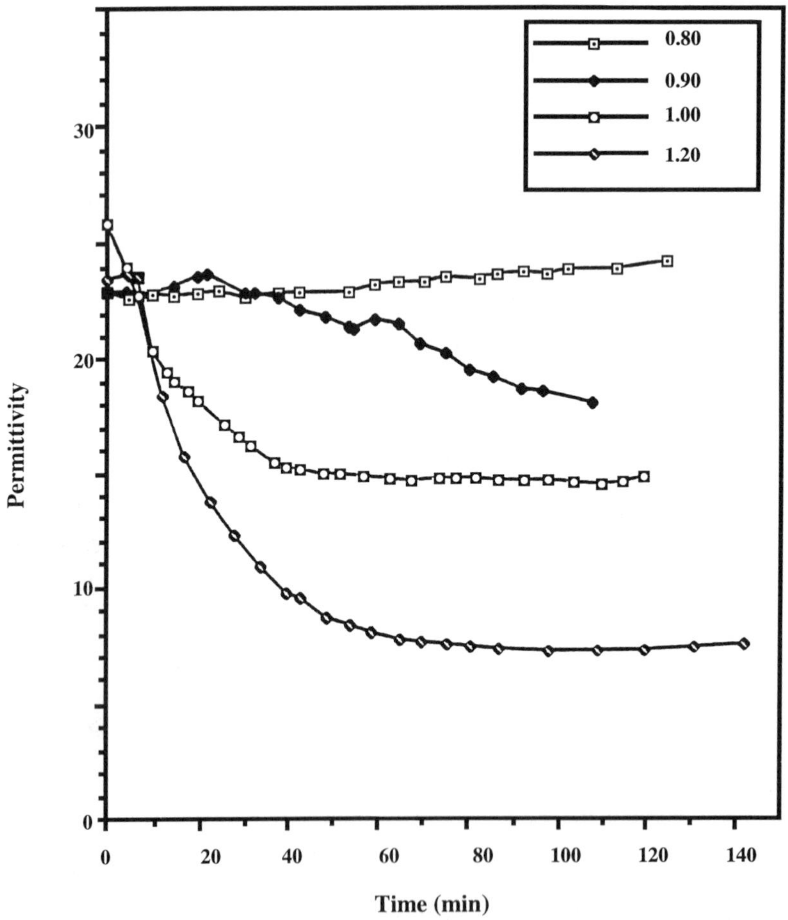

Figure 9. Static permittivity (ε_s) as a function of time for chlathrate hydrate formation in model w/o emulsions. The relative concentrations of CCl_3F:H_20 were 0.80, 0.90, 1.00 and 1.20, respectively.

Gas hydrates (clathrates) are a technically alternative form of ice, which can entrap relatively large volumes of gas within cavities in the ice crystal matrix. The entrapped guest molecules (gas) stabilize the structures by means of van der Waals interactions, and combinations of the different unit cells give rise to structures I, II [49-51] and H.[52] The most common gas to form gas hydrate is methane, but ethane, propane, butane, carbon dioxide, nitrogen and many other types of gases also give rise to gas hydrates.

When gas hydrate formation is taking place in a w/o emulsion, the emulsified water is converted to clathrate structures and decreases the volume fraction of free water in the emulsion droplets. The formation of new structures will change the dielectric properties, which can be used to follow the gas hydrate formation. See Figure 9[53,54] where CCl_3F is used as the gas hydrate forming species in a nonionic w/o emulsion.

The relaxation modes in these studies are due to the Maxwell-Wagner-Sillars effect. Electrolyte is not incorporated in the clathrate structures, and this results in an increased electrolyte concentration in the remaining free water, which changes the dielectric relaxation mode. The relaxation time (τ) decreases from about 1000 ± 100 ps to a final level of 200 ± 20 ps. The experimental value of 200 ps corresponds to a 3 % (weight/volume) NaCl solution,

CCl_3F is soluble in the model oil and has a low solubility in the aqueous phase as seen from the low concentration of CCl_3F inside the emulsion droplet. The surfactant membrane acts as the diffusion barrier and this barrier depends of the hydrophilic/hydrophobic properties of the surfactant, the concentration of the surfactant, the hydrophobic property of the oil phase and the hydrophilic property of the aqueous phase. The CCl_3F concentration in the water phase is very low after the autocatalytic reaction has started, and it is obvious that the diffusion through the interfacial membrane appears to be the rate limiting step in the overall process.[53,54]

6. Sedimentation in Particulate Systems as Followed by Means of Dielectric Spectroscopy

The experimental set up for the sedimentation experiments is shown in Figure10 where the top and bottom sensors give identical values during mixing. When the mixing is stopped we follow the sedimentation by means of the bottom sensor.[55] In order to verify the appropriateness of the TDS method for sedimentation studies, we started with two model systems, i.e. SiO_2 and Al_2O_3. Figure11 shows the permittivity spectra of alumina in water. The most significant change is in static permittivity, which is the reason why we have concentrated on this parameter in the following figures. Figure 12 (a) shows the variation in static permittivity as a function the volume fraction of silica. The dependence follows a typical model for heterogeneous systems in this case the Boned-Peyrelasse simplified to the Boyle or Hanai equation. Figure 12 (b) gives the corresponding data for Al_2O_3 and similar conclusions can be drawn based on these data.[55]

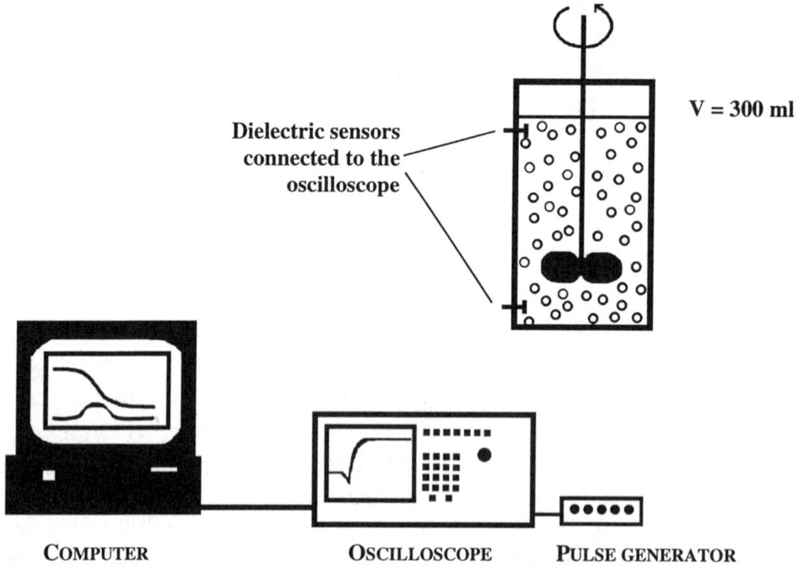

Figure 10. Experimental set-up for dielectric measurements of sedimentation rates in particulate suspensions.

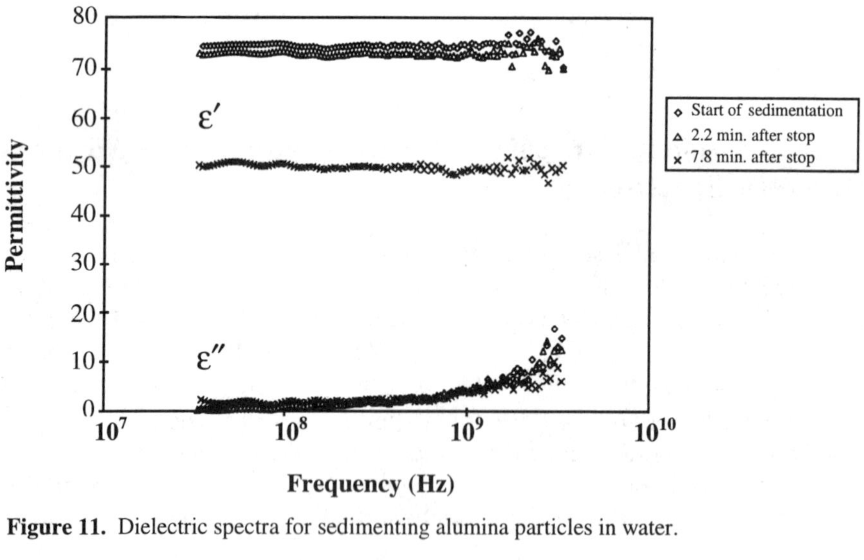

Figure 11. Dielectric spectra for sedimenting alumina particles in water.

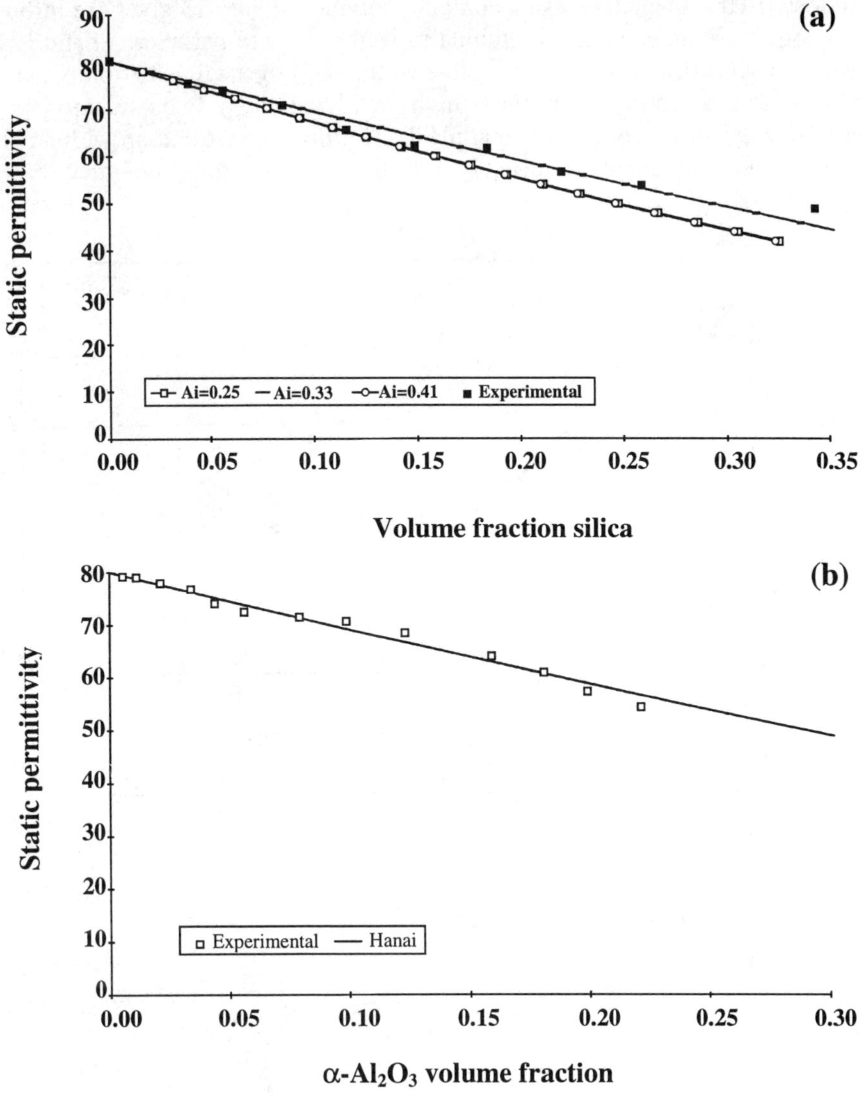

Figure 12. Static permittivity for particle suspensions in water. The solid line is a fit according to the Boned-Peyrelasse model: (a) silica particles and (b) alumina particles.

In order to alter the sedimentation rate we have surface modified the solid particles SiO_2 and Al_2O_3. The surface modification includes surfactants, and polymers. We have used the nonionic surfactant C_9PhEO_6 and ethyl hydroxoethyl cellulose (EHEC) which is also a nonionic polymer. Figure 13 gives the influence of pH on the sedimentation of alumina in water for pure particles (a) and EHEC coated particles (b). At pH 8.5 (i.e. close to the IEP) the sedimentation is fast and gives rise to a porous sediment (i.e. a high final level of ε). In contrast to this one can follow the slow sedimentation at pH 3.5. In this case we end up with a denser sediment and consequently a lower ε. For the coated particles the effect of pH is rather small.[57]

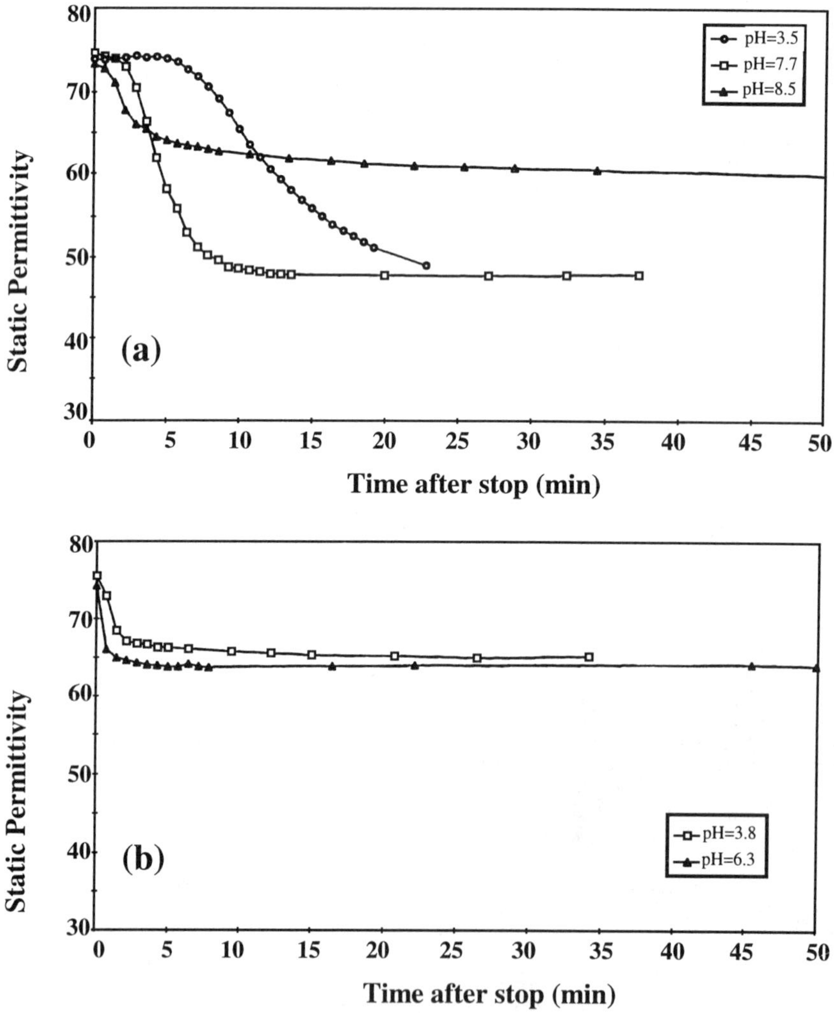

Figure 13. Change in static permittivity due to sedimentation at different pH's: (a) alumina in water and (b) EHEC-coated alumina in water.

In the following we focus on SiO_2 that we have synthesized according to the sol-gel technique. For this purpose we undertook a multivariate screening design with the variables defined in Ref. (56). For our purposes we have chosen a particle size of 1020 nm.

In order to visualize the parameters responsible for the sedimentation of precipitated SiO_2 we undertook a multivariate analysis where we defined surface and bulk properties responsible for stabilization of the silica particles. Figure 14 shows our response parameter, i.e. the sedimentation rate expressed as the derivative of permittivity vs. time. Figure 15 defines the variables and their high and low levels together with the regression coefficients for the variables and their cross-terms.

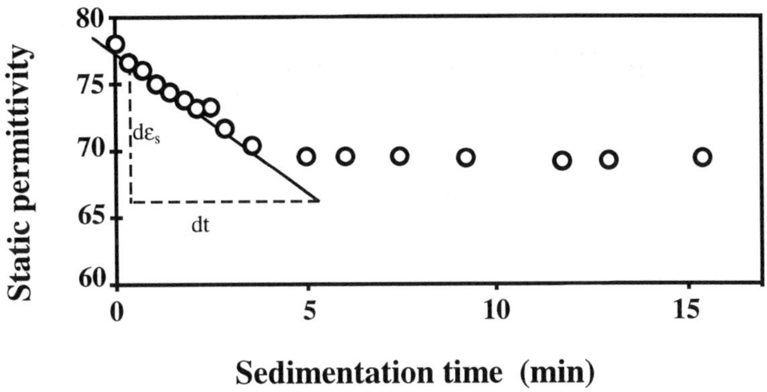

Figure 14. Static permittivity *vs.* time as response variable in the reduced factorial design for the sedimentation tests.

Since this investigation involves monodisperse particles with equal sizes and densities, the different sedimentation rates will mainly reflect changes in the intermolecular forces between the particles. The following was found for the variables we investigated:[58]

Volume fraction disperse phase (variable 1)

The value of $d\varepsilon_S/dt$ increases as the volume fraction of solid particles increases. Due to volumetric reasons the particle-particle distance will be lower at higher ϕ. This can result in for instance a higher degree of flocculation, which will accelerate the sedimentation.

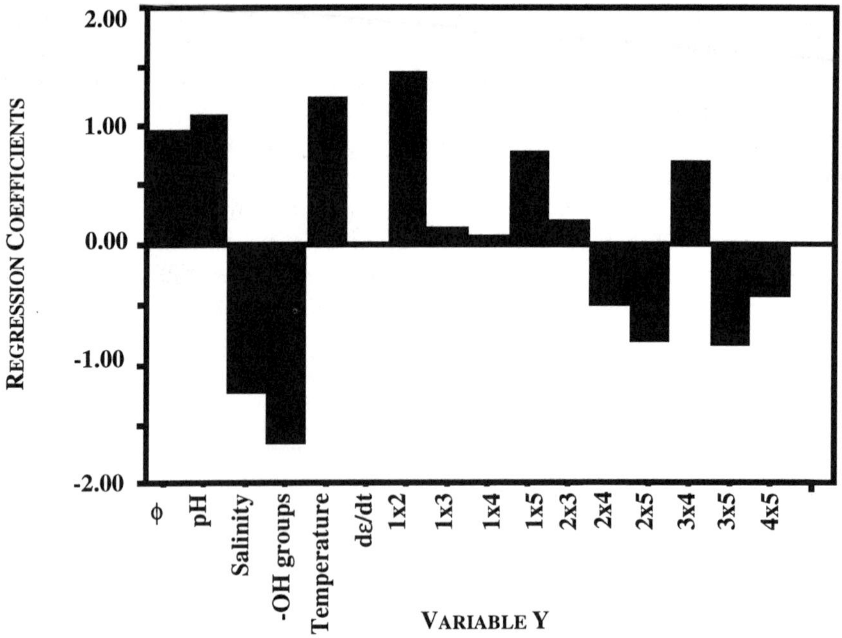

Variables		Levels -	+	Centerpoints
Volume fraction disperse phase	X_1	0.02	0.06	0.04
pH	X_2	3	11	7
Dehydroxylation temperature (°C)*	X_3	800	120	400
Salinity (NaCl)	X_4	0.001	0.1	0.01
Temperature (°C)	X_5	4	25	15

*Reflects the amount of –OH groups at the surface of the mica particles.
X_3 = 120 gives a high amount of –OH groups at the surface. X_3 = 800 gives a low amount of –OH groups.

Figure 15. Definition of variables and their levels together with regression coefficients for sedimentation of precipitated silica based on reduced factorial design.

pH (variable 2)

A higher pH tends to increase $d\varepsilon_S/dt$, indicating a less stable suspension against sedimentation. This is not in agreement with the classical DLVO theory concerning zeta potential and stability. Silica has a high negative zeta potential at high pHs and a low value at low pHs. One would therefore expect an increase in stability at high pH values. Allen and Matijevic[60] found that Ludox™ HS precipitated silica does not coagulate in acidic pH where the electrophoretic mobility is minimum. Wattilon and Gèrard[61] attributed the exceptional stability of silica suspension to the presence of immobilized surface water layer. The first evidence of this hydration layer was given by Derjaguin and Zorin[62] and several authors[63,64] have given further evidence of the existence of this hydration force. Hence the pH dependence is not entirely correlated with an electrostatic stabilization of the silica particles.

Level of salinity, [NaCl] (variable 3)

A higher level of salinity in the continuous phase decreases the value of $d\varepsilon_S/dt$. For electrostatically stabilized systems one would expect that an increase in salinity would decrease the stability of the suspensions. Iler[65] stated that for dilute silica sols at pH 2, where there is little ionic charge on the particles, no coagulation is observed when electrolyte is added presumably because of the hydration layer. The mechanism of coagulation with salts has been investigated for half a century and the mechanism is not yet fully understood. Dumont[66] calculated the Hamaker constant for a series of oxide particles based on an approximate expression derived by Israelachvili[67] (Eq.10).

$$A_{13} = \frac{3}{4} kT \left[\frac{\varepsilon_1 - \varepsilon_3}{\varepsilon_1 + \varepsilon_3} \right]^2 + \frac{3h\nu_e}{16\sqrt{2}} \frac{\left(n_1^2 - n_3^2\right)^2}{\left(n_1^2 + n_3^2\right)^{3/2}} \qquad (10)$$

where A_{13} is the Hamaker constant for particles (medium 1) dispersed in medium 3. ε_1 and ε_3 are the static dielectric constants for the media, n_1 and n_3 are their refractive indexes in the visible range. ν_e is the adsorption frequency of the media, assumed to be the same for both of them. k is the Boltzmann constant, T is the temperature, and h is Planck's constant. In this expression one can identify a static term (depending on the dielectric constant) and a dispersive term (depending on the refractive index and the frequency.) For SiO_2 the total Hamaker is of the order 0.65x 10^{-20} J while TiO_2 has a value of 23x 10^{-20} J. As a consequence the attraction between two SiO_2 particles is about 35 times smaller than the similar attraction between TiO_2 particles. Due to the small value of the attraction force the total interaction energy between two SiO_2 particles will be depending on structural forces like hydration forces to a higher extent than for instance the TiO_2 system. For SiO_2 suspended in water an electrolyte addition will mainly alter the solvent properties (dielectric constant and refractive index). For the Hamaker constant this will result in a substantial decrease in the static part and a smaller decrease in the dispersive part. All- in-all the electrolyte addition will decrease the attraction energy and hence stabilize the system.

Qualitatively our findings are in accordance with these literature results. It should also be noted that we are limited in the concentration of electrolyte due to the experimental setup. Using a higher salinity will cause problems in resolving the dielectric spectra.

Level of OH-groups, (dehydroxylation temperature, variable 4)

A higher level of this variable tends to reduce the value of $d\varepsilon_S/dt$, i.e. increase the stability against sedimentation. This is explained on the basis of the increased hydrophilicity of the silica particles with increasing amount of OH-groups at the solid surface.[68] Studies of dehydroxylation and OH coverage of silica particles have been conducted by different authors, reporting that the coverage of OH-groups is almost independent of silica type and dependent solely on the dehydroxylation temperature.[70-72] The degree of rehydroxylation is small for silica dehydroxylated at 800°C when the particles are dispersed in water at ambient temperature and below. The amount of OH groups will dictate the level of hydration of the particles, and the strength of the hydration forces which is the source of the deviation from classical DLVO theory. Hence our result that an increase in the amount of OH groups will stabilize the system is logical.

Level of T, temperature (variable 5)

A higher level of temperature in the suspensions increases $d\varepsilon_S/dt$. An increase in T will most likely influence the diffusion properties (Brownian motions) of the particles and the viscosity of water. A higher Brownian motion will increase the collision frequency between the particles and enhance the possibility of a floc formation. Especially so when the repulsive hydration force between the SiO_2 colloids will decline as a function of temperature. The viscosity of water at 5°C is slightly higher than at 25°C and may thus contribute to the increased value of $d\varepsilon_S/dt$ at high temperature.

Our final experiments deal with magnetic particles.[59] For these tests we have utilized the experimental design presented in Figure16. The particles are monodisperse polystyrene particles (2.8 - 4.5 ηm) containing 25 % magnetic oxides. When these particles are introduced to a magnetic field an induced flocculation will take place (Fig. 17). The next figure (18) shows how the static permittivity changes with time (due to sedimentation) and with applied magnetic field. The strength of the magnetic field will dictate the degree of flocculation. For single particle sedimentation with a linear decrease in as a function of concentration, one would expect proportionality between the derivatives of

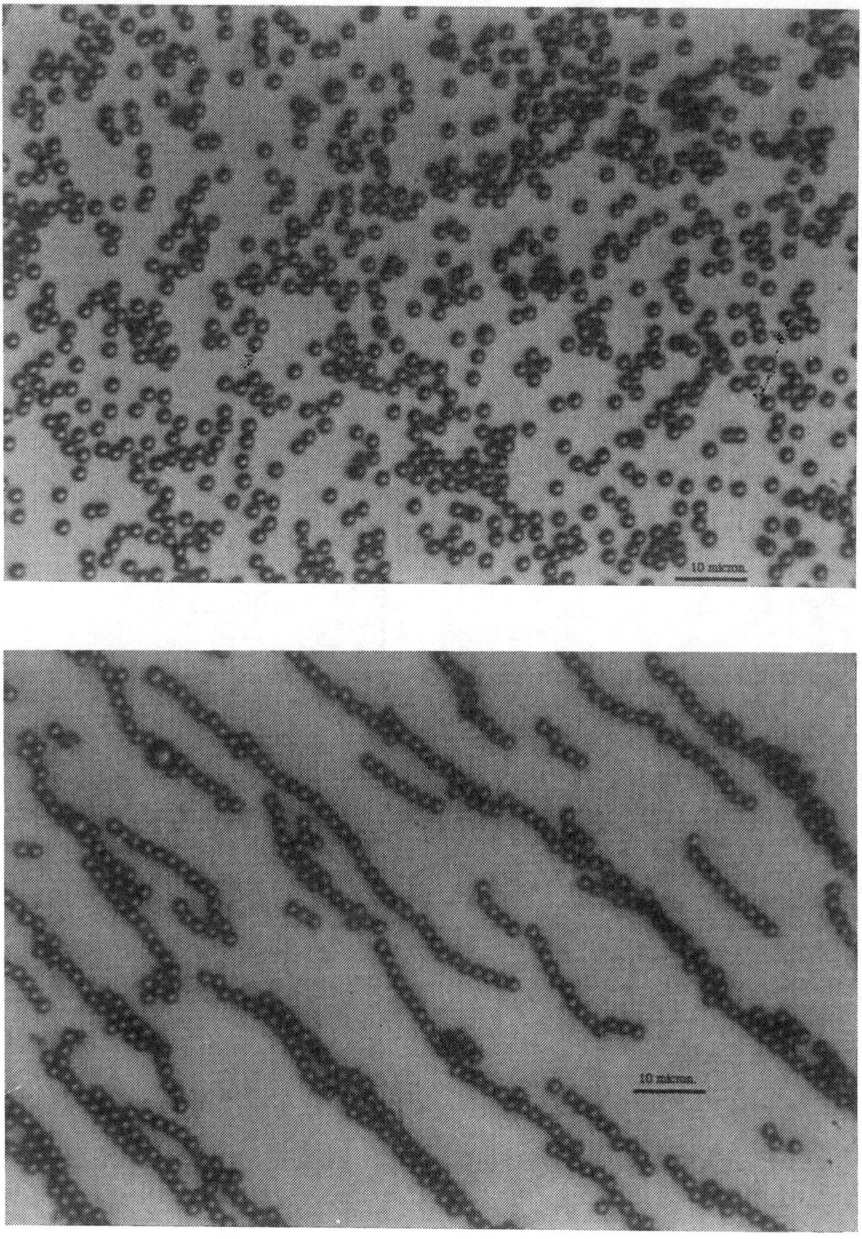

Figure 17. (Top) Polystyrene based magnetic monodisperse colloids. The particles are 2.8μm. (Bottom) Induced flocculation due to an external magnetic field.

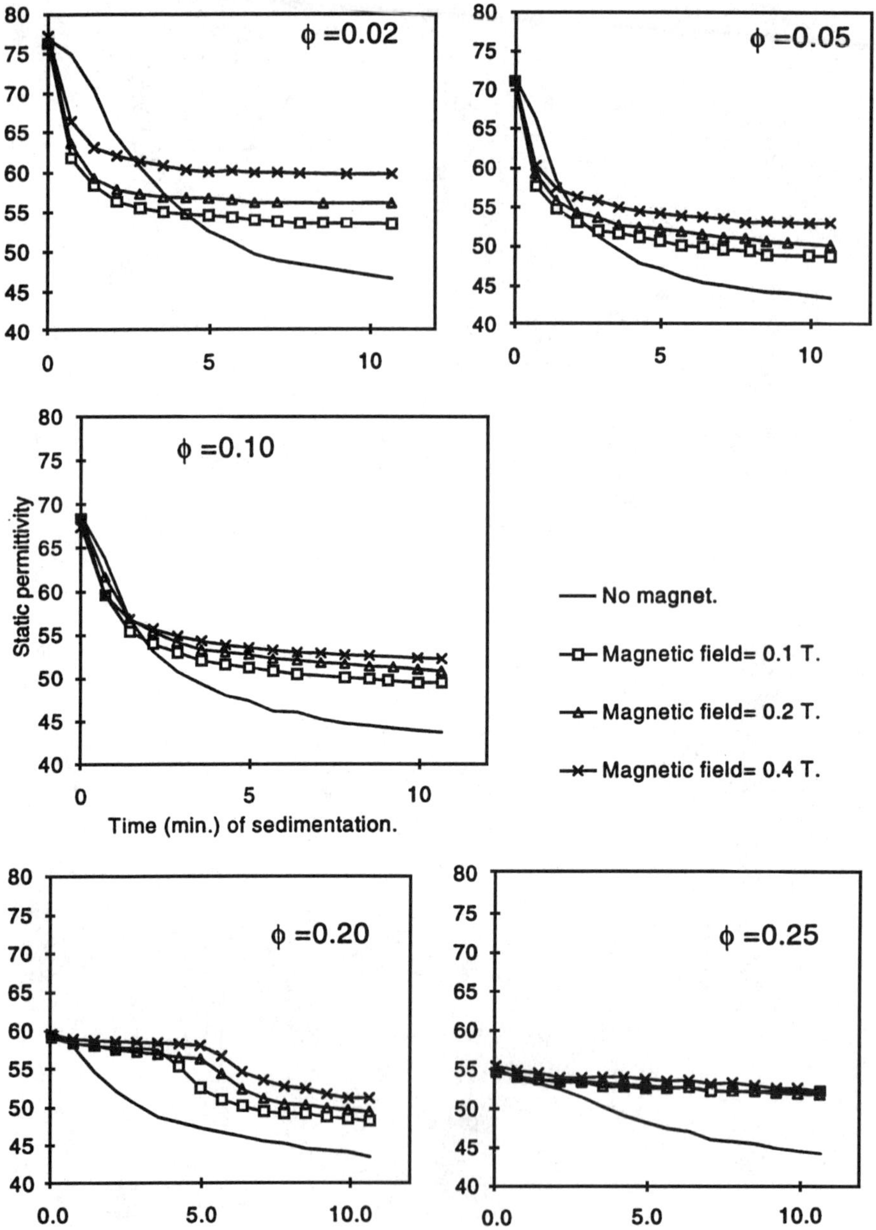

Figure 18. Static permittivity for sedimenting magnetic particles at different magnetic fields and volume fractions.

volume fraction and permittivity vs. time. From Figure 18 we can observe that this is more or less always the case for sedimentation without a magnetic field, at least up to a time period between 0 and maximally 5 minutes. In the presence of an external field this period is substantially shorter for low volume fractions and substantially extended for high fractions.

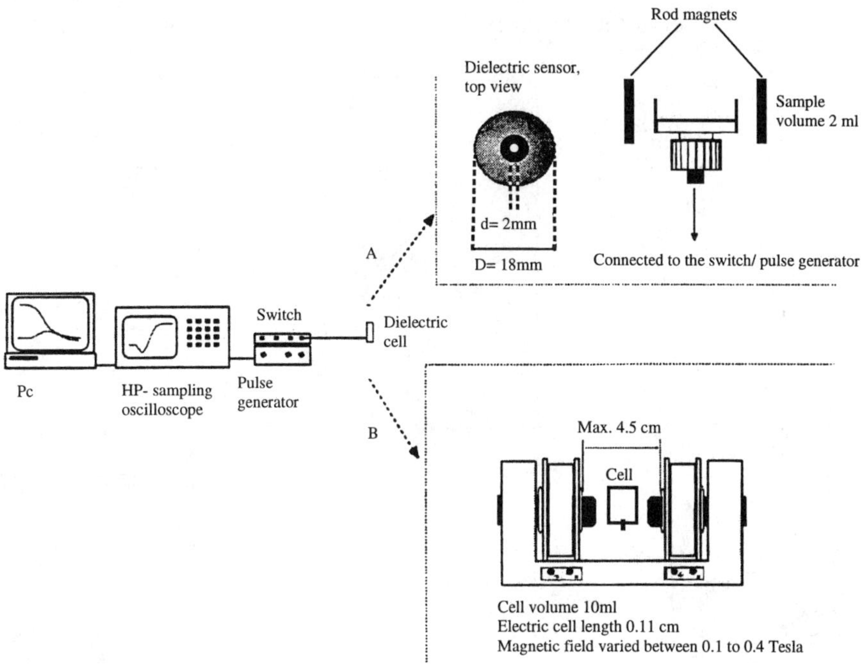

Figure 16. Experimental set-up for dielectric measurements on suspensions containing magnetic particles.

Two major effects will dictate the suspension behavior, i.e. a gravity-induced sedimentation and a bridging due to the onset of the magnetic field. At lower volume fractions the distance between the particles is rather large, and the hindrance of settling is low. When a magnetic field is applied across the suspension there will be an instantaneous sedimentation of small flocs (as seen from Figure 18, Φ=0.02 and 0.05). At higher Φ values the distance between the colloids is much smaller and the applied magnetic field will bridge the particles and there will be a hindered sedimentation. This is observed for Φ=0.20 and 0.25. For a field of 0.4 T it takes about 5 minutes before sedimentation sets in while the weaker field (0.1 T) can hinder the sedimentation for 3.5 minutes. For the highest volume fraction (Φ =0.25) no significant sedimentation is observed when the magnetic field is applied.

When the external magnetic field is applied the particles will to a higher degree flocculate and settle as flocs giving a more porous bottom layer, as illustrated in Figure 19. When the sedimentation takes place as individual particles the result is a dense sediment. The porosity of the sediment is also reflected in the level of the permittivity.

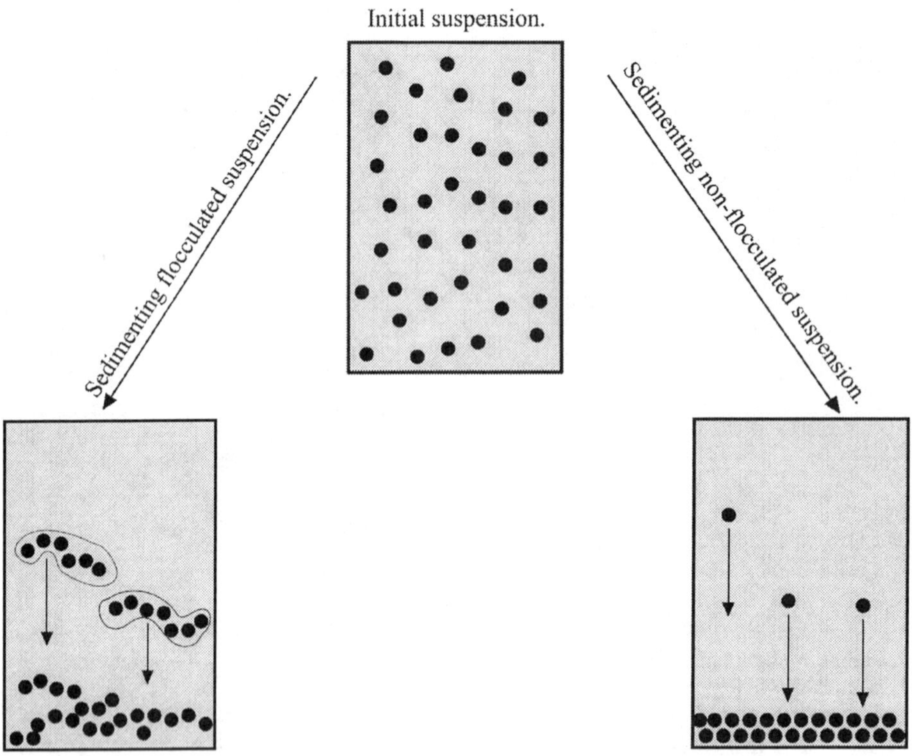

Figure 19. Influence of flocculation on the nature of the sediment according to dielectric measurements.

Acknowledgement

The Norwegian Research Council (NFR) is acknowledged for financial support of the Flucha program together with industry. B.P. acknowledges a Ph.D. grant from NFR.

References

1. *Encyclopedia of Emulsion Technology*, Vol. 1. Edited by P. Becher. Marcel Dekker, New York, 1983.
2. *Emulsions - A Fundamental and Practical Approach*, **C 363**. Edited by J. Sjöblom., Kluwer, Dordrecht, 1992.
3. *Food Emulsions*. Edited by K. Larsson and S. E. Friberg. Dekker, New York, 1996.
4. N.E. Hill, W.E. Vaughan, A.H. Price, and M. Davies, *Dielectric Properties and Molecular Behavior*, Van Norstrand, London, 1969.
5. J. B. Hasted, *Aqueous Dielectrics*, Chapman & Hall, London, 1973.
6. E.H. Grant, R. J. Sheppard, and G.P. South, *Dielectric Behavior and Biological Molecules in Solution*, Clarendon Press, Oxford, 1978.
7. P. Debye, *Polar Molecules*, Reinhold, New York, 1929.
8. K.S. Cole and R.H. Cole, *J. Phys. Chem.*, **9**, 341 (1941).
9. B. Gestblom and E. Noreland, *J. Chem. Phys.*, **81**, 782 (1977).
10. J.C. Maxwell, *Electricity and Magnetism*, Vol. 1, Clarendon, Oxford, 1892.
11. G. Banhegyi, *Colloid Polym. Sci.*, **266**, 11 (1988).
12. K.W. Wagner, *Arch. Elctrotechn.*, **2**, 371 (1914).
13. R.W. Sillars, *J. Inst. Electr. Engrs.*, **80**, 378 (1937).
14. H. Fricke, *J. Phys. Chem.*, **57**, 934 (1953).
15. D.A.G. Bruggeman, *Ann. Phys. Leipzig*, **24**, 636 (1935).
16. T. Hanai, *Kolloid-Z.*, **171**, 23 (1960).
17. T. Hanai, *Kolloid-Z.*, **175**, 62 (1961).
18. M.H. Boyle, *Colloid Polym. Sci.*, **263**, 51 (1985).
19. C. Boned and J. Peyrelasse, *Colloid Poym. Sci.*, **261**, 500 (1983).
20. J. Sjöblom, T. Skodvin, T. Jakobsen, and S.S. Dukhin, *J. Disp. Sci. Techn.*, **15**, 401, (1994).
21. T. Skodvin , J. Sjöblom, J.O. Saeten, O. Urdahl, and B. Gestblom, *J. Colloid Interface Sci.*, **166**, 43 (1994).
22. T. Skodvin and J. Sjöblom, *J. Colloid Interface Sci.*, **182**, 190 (1996).
23. F.G. Cottrell, US Patent No. 987114 (1911).
24. F.G. Cottrell and J.B. Speed, US Patent No. 987115 (1911).
25. S.E. Taylor, *Inst. Phys. Conf. Ser.*, **118**, 185 (1991).
26. T.J. Williams and A.G. Bailey, *IEEE Trans. Ind. Appl.*, **IA-22**, 536 (1986).
27. S.E. Taylor, *Colloids Surfaces*, **29**, 29 (1988).
28. C.A.R. Pearce, *British J. Appl. Phys.*, **5**, 136 (1954).
29. U. Zimmermann, G. Pilwat, and F. Riemann, *Biophys. J.*, **14**, 881 (1981).
30. U. Zimmermann, *Biochim. Biophys. Acta*, **694**, 227 (1982).
31. A.H. Brown and C. Hanson, *Trans. Faraday Soc.*, **61**, 1754 (1965).
32. B. Gestblom, H. Førdedal, and J. Sjöblom, *J. Dispersion Sci. Technol.* **15**, 449 (1994).
33. J. Sjöblom, O. Urdahl, H. Høiland, A.A. Christy, and J. Johansen, *Progr. Colloid Polym. Sci.*, **82**, 131 (1990).
34. J. Sjöblom, L. Mingyuan, T.Gu, and A.A. Christy, *Colloids Surfaces*, **66**, 55 (1992).
35. K.G. Nordli, J. Sjöblom, J. Kizling, and P. Stenius, *Colloids Surfaces*, **57**, 83 (1991).
36. K.G. Nordli, J. Sjöblom, and P. Stenius, *Colloids Surfaces*, **63**, 241 (1992).
37. J. Sjöblom, L. Mingyuan, A.A. Christy, and H.P. Rønningsen, *Colloids Surfaces A*, **96**, 261 (1995).
38. H.P. Rønningsen, J. Sjöblom, and L. Mingyuan, *Colloids Surfaces A*, **97**, 119 (1995).

39. S.E. Friberg, p. 1 in *Emulsions - A Fundamental and Practical Approach.* Edited by J. Sjöblom. NATO ASI Series C 363, Kluwer, Dordrecht, 1992.

40. H. Førdedal, E. Nodland, J. Sjöblom, and O.M. Kvalheim, *J. Colloid Interface Sci.*, **173**, 396 (1995).

41. H. Førdedal, Y. Schildberg, J. Sjöblom, and J-L.Volle, *Colloids Surfaces A*, **106**, 33 (1996).

42. H. Førdedal, Ø. Midttun, J. Sjöblom, O.M. Kvalheim, Y. Schildberg, and J-L.Volle, *J.Colloid Interface Sci.*, **182**, 117 (1996).

43. H. Førdedal and J. Sjöblom, *J. Colloid Interface Sci.*, **181**, 589 (196).

44. Y. Feldman, N. Kozlovich, I. Nir, and N. Garti, *Phys. Rev. E.*, **51**, 478 (1995).

45. J. Peyrelasse, M. Moha-Ouchane, and C. Boned, *Phys. Rev. A*, **38**, 904 (1988).

46. C. Thomas, J.P. Perl, and D.T. Wasan, *J. Colloid Interface Sci.*, **139**, 1 (1990).

47. J.P. Perl, H.W. Bussey, and D.T. Wasan, *J. Colloid Interface Sci.*, **108**, 528 (1985).

48. T. Skodvin, PhD Thesis, University of Bergen, Norway, 1995.

49. M. von Stackelberg, *Naturwissenchaften*, **11**, 327 (1949).

50. M. von Stackelberg and H. R. Müller, *Z. Elektrochem.*, **58**, 25 (1954).

51. M. von Stackelberg and W. Jahns, *Z. Elektrochem.*, **58**, 162 (1954).

52. J.A. Ripmeester, J. S. Tse, C.I. Ratcliffe, and B.M. Powell, *Nature*, **325**, 135 (1987).

53. T. Jakobsen, J. Sjöblom, and P. Ruoff, *Colloids Surfaces A*, **112**, 73 (1996).

54. J. Sjöblom, H. Førdedal, T. Jakobsen, and T. Skodvin, in *Handbook of Surfactants*, Edited by K.S. Birdi, CRC Press, in press, 1997.

55. B. Pettersen and J. Sjöblom, *Colloids Surfaces A*, **113**, 175 (1996).

56. R. Lindberg, G. Sundholm, B. Pettersen, J. Sjöblom, and S.E. Friberg, *Colloids Surfaces A*, **123-124**, 549 (1997).

57. B. Pettersen, L. Bergflödt, and J. Sjöblom, *Colloid Surfaces A*, **127**, 175 (1997).

58. B. Pettersen and J. Sjöblom, submitted to *J. Colloid Interface Sci.*, (1997).

59. B. Pettersen, T. Skodvin, and J. Sjöblom, submitted to *Colloids Surfaces A*, (1997).

60. L.H. Allen and E. Matijevic, *J. Colloid Interface Sci.*, **31**, 287 (1969).

61. A. Watillon and P.Gérard, *Proc. Int. Congr. Surface Activ.*, **4**, 1261 (1964).

62. B.V. Derjaguin and Z.M. Zorin, p. 14 in *Proceedings of the 2nd International Congress of Surface Activity*, **2**, Butterworths, London, 1957.

63. R.M. Pashley and J.A. Kitchener, *J. Colloid Interface Sci.*, **71**, 491 (1979).

64. Ya.I. Rabinovich, B.V. Derjaguin, and N.V. Churaev, *Adv. Colloid Interface Sci.*, **16**, 63 (1982).

65. R.K. Iler, *The Chemistry of Silica*, John Wiley & Sons, 1979.

66. F. Dumont, "Stability of Sols", pp.143-46 in *The Colloid Chemistry of Silica*, American Chemical Society, Washington D.C., 1994.

67. J.N. Israelachvili, *Intermolecular and Surface Forces*, Academic Press, New York, 1979.

68. *The Colloid Chemistry of Silica*, Chapt. 1. Edited by H.E. Bergna. American Chemical Society, Washington D.C., 1994.

69. M.L. Hair, *Infrared Spectroscopy in Surface Chemistry*, Dekker, New York, 1967.

70. L.T. Zhuralev, *Langmuir*, **3**, 316 (1987).

71. L.T. Zhuralev, *Colloids Surfaces*, **74**, 71 (1993).

72. M.L. Hair and A.T. Bell, ACS Symposium Series 137, American Chemical Society, Washington DC 1980.

Measurements in Concentrated Systems

Acoustic and Electroacoustic Spectroscopy

Andrei S. Dukhin and Philip J. Goetz
Dispersion Technology Inc., 3 Hillside Avenue, Mount Kisco, NY 10549 USA

Abstract

Both acoustic and electroacoustic spectroscopy are related to sound propagation through a heterogeneous system such as a suspension or an emulsion. Acoustic spectrometers measure only changes of sound wave properties, whereas electroacoustic spectrometers deal with the coupling between electrodynamic phenomena and the sound wave pressure field. We have suggested in previous papers that the combination of both techniques provides the most reliable and complete characterization of the dispersed system. This paper describes experimental performance of this new technique using various stable and unstable concentrated dispersed systems.

1. Introduction

Acoustic and electroacoustic spectroscopy are developing rapidly as an alternative to light scattering methods. The ability to characterize concentrated disperse systems provides much of the impetus for these developments.

Both techniques are based on well-established scientific background.[1-19] In both methods, the interaction of sound with the dispersed particles provides useful information; however, the set of measured parameters is different. In acoustic spectroscopy the parameter is attenuation and/or sound speed,[3-5] whereas for electroacoustic spectroscopy it is either colloid vibration potential/current (CVP/CVI)[6] or electrokinetic sonic amplitude (ESA).[8,9]

Electroacoustic spectra depend on two characteristics: particle size distribution (PSD) and zeta (ζ) potential. This type of spectroscopy is very attractive for providing complex characterization of both PSD and ζ-potential. However, there are several disadvantages of this simple approach. We have analyzed and listed these problems in our previous paper.[10] Here we present them again:

- **Calibration with a known colloid required**

Such calibration is an essential part of electroacoustic spectroscopy. It is a possible source of error, not only for ζ-potential, but for PSD as well.

• Absence of electroacoustic theory which is valid for concentrated systems
Electroacoustic theory is another essential part of electroacoustic spectroscopy. It is the basis for conversion of the spectra to PSD and ζ-potential. Absence of an electroacoustic theory for concentrated systems makes it hard to explore the main advantage of acoustic-based spectroscopy over light scattering: namely, the ability to characterize concentrated dispersed systems.

• Constant ζ-potential restriction
All electroacoustic theories assume that all particles have the same ζ-potential. This assumption might be valid for single component systems if the thickness of the electrical double-layer, κ^{-1}, is much less than the particle radius a. This further assumption might indeed be correct for many aqueous disperse systems, but it still introduces uncertainty in the measurement, even more so in the case of very small particles or non-aqueous media. More important perhaps is the fact that this assumption makes it impossible to handle the case of mixed component systems, as for example ceramic slurries.

• Double-layer (DL) model required
Current electroacoustic theory requires certain simplifying assumptions concerning the nature of the DL. The uncertainties with the DL model affect not only the calculated value of the ζ-potential but the PSD as well.

• Conductivity restrictions
Electroacoustic spectroscopy works within a restricted conductivity range, usually between 0.0001 S/m and 1 S/m. Therefore this type of spectroscopy is not suitable for particle size characterization in many non-aqueous systems.

In order to eliminate these disadvantages of electroacoustics we have suggested combining it with acoustics.[10] This new approach requires more sophisticated hardware; nevertheless, we believe that it is worthwhile because the reliability of the data improves significantly. This paper illustrates experimentally the advantage of this combined approach.

Acoustic spectroscopy based on the attenuation measurement does not require any assumptions about electro-surface properties. As a result the PSD obtained with acoustic spectroscopy is more reliable than the PSD extracted from electroacoustic spectra. Therefore, acoustic spectroscopy is preferable for characterization of the particle size distribution.

The more reliable PSD based on acoustic spectroscopy can then be used in the interpretation of the electroacoustic spectra. As a result, the improvement in PSD reliability leads to a more reliable estimate of ζ-potential.

We are going to characterize the accuracy and precision of acoustics and electroacoustics using stable, well-defined, concentrated dispersed systems. We will show that this combined acoustic-electroacoustic technique is suitable for characterizing unstable dispersed systems as well. In order to do this we perform titrations starting with a concentrated dispersed system. The evolution of measured parameters (attenuation spectra and CVI) reflect changes of PSD and ζ-potential. Comparison of these two parameters allows us to determine the value of ζ-potential at which systems lose their stability.

2. Theoretical Background

2.1 Acoustic Spectroscopy

Acoustic spectroscopy deals with the measurement of attenuation, sound speed, or acoustic impedance. The sound attenuation coefficient α is a convenient parameter characterizing total energy losses. It is equal to:

$$\alpha = \frac{10 \log_{10} \frac{I_{ini}}{I_{end}}}{L} \frac{db}{cm} \tag{1}$$

where I_{ini} is initial sound intensity, and I_{end} is the intensity after traveling a distance L. The ratio of sound intensities over some specific distance can be experimentally measured. The computed value of α reflects the total acoustic energy loss.

In acoustic spectroscopy we will speak mainly about attenuation and sound speed. Nevertheless, it is convenient to define a complex acoustic impedance *Z* in order to compare acoustic and electroacoustic spectroscopy. This impedance is the ratio of the pressure in the acoustic wave to the velocity of the medium displacement. Acoustic impedance is equal to:

$$Z = V\rho\left(1 - j\frac{V\alpha}{\omega}\right) = V\rho\left(1 - j\frac{\alpha\lambda}{2\pi}\right) \tag{2}$$

where *j* is $\sqrt{-1}$, ω is the sound frequency in radians/sec, λ is the wavelength of the sound, and *V* is the sound speed.

A sound wave interacts with the dispersed particles, and to a lesser extent with the dispersion medium. As a result of this interaction, the properties of the propagating sound wave change. A part of the acoustic energy fails to reach the receiving transducer. This part of the acoustic energy is dissipated by conversion to some other form of energy or diverted from the path between the transmitter and receiver transducers by scattering or other geometrical effects. Altogether, there are six known loss mechanisms produced by this interaction between the sound wave and the dispersed phase: 1) viscous losses; 2) thermal losses; 3) scattering losses; 4) intrinsic losses; 5) electrokinetic losses and 6) structural losses. Viscous, thermal, scattering, and intrinsic losses are included in most analyses of acoustic spectroscopy, whereas the electrokinetic and structural losses are either ignored or assumed to be negligibly small. However, structural losses can be quite significant, especially in concentrated systems. Nevertheless, spectroscopy based only on these first four mechanisms has been applied successfully for many real systems.

Acoustic spectroscopy is successful, first of all, in dilute systems. The theory of sound propagation is well established only for dilute systems when particle-particle interaction is negligibly weak. This theory has been in principle created by Epstein and Carhart[1] and then generalized for emulsions and suspensions by Allegra and Hawley.[2] This theory takes into account only the first four above mentioned mechanisms (viscous, thermal, intrinsic, and scattering losses).

For dense rigid sub-micrometer particles within the frequency range from 1 MHz to 100 MHz the viscous losses are dominant. The viscous losses of acoustic energy occur due to the shear wave generated by the particle oscillating in the acoustic pressure field. The amplitude of the shear wave decays exponentially with distance from the particle surface. The dimension characteristic of this decay is the so-called "viscous depth" δ_v. It is equal to the distance from the particle surface at which the amplitude drops by a factor of *e* :

$$\delta_v = \sqrt{\frac{2\nu}{\omega}} \tag{1}$$

where ν is kinematic viscosity and ω is angular frequency.

The "viscous depth" is frequency dependent. It decreases when the frequency increases and eventually becomes comparable with the particle radius *a*. Attenuation expressed in *dB/cm/MHz* reaches a maximum at that frequency range. The frequency corresponding to the maximum attenuation is the so-called "critical frequency".

The mechanism of scattering loss is quite different compared to the viscous mechanism. The acoustic scattering mechanism is similar to light scattering. Acoustic scattering does not produce dissipation of acoustic energy; particles simply redirect part of the acoustic energy flow so that this portion of the sound wave does not reach the receiver.

As mentioned above, the acoustic theory for dilute dispersed systems is complete, whereas the theory for concentrated systems, especially polydisperse systems, is far from being done. There are three different approaches to the concentration-dependent effects reviewed by Harker and Temple[11]: (i) phenomenological, (ii) coupled-phase model, and (iii) multiple scattering treatment. The goal of these approaches is to incorporate particle-particle interactions into the acoustic theory.

Dispersed particles oscillating in the acoustic pressure field generate several waves: long scale compression waves and short scale shear and temperature waves. These waves could interfere with the motion of other particles, which could be effectively described as particle-particle interactions.

Particle-particle interaction through the long scale compression waves is a part of the scattering mechanism of sound attenuation in the concentrated system. The multiple scattering treatment is the most suitable to describe this effect. The theory of multiple scattering is reviewed by Harker and Temple.[11] This aspect of concentration effects in acoustic spectroscopy is less important for colloidal systems because these scattering losses are exhibited by large particles and at high frequencies.

Particle-particle interaction through shear waves is much more important for colloidal systems. It affects the viscous loss mechanism which is dominant in dispersions of small particles. Particle-particle interaction causes a shift in acoustic spectra to higher frequencies. This shift can be very significant. For instance, our dilution experiment with a rutile dispersion[3] shows a shift of the characteristic frequency from 15 MHz for 4 wt % to above 100 MHz for 74.6 wt %. The theory of sound propagation must take this effect into account, otherwise the error in particle size calculation can be enormous.

There have been several attempts to adjust the viscous loss theory for concentrated systems.[11-15] The last one[3] combines the achievements of the previous works. It explores the "coupled phase approach", generalizing it for the polydisperse system. The "cell model" provides an expression for the particle drag following Pendse and Strout[14,15] but using a different mathematical presentation. The "cell model" concept has been generalized for the polydisperse system as well.

The new theory of viscous loss has been successfully tested experimentally with concentrated rutile dispersions.[3] Therefore the existing theory of acoustic attenuation makes it possible to calculate PSD in concentrated dispersions.

2.2 Electroacoustic Spectroscopy

Whereas acoustic spectroscopy describes the combined effect of the six separate loss mechanisms, electroacoustic spectroscopy (as it is presently formulated) emphasizes only one of these interaction mechanisms, i.e. the electrokinetic losses.

In acoustic spectroscopy sound is utilized as both the excitation and the measured variable, and therefore there is but one basic implementation. In contrast, electroacoustic spectroscopy deals with the coupling of electric and acoustic fields and therefore there are two possible implementations. One can apply a sound field and measure the resultant electric field, which is referred to as the colloid vibration potential (CVP), or conversely one can apply an electric field and measure the resultant acoustic field, which is referred to as the electrokinetic sonic amplitude (ESA).

First let us consider the measurement of CVP. When the density of the particles ρ_p differs from that of the medium ρ_m, the particles move relative to the medium under the influence of an acoustic wave. This motion causes a displacement of the internal and external parts of the DL. This phenomenon is usually referred to as polarization of the DL.[16] This displacement of opposite charges gives rise to a dipole moment. The superposition of the electric fields of these induced dipole moments over the collection of particles gives rise to a macroscopic electric field, the CVP.[6] Thus, some of the acoustic energy is converted to electric energy. This electric energy may then be dissipated if the opportunity for electric current (CVI) exists.

The ESA phenomenon occurs when an alternating electric field is applied to the disperse system.[8,9] If the ζ-potential of the particle is greater than zero, then the oscillating electrophoretic motion of the charged dispersed particles generates a sound wave.

Both electroacoustic parameters CVP and ESA can be experimentally measured. The following expressions give their relationship with the system parameters according to O'Brien[9]:

$$ESA = C\varphi\mu_e \frac{\rho_p - \rho_m}{\rho_m} \frac{Z_g Z_s}{Z_g + Z_s} \qquad (2)$$

$$CVP = C'\varphi\mu_e \frac{\rho_p - \rho_m}{K^* \rho_m} \frac{Z_g Z_s}{Z_g + Z_s} \qquad (3)$$

where C and C' are calibration constants independent of the properties of the dispersed system, φ is the volume fraction, μ_e is the dynamic electrophoretic mobility, K^* is the complex conductivity of the dispersed system, Z is the acoustic impedance, and indexes g and s correspond to the sound transducer and dispersed system.

The CVP or ESA spectrum is the experimental output from electroacoustic spectroscopy. Both of these spectra depend on the dynamic electrophoretic mobility, which contains information about ζ-potential and PSD. The first theory of dynamic electrophoretic mobility, created by Dukhin,[17] works at low frequency. O'Brien has created a new theory[8] which is valid in the frequency range important for electroacoustic spectroscopy.

Only one of the electroacoustic spectra is required because both of them contain essentially the same information about the dispersed system. Both CVP and ESA are complex parameters. The phase spectra is more suitable for the PSD calculation because it is independent of the ζ-potential. However, the dependence of the phase on the surface conductivity might still create some problems. The magnitude spectra provides information about ζ-potential for a known PSD.

The conversion of electroacoustic spectra into the PSD requires a theoretical model of the electroacoustic phenomena. This conversion procedure is much more complicated for electroacoustics compared to the relatively simple acoustics. The theory becomes much more complicated because of the additional (electric) field. As a result even the most sophisticated existing theory, created by O'Brien[8], includes additional assumptions. However, even this dilute case theory has not been completed.[10]

The electroacoustic theory available now is valid only for the dilute case. The first attempts to generalize this theory for the concentrated case[18,19] reveals a great deal about the complicated theoretical problems. As a result, electroacoustic spectroscopy is currently applicable for the characterization of dilute dispersed systems only.

3. Measuring Technique

We use the acoustic and electroacoustic spectrometer DT-1200, developed by Dispersion Technology Inc. This instrument has two separate sensors for measuring acoustic and electroacoustic signals independently. Both sensors use a pulse technique.

The Acoustic sensor has two piezo crystal transducers. The gap between transmitter and receiver is variable in steps. In default it changes from 0.15 mm up to 20 mm in 21 steps. The frequency changes in steps as well. By default it changes from 1 to 100 MHz in 18 steps. The number of pulses collected at each gap and each frequency is automatically adjustable in order to reach the target signal-to-noise ratio. The acoustic sensor also measures sound speed at one chosen frequency. It makes this measurement using a time-of-pulse detection by the receiver. The instrument automatically adjusts pulse sampling depending on the value of the sound speed. This is necessary for eliminating possible artifacts, like excess attenuation at low frequencies.

The electroacoustic sensor measures the magnitude and phase of the CVI at a chosen frequency. It has a piezo crystal transducer and a specially designed electric antenna. The distance between transmitter and antenna is 5 mm. There is provision for automatic correction for sound speed and attenuation measured with the acoustic sensor.

A special program calculates PSD from the attenuation spectra and ζ-potential from CVI. This program tests log-normal, bimodal, and modified log-normal[20] particle size distributions. It uses an error analysis in order to search for the best PSD. The goal of the optimization procedure is to minimize the error of theoretical fit to the experimental attenuation spectra. The analysis program takes into account several correction factors when calculating the ζ-potential, including:

1. Particle size correction. It uses either the PSD calculated from the attenuation spectra or *a priori* known PSD.

2. Attenuation of the sound pulse.

3. Back flow caused by the moving dispersed particles.

The instrument also has a special prediction program. This is an educational tool which gives an opportunity to calculate attenuation and CVI spectra for any system specified by the user. This program is suitable also for a manual search of the best PSD for the given experimental attenuation spectra. It helps to overcome problems of a local minimum which might be an obstacle for an automatic optimization program.

The total required sample volume is about 100 cm^3. There is a special magnetic stirrer to prevent sedimentation and facilitate mixing of chemicals during titration. The instrument has two burettes and appropriate software for automatic titration. The instrument has also conductivity and temperature probes. One attenuation spectra measurement using a default setup takes about 5 minutes. The user can speed up measurement by changing the setup parameters. One CVI measurement takes from 10 seconds to a minute depending on the system properties.

3.1 Precision

Precision is a measure of the reproducibility of measurements. This section presents results concerning precision of the acoustic and electroacoustic sensors. Attenuation spectra shown in Figure 1 illustrate the acoustic sensor precision. These attenuation spectra were measured for Sumitomo AA-2 alumina and Ludox™ silica. The alumina sample and silica sample were measured 10 and 11 consecutive times, respectively. The corresponding median particle size is given in Table I.

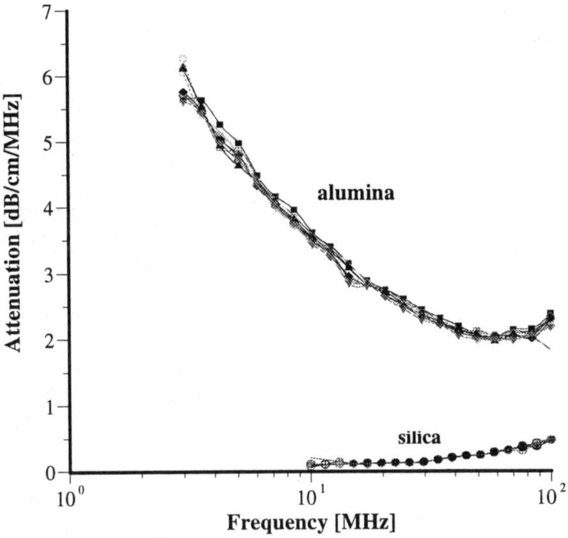

Figure 1. Attenuation spectra of Sumitomo AA-2 alumina and Ludox™ silica.

Table I. Median particle size

alumina	2.015	2.076	2.057	2.092	2.065	2.047	2.035	2.075	2.039	2.085	
silica	0.03	0.029	0.029	0.029	0.029	0.03	0.03	0.029	0.03	0.03	0.03

It is seen that absolute variation of the median particle size is 0.9 % for alumina and 1.5 % for silica. These numbers illustrate the precision of the acoustic sensor.

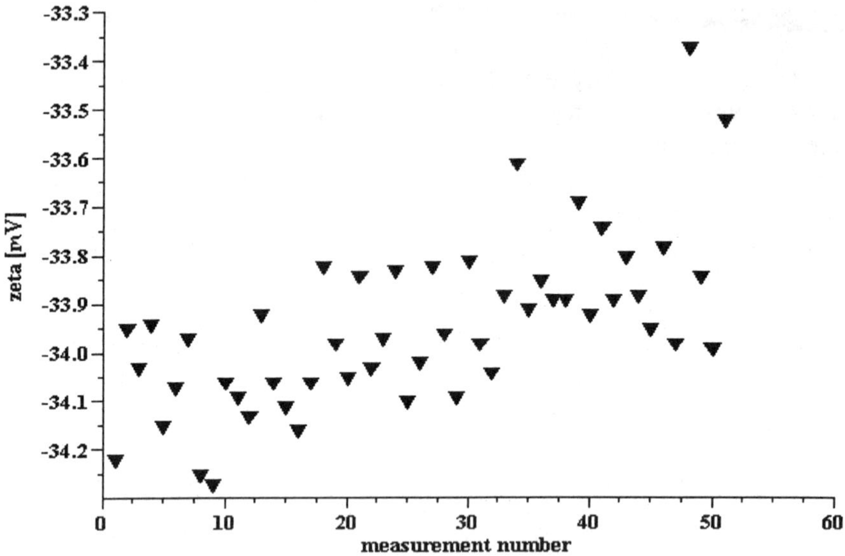

Figure 2. Continuous measurement of zeta potential for Ludox™ silica, 32 vol %.

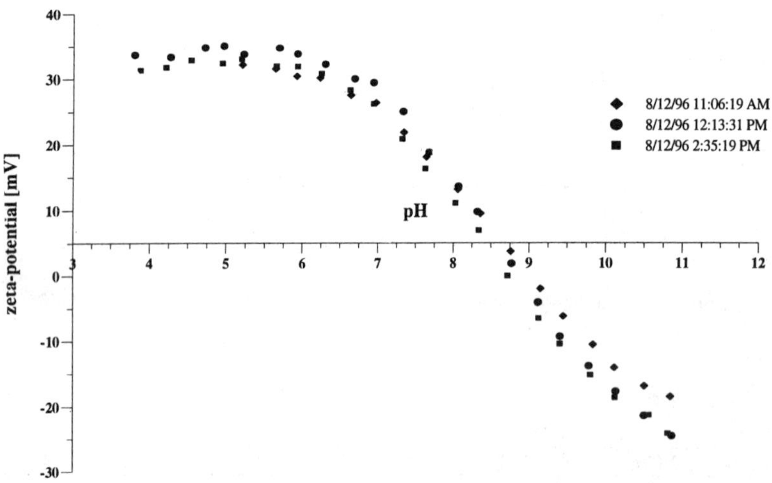

Figure 3. Titration of Sumitomo AKP-30 alumina at 5 vol %.

Figure 2 and Figure 3 illustrate the precision of the electroacoustic sensor. Figure 2 shows the result of 51 continuous CVI measurements for silica. It is seen that the precision of the absolute value of the ζ-potential measurement is a fraction of a mV. Figure 3 shows titration curves for three identical samples of 6 wt % alumina with particles about 100 nm in size. It is seen that precision of the isoelectric point determination is about 0.1 pH unit.

3.2 Accuracy

Accuracy characterizes the correlation between actual and measured values. In order to determine accuracy of PSD one needs a standard system with a known PSD. Such systems have been developed in Germany, and we used BCR silica quartz with a median size of about 3 µm.

Figure 4 and Figure 5 show the measured attenuation spectra and corresponding PSD. The difference between the nominal median particle size and that measured with the DT-1200 is less than 1 %. At the same time there is some difference in the PSD width. This means that the acoustic sensor determines median size with an accuracy of at least 1 % and PSD width with an accuracy of about 5 %.

It is also important to illustrate accuracy for different materials. For instance, we measured several aluminas produced by Sumitomo. Attenuation spectra measured at 5 vol % are shown in Figure 6. Corresponding particle sizes are given in Table II. It is seen that accuracy is quite high for these alumina samples as well.

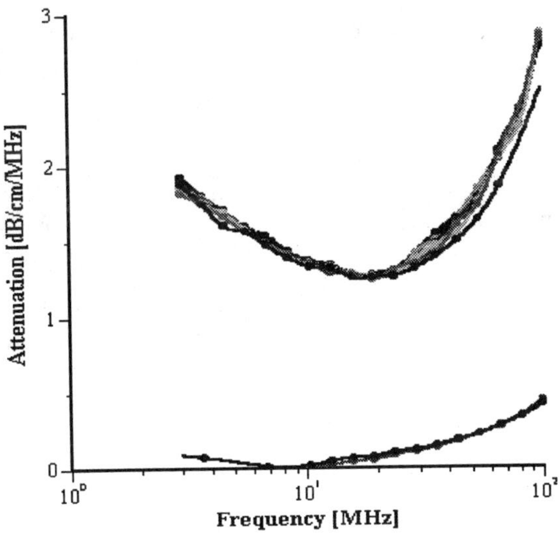

Figure 4. Attenuation spectra for BCR quartz at 10.98 wt % in ethanol.

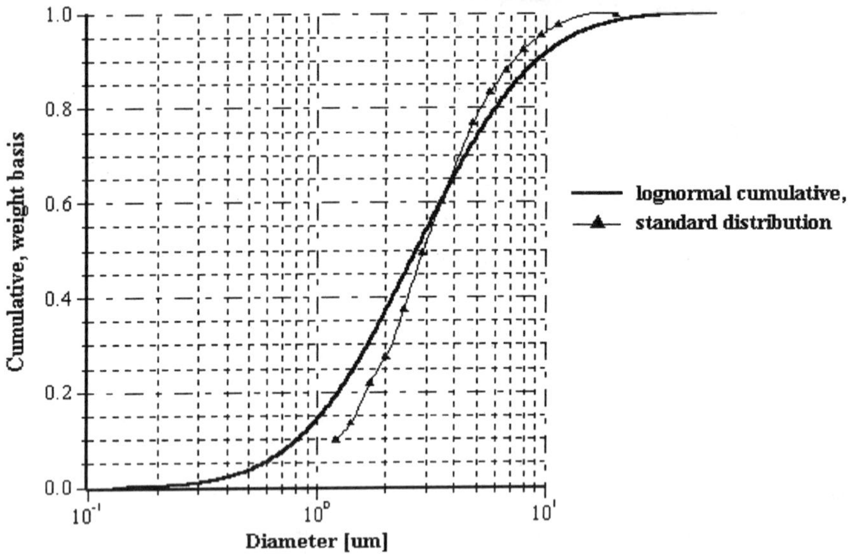

Figure 5. Comparison of particle size distribution calculated from viscous losses with BCR standard distribution.

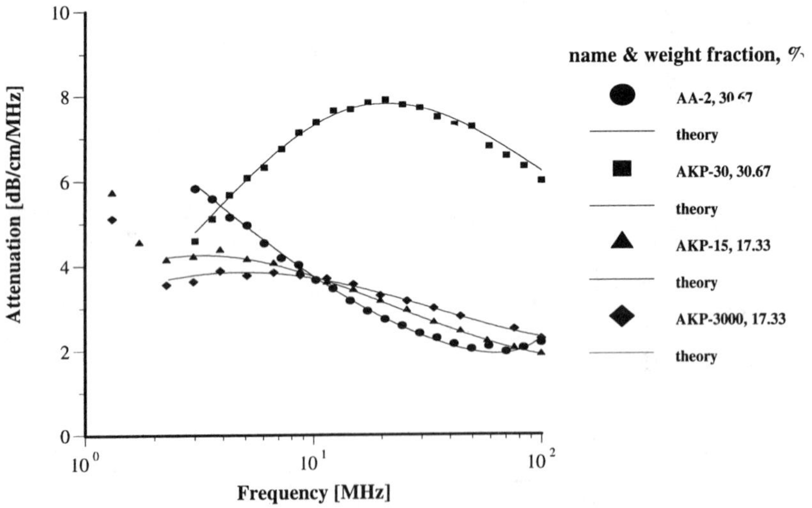

Figure 6. Experimental and theoretical attenuation spectra of Sumitomo alumina 5 vol % slurries.

Table II. Median particle size in microns.

	acoustics, DT-1200	manufacturer
AKP-15	0.684	0.7
AKP-30	0.319	0.3
AKP-3000	0.520	0.5
AA-2	1.956	2

Test of the ζ-potential measurement accuracy is much more complicated because there is no ζ-potential standard for concentrated systems. Absence of electroacoustic theory for concentrated systems creates additional complexity. Our experience is that CVI makes it possible to measure ζ with almost the same accuracy as microelectrophoresis.

4. Titration Experiments

4.1 Materials

A titration experiment has been performed with two different materials: alumina and rutile. We chose these materials because of the different expected values of their isoelectric points. The isoelectric point of rutile is about pH 4, whereas the isoelectric point of alumina is around pH 9.

The concentrated aqueous rutile slurry was obtained from E.I. DuPont de Nemours. It was commercially referred to as R746 and was normally supplied at a weight fraction of 76.5 % (44.5 % by volume). The density of the particles was 4.06 g/cm^3. It was slightly less than for typical rutile because of the various surface modifiers used to stabilize the slurry.

The pH titration was performed with a sample diluted to 7 vol %. The sample was diluted using distilled water and adjusted to pH 8.5 with potassium hydroxide. The median particle size of this rutile was about 0.3 μm according to manufacturer data.

The weight fraction of the alumina slurry is 11.6 wt %. This is equivalent to a volume fraction of 4 % for an alumina density of 3.4 g/cm^3 . The initial slurry pH was 4. This sample was used in titration experiments as is without additional dilution. The median particle size of the alumina particles was unknown.

4.2 Results and Discussion

Electroacoustic measurement of ζ-potential confirms the expected values of the isoelectric points for both alumina and rutile. Corresponding titration curves are shown on

Figure 7 and Figure 8. A dispersed system loses stability at pH values close to the isoelectric point if the stability of this system depends on DL interactions. Aggregation of the particles changes the PSD, which affects the measured attenuation spectra. Experiments indeed show (see Figure 9 and Figure 10) that the attenuation spectrum changes with pH for both alumina and rutile.

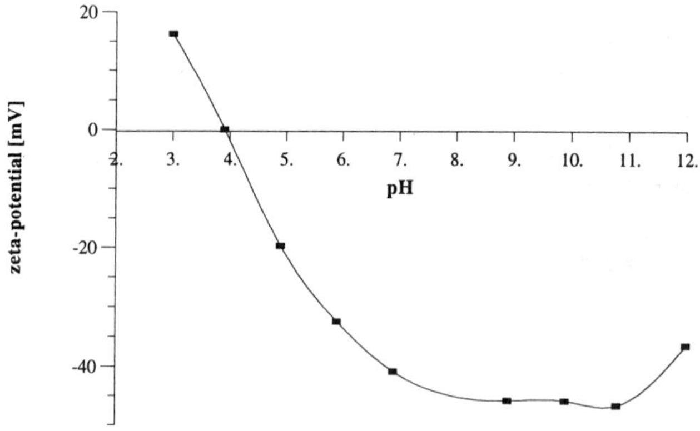

Figure 7. ζ-potential versus pH for rutile at 7 vol %.

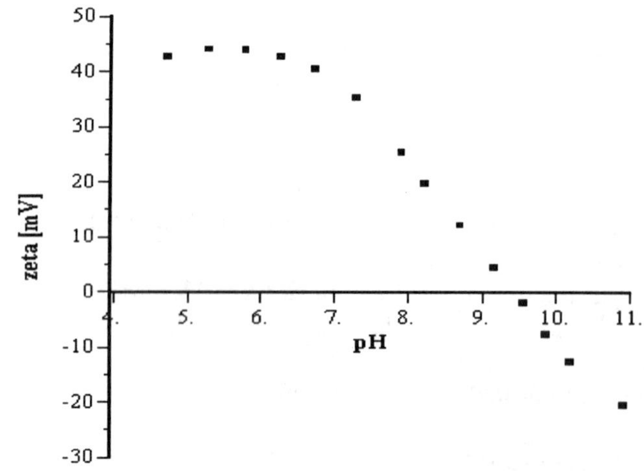

Figure 8. ζ-potential versus pH for alumina at 4 vol %.

It is interesting that in both cases the attenuation spectrum remains stable up to a certain pH. In the case of alumina it is stable up to pH 7.3. In the case of rutile this critical pH is 5.9. Attenuation spectra change beyond this critical pH, which marks the range of aggregation stability. Alumina is stable for pH values below 7.3 whereas rutile is stable for pH values above 5.9. It is noteable that in both cases this critical pH corresponds to the same value of the ζ-potential (about 30 mV). It looks like both dispersed systems lose their stability when the ζ-potential drops to a level of 30 mV.

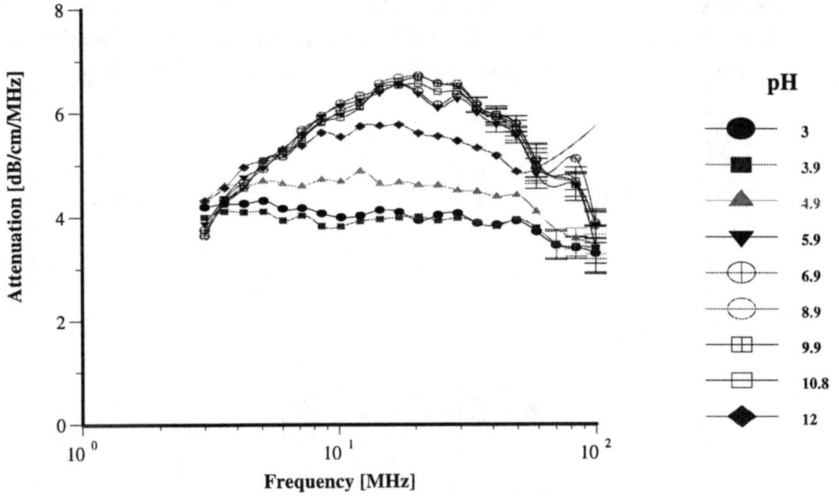

Figure 9. Attenuation spectra as a function of pH for rutile at 7 vol %.

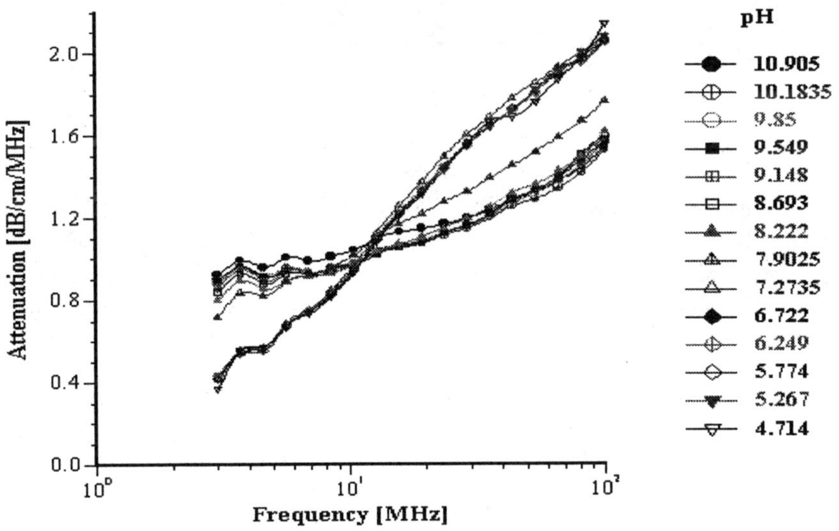

Figure 10. Attenuation spectra as a function of pH for alumina at 4 vol %.

Continuous change of pH beyond the critical point causes a fast change in the attenuation spectra, although it reaches another stable level in both cases. Existence of this second pH-independent attenuation spectra is an interesting feature which is common to both systems. In the case of alumina, the attenuation spectrum remains the same over a wide pH range from 8.2 to11. Attenuation spectra of rutile at pH values below at least 4 also do not change. For some reason the particle size distribution does not change in the vicinity of the isoelectric point. The system is certainly unstable at these pH values; however, there is another factor which prevents particles from collapsing into large aggregates.

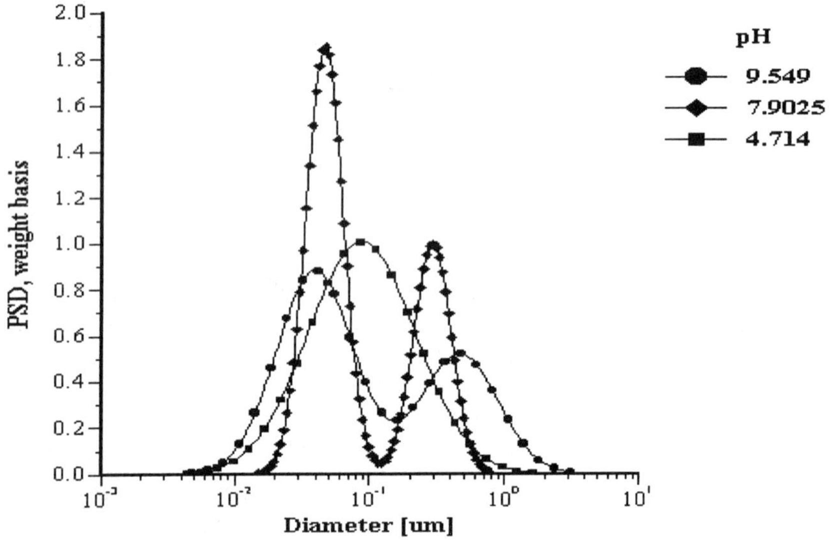

Figure 11. Particle size distribution as a function of pH for alumina at 4 vol %.

Particle size distributions are shown in Figure 11 and Figure 12 for various values of pH. It is seen that the distribution becomes bimodal beyond a critical pH. However, the size of the aggregates does not exceed several microns. We believe that these bimodal particle size distributions are stable in time because of stirring. The titration experiment requires the sample to be stirred for homogeneous mixing of added chemicals. Shear stresses caused by this stirring may break apart large aggregates. This shear can be a factor which restricts the maximum size of the aggregates. This idea can be tested in the future by making measurements at different stirring rates.

There is another interesting unsolved question; we don't know if the aggregation that we observe is reversible. The systems which we use do not allow us to reach ζ-potentials above the critical 30 mV level after crossing through the isoelectric point. The value of ζ-potential is restricted by increasing ionic strength at high pH values for alumina and at low pH values for rutile. We will need another material with an isoelectric point in the intermediate pH range in order to answer this question definitively.

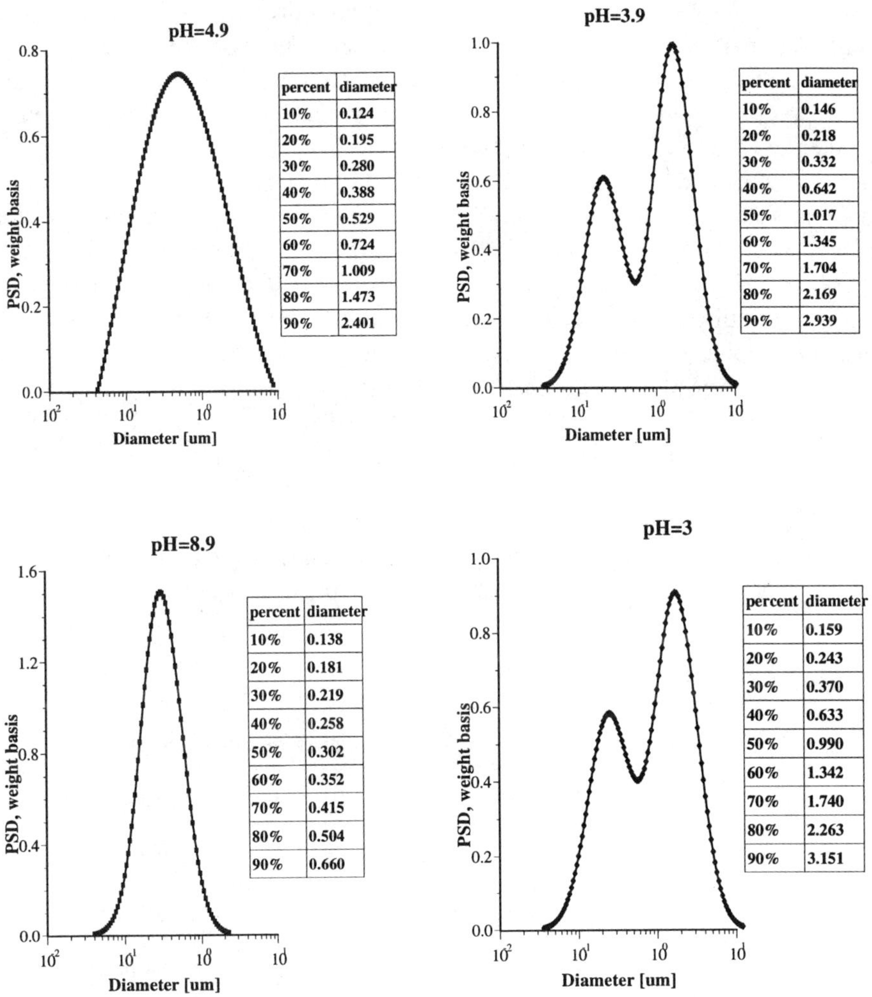

pH=4.9

percent	diameter
10%	0.124
20%	0.195
30%	0.280
40%	0.388
50%	0.529
60%	0.724
70%	1.009
80%	1.473
90%	2.401

pH=3.9

percent	diameter
10%	0.146
20%	0.218
30%	0.332
40%	0.642
50%	1.017
60%	1.345
70%	1.704
80%	2.169
90%	2.939

pH=8.9

percent	diameter
10%	0.138
20%	0.181
30%	0.219
40%	0.258
50%	0.302
60%	0.352
70%	0.415
80%	0.504
90%	0.660

pH=3

percent	diameter
10%	0.159
20%	0.243
30%	0.370
40%	0.633
50%	0.990
60%	1.342
70%	1.740
80%	2.263
90%	3.151

Figure 12. Particle size distributions for 7 vol % rutile at different pH values.

5. Conclusions

Simultaneous acoustic and electroacoustic measurements make it possible to characterize, independently, the particle size distribution and ζ-potential in concentrated dispersed systems. Acoustic spectra are a source of particle size data which is independent in this case of all uncertainties about electrical double-layer structure and properties. Electroacoustic measurement of the colloid vibration current (CVI) provides data for ζ-potential calculation with a correction for the independent particle size known from the acoustic attenuation measurement.

Acoustic measurements are able to characterize the median particle size with a precision and accuracy of roughly 1 %. The distribution width can be characterized with a precision and accuracy of roughly 1 % and 5 %, respectively. The value of ζ-potential can be determined with a precision equal to a fraction of a mV and with an accuracy of several mV, depending on particle volume fraction. Acoustic attenuation spectra and CVI measured for concentrated 7 vol % rutile and 4 vol % alumina dispersions show pronounced changes caused by pH titration. Both dispersions lose their kinetic stability towards aggregation within ± 2 pH units vicinity of the isoelectric point, and both become unstable when the absolute value of ζ-potential drops below 30 mV. The particle size distribution of both dispersions becomes bimodal near the isoelectric point. This bimodal distribution is stable and pH-independent within ± 2 pH units of the isoelectric point.

We believe that this work provides sufficient evidence to conclude that combined acoustic and electroacoustic spectroscopy is a valuable tool for characterizing stable and unstable concentrated dispersed systems.

References

1. P.S. Epstein and R.R. Carhart, "The Absorption of Sound in Suspensions and Emulsions", *J. Acoust. Soc. Amer.*, **25** [3] 553-565 (1953).
2. J.R. Allegra and S.A. Hawley, "Attenuation of Sound in Suspensions and Emulsions: Theory and Experiments", *J. Acoust. Soc. Amer.*, **51** [5] 1545-1564 (1972).
3. A.S. Dukhin and P.J. Goetz, "Acoustic Spectroscopy for Concentrated Polydisperse Colloids with High Density Contrast", *Langmuir*, **12** [21] 4987-4997 (1996).
4. A.S. Dukhin, P.J. Goetz, and C.W. Hamlet, "Acoustic Spectroscopy for Concentrated Polydisperse Colloids with Low Density Contrast", *Langmuir*, **12** [21] 4998-5004 (1996).
5. D.J. McClements, "Ultrasonic Characterization of Emulsions and Suspensions", *Adv. Colloid Interface Sci.*, **37,** 33-72 (1991).
6. B.J. Marlow, D. Fairhurst, and H.P. Pendse, "Colloid Vibration Potential and the Electrokinetic Characterization of Concentrated Colloids", *Langmuir*, **4** [3] 611-626 (1983).

7. F. Booth and J. Enderby, "On Electrical Effects due to Sound Waves in Colloidal Suspensions", *Proc. Amer. Phys. Soc.*, **208A**, 32 (1952).

8. R.W. O'Brien, W.N. Rowlands, and R.J. Hunter, "Determining Charge and Size with the Acoustosizer", pp.1-22 in *Electroacoustics for Characterization of Particulates and Suspensions*, Special Publication 856. Edited by S.G. Malghan. National Institute of Standards and Technology, US Department of Commerce, Gaithersburg, MD, 1993.

9. R.W. O'Brien, "Electro-Acoustic Effects in a Dilute Suspension of Spherical Particles", *J. Fluid Mech.*, **190**, 71-86 (1988).

10. A.S. Dukhin and P.J. Goetz, "Acoustic and Electroacoustic Spectroscopy", *Langmuir*, **12** [19] 4334-4344 (1996).

11. A.H. Harker and J.A.G. Temple, "Velocity and Attenuation of Ultrasound in Suspensions of Particles in Fluids", *J. Phys. D.:Appl. Phys.*, **21**, 1576-1588 (1988).

12. A.H. Harker, P. Schofield, B.P. Stimpson, R.G. Taylor, and J.A.G. Temple, "Ultrasonic Propagation in Slurries", *Ultrasonics*, **29**, 427-439 (1991).

13. R.L. Gibson and M.N. Toksoz, "Viscous Attenuation of Acoustic Waves in Suspensions", *J. Acoust. Soc. Amer.*, **85**, 1925-1934 (1989).

14. H.P. Pendse, T.A. Strout, and A. Shanna, "Theoretical Consideration in Acoustophoretic Analysis of Concentrated Colloids" pp. 23-39 in *Electroacoustics for Characterization of Particulates and Suspensions*, Special Publication 856. Edited by S.G. Malghan. National Institute of Standards and Technology, US Department of Commerce, Gaithersburg, MD, 1993.

15. T.A. Strout, "Attenuation of Sound in High-Concentration Suspensions: Development and Application of an Oscillatory Cell Model", PhD Thesis, The University of Maine, 1991.

16. S.S. Dukhin and B.V. Derjaguin, "Electrokinetic Phenomena" in *Surface and Colloid Science*, vol. 7. Edited by E. Matijevic. John Wiley & Sons, NY, 1974.

17. T.A. Vorobyeva, I.N. Vlodavets, and S.S. Dukhin, *Colloid J. USSR*, **32** [189] (1970).

18. P.F. Rider and R.W. O'Brien, "The Dynamic Mobility of Particles in a Non-Dilute Suspension", *J. Fluid. Mech.*, **257**, 607-636 (1993).

19. P.F.Rider, "Sound Wave Mobilities in a Nondilute Suspension of Spheres", *J. Colloid Interface Sci.*, **172,** 1-13 (1995).

20. R.R. Irani and C.F. Callis, *Particle Size: Measurement, Interpretation and Application*, John Wiley & Sons, NY-London, 1971.

Ultrasonic Extinction and Velocity in Dense Suspensions

Ulrich Riebel

Lehrstuhl Mechanische Verfahrenstechnik, Brandenburgische Technische Universität Cottbus, 03013 Cottbus, Germany

Udo Kräuter

BAYER AG, 51368 Leverkusen, Germany

Abstract

Theoretical predictions, measurement methods, and experimental results for ultrasonic extinction and velocity in suspensions are discussed for the intermediate and short wavelength regime. Special attention is given to the various mechanisms of particle-particle interaction in dense suspensions.

1. Introduction

Frequency-dependent extinction and velocity of ultrasound have become the basis of a number of measurement techniques, which are applicable to suspensions and emulsions even at high concentrations.

The present paper is focused on the range of intermediate and short wavelengths, and has the intention to answer two basic questions:

- How good are the theoretical approaches for describing ultrasonic extinction and velocity in dilute suspensions?

- What is the physical background of particle-particle interactions in dense suspensions and how are they modelled properly?

2. Ultrasonic Extinction and Velocity in Dilute Suspensions

In dilute suspensions, ultrasonic extinction is described phenomenologically by the Bouguer-Lambert-Beer-Law (BLBL), which is derived from an intensity balance for an infinitesimally thin layer of suspension:

$$dI = -\text{const.} \cdot I \cdot dz \qquad (1)$$

Integration yields the transmission

$$T = I/I_0 = \exp\{-\text{const.}\ \Delta z\} \qquad (2)$$

or the extinction, respectively:

$$E = -\ln I/I_0 = \text{const.} \cdot \Delta z \qquad (3)$$

In the above equations, the constant is a measure of concentration multiplied by the corresponding measure for the particle extinction properties:

$$\text{const.} = C_N \cdot \int A_{ext} < x > \cdot\, q_0 < x > dx \qquad (4\text{ a})$$

$$= C_{PA} \cdot \int K_{ext} < x > \cdot\, q_2 < x > dx \qquad (4\text{ b})$$

$$= C_V \cdot \int \alpha_{ext} < x > \cdot\, q_3 < x > dx \qquad (4\text{ c})$$

The above equations are perfectly equivalent, and it is, on a first sight, just a question of convenience which one to use. The differences may be important, however, when measurement errors or particle-particle interactions in highly concentrated media are discussed.

In the short wavelength range, it is most appropriate to present the extinction efficiency K_{ext} as a function of the particle size parameter $\pi x / \lambda$. As far as scattering is the dominant mechanism of extinction, this yields a size independent representation of the extinction behaviour,[1] (see Fig. 1). In the range of lower $\pi x / \lambda$, inertial and viscous entrainment and thermal effects become important, and the curves split up.

The presence of particles in a suspension also leads to a change in ultrasonic velocity. This is described by the equation:

$$c_{susp} = c_0\ (1 - \Phi \cdot C_V) \qquad (5)$$

where Φ is the phase function. Just as K_{ext}, c_{Susp}, and Φ can be represented as functions of $\pi x / \lambda$, independently from the particle diameter; compare Figures 3, 4 and 9.

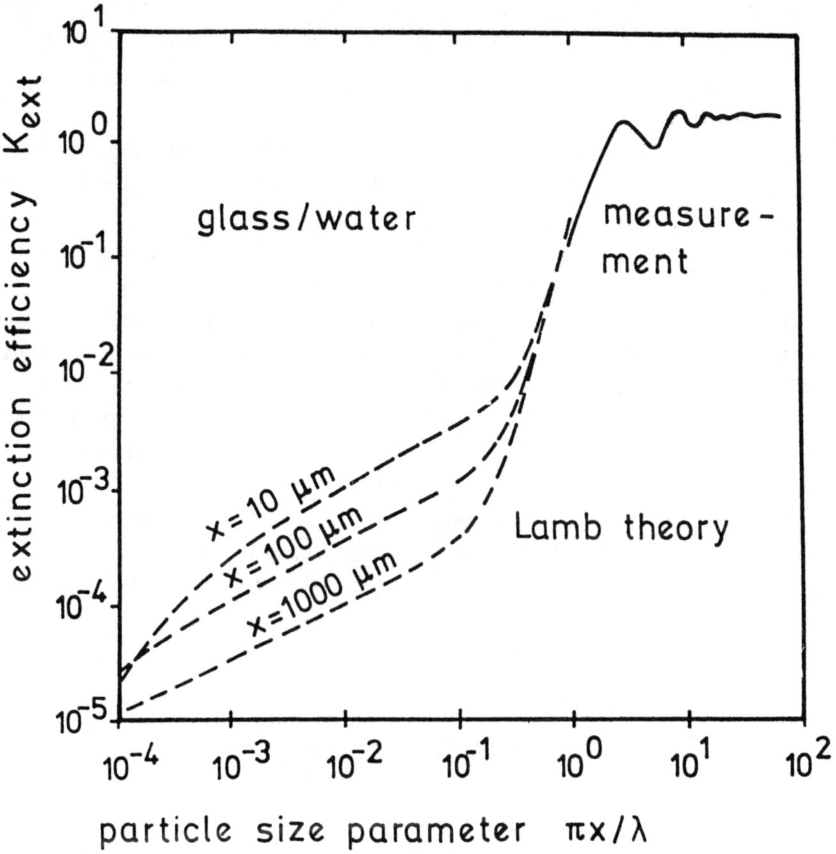

Figure 1. Extinction efficiency as a function of the dimensionless particle size parameter $\pi x / \lambda$.

The theoretical prediction of extinction efficiency and phase function in the scattering region is done in analogy to Mie theory.[2] Both the extinction efficiency and the phase function can be presented by series of complex coefficients:

$$K_{ext} = -\frac{4\lambda^2}{\pi^2 x^2} \sum_{n=0}^{\infty} (2n+1) \, \mathrm{Re}\{An\} \tag{6}$$

$$\Phi = \frac{3\lambda^3}{2\pi^3 x^3} \sum_{n=0}^{\infty} (2n+1)\, \mathrm{Im}\{An\} \tag{7}$$

whereby in the following examples, the A_n were calculated using Hay and Mercer´s theory.[3]

Measurements of ultrasonic extinction efficiencies require a broad frequency range, an extreme dynamic range and a high resolution. For applications both in basic research and in industrial on-line measurements, the following set-up[1] was used: Frequencies from 0.1 to 150 MHz are fed to a pair of piezoelectric transducers, whereby both the frequency and the intensity are selected by a computer. With the feedback from the receiving circuitry the intensity is adjusted so that the received signal remains almost constant within a narrow bandwidth. This allows to combine a dynamic range of 120 dB with a resolution of 0.1 dB. At the same time, ultrasonic intensities are always kept inferior to 10^{-3} W/cm², so that cavitation effects, destruction of materials, and even changes in agglomerates of low strength are prevented.

For measurements of ultrasonic phase velocity, two different set-ups are used. For the operation with continuous waves, a signal transmitted through the measurement section and a reference signal are fed into a phase comparator. The phase shift obtained with the suspension is compared to a reference measurement with the pure liquid, so that all phase shifts resulting from the electronic circuit are cancelled out.

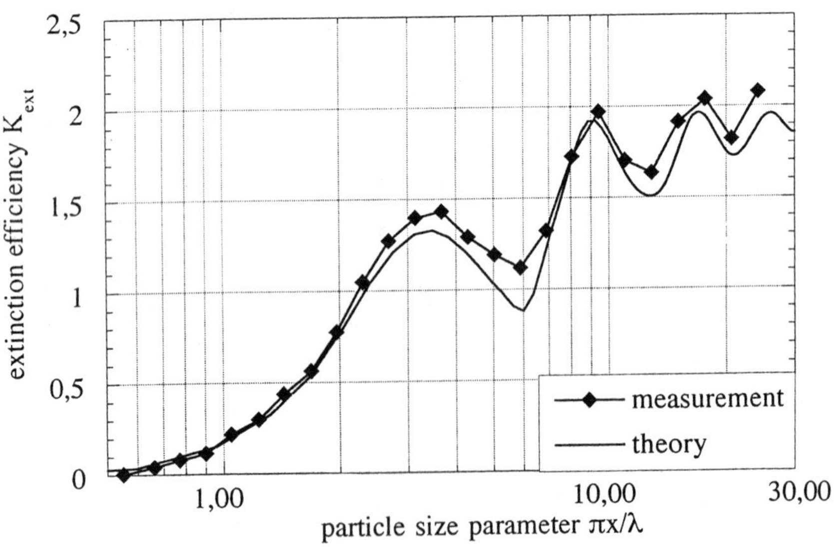

Figure 2. Extinction efficiency of glass beads in water, theory and experiment in comparison.[2]

For measurements of ultrasonic velocity by burst transmission,[2] an ultrasonic burst is transmitted through the suspension and recorded digitally. Bursts transmitted through the suspension and through the pure suspending liquid are correlated to obtain the phase velocity.

For homogeneous, isotropic, spherical particles with well-known material properties, experimental values and theoretical predictions of extinction efficiency and phase velocity are in good agreement; see Figures 2 and 3 with results for glass beads.

On the other hand, with materials like sharp-edged SiC or bronze spheres having a slightly rough surface, the theoretical prediction fails completely.[2] At the present state of knowledge, it is not clear however, if the failure of theory is to be attributed mainly to shape, to material anisotropy and inhomogeneity, or to a general failure of the theory which might be due to the much larger mismatch of particle and fluid impedances.

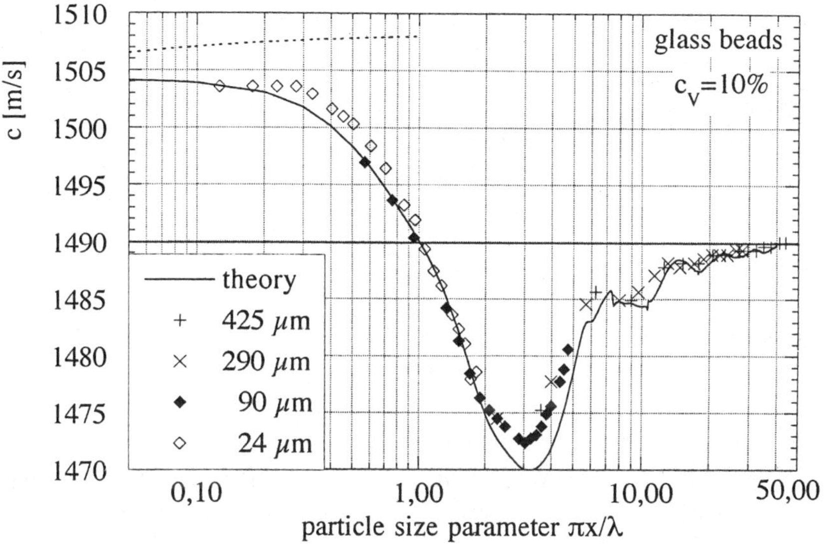

Figure 3. Phase velocity for suspension of glass beads in water, theory and experiment (burst transmission) in comparison.

An additional aspect to this is shown in Figure 4, where phase velocities measured by burst transmission and by continuous wave transmission are compared. Both measurements, even though done with great care and with the same materials, give quite different results, while the extinction efficiency determined by burst transmission does not differ from the result from continuous wave measurements. As the theory coincides with the result from burst transmission, this may lead to the conclusion that additional, weakly coupled modes of resonance exist, which are not excited by a short signal, and which are not included in the theory by Hay and Mercer.

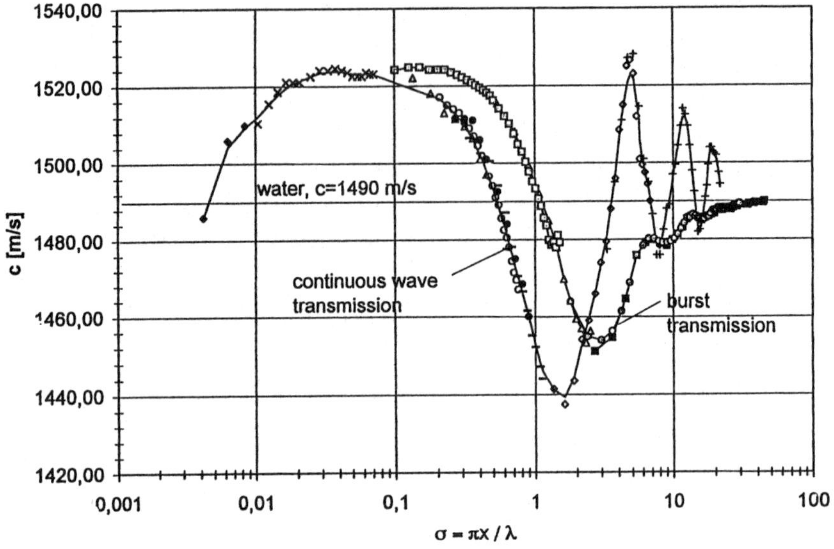

Figure 4. Phase velocity for suspension of glass beads in water, measurements by burst transmission and continuous waves in comparison.

3. First Conclusion

As a first conclusion, it may be stated that a theoretical prediction of extinction efficiency and phase function in the range of larger $\pi x / \lambda$ is possible in some cases, but the reliability of theories is not sufficient. When extinction spectra are inverted to obtain particle size distribution, data confirmed by experiment should be used.

4. Particle-Particle Interactions in Dense Systems

In concentrated suspensions, particle-wave interactions are modified by particle-particle interactions. According to the different physical mechanisms, at least three different types of particle-particle interaction must be discerned: multiple scattering, dependent scattering, and steric interactions are important in the range of intermediate and short wavelengths.[4] The thermal interactions, which can be important in the long wavelength regime, will not be discussed here.

The notion of multiple scattering signifies that parts of the ultrasonic radiation undergo interactions with different particles consecutively without being absorbed. While the ultrasonic radiation emitted from a common transducer is coherent and has a high degree of directivity (depending on transducer diameter and wave-length), multiple scattering events lead to a complete loss of coherence and directivity. Since piezoelectric receivers of sufficient diameter are very little sensitive to incoherent radiation, ultrasonic extinction measurements will, in general, not be disturbed by multiple scattering effects.

In extreme cases, where the level of multiply scattered ultrasound may be many orders of magnitude higher than the remaining level of coherent ultrasound, a double measurement with differently adjusted transducers may be used to quantify a possible disturbance:[5] When the transducers are bonded to wedge shaped windows, the emitter-receiver arrangement can be readjusted by rotating the windows by 180°. In the readjusted mode, the sensitivity of the receiver for directly transmitted, coherent, and collimated radiation will drop dramatically, while the sensitivity to the non-directive, multiply scattered radiation will remain unchanged. The evaluation of the double measurement hence allows to quantify coherent and incoherent contributions to the signal received.

Figure 5 shows an early result on the concentration-dependence of transmission, where a correction with double measurements had been necessary. It is evident, that even after correction for multiple scattering, strong deviations from Bouguer-Lambert-Beer´s law persist, which must be accounted to dependent scattering and steric interactions.

The notion of dependent scattering includes all effects where the mechanism of interaction between particle and wave is modified by the presence of neighbouring particles. Two cases are possible: When the particles are much smaller than the wavelength, the suspension as a whole may be treated as a homogeneous medium with different density, viscosity etc. Approaches of this type are known as "effective medium theories".

Dependent scattering is also possible with particles larger than the wavelength. When the gap between particles is smaller than about one third of the wavelength, the tunnelling effect leads to a coupling of the wave fields within neighbouring particles. This condition leads to a relation for the onset of dependent scattering,[4] where the critical C_V is a function of $\pi x / \lambda$ (see Fig. 6).

Steric interactions between particles lead, with increasing volume concentration, to the formation of structure in the suspension. In the limit of high concentration, a suspension of monodisperse spheres would evolve into a dense packing, and no more radiation could be transmitted along a line of sight beyond $C_V = 0.7$.

A rigorous treatment of steric interactions would have to use the Percus-

Yevick theory for the particle pair correlation function, together with modifications for hydrodynamic (flow-induced) and electrostatic effects. As for now, it seems impossible to implement this exact theory for the short wavelength regime, especially since the theories on scattering are not yet satisfactory.

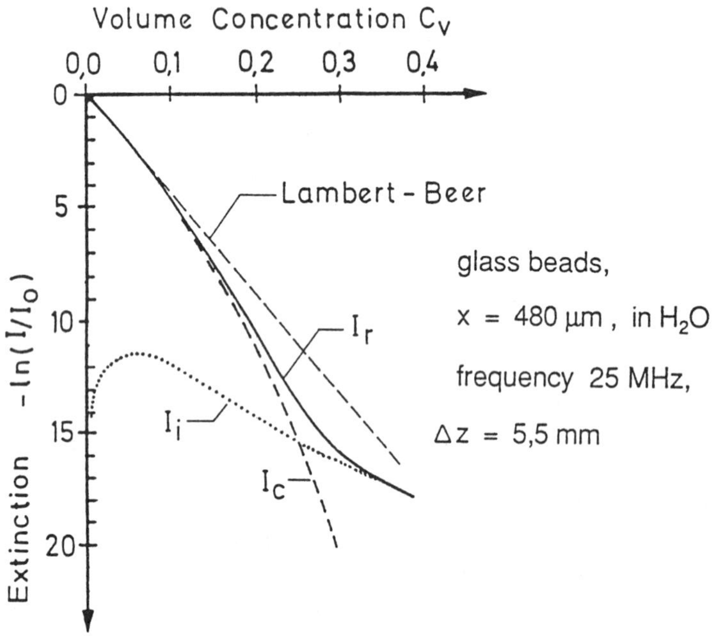

Figure 5. Concentration-dependent transmission with the contributions of coherent (I_C) and incoherent, multiply scattered (I_i) radiation. Experimental data obtained with transducers of approximately 10 mm in diameter.

Instead, a very simple approach[4] has proven to be successful as well. Looking back to the derivation of the BLBL, Equation (1), one finds the intensity balance for an infinitesimally thin layer of suspension. In a physically sound interpretation the infinitesimally thin suspension layer would have at least the thickness of one particle diameter, since particles are, physically, a unit which must not be subdivided. To allow integration, the intensity change within a layer of particle thickness would still have to be small relative to the local intensity itself. This is the case for small concentrations only. For a mathematical treatment of higher concentrations, finite changes of intensity have to be treated somehow. A number of different models have been developed to do so, including a model with particle monolayers, a statistical model, and numerical simulations. All results agree to give extinction as:

$$E = -\ln\left(\frac{I}{I_0}\right) = \frac{\Delta z}{x} \cdot \frac{1{,}5}{P} \cdot K_{ext} \cdot \ln(1 - C_V)$$

$$= E_{BLBL} \cdot \frac{\ln(1 - PC_V)}{PC_V} = E_{BLBL} \cdot F < C_V > \qquad (9)$$

This equation incorporates steric interactions without assuming a certain type of suspension structure. Empirically, the parameter P is found to be dependent on the suspension structure and may vary between 1.7 and 2.9, depending on the flow conditions. The equation has proven to be valid in the short wavelength regime with concentrations up to at least about C_V=0.4.

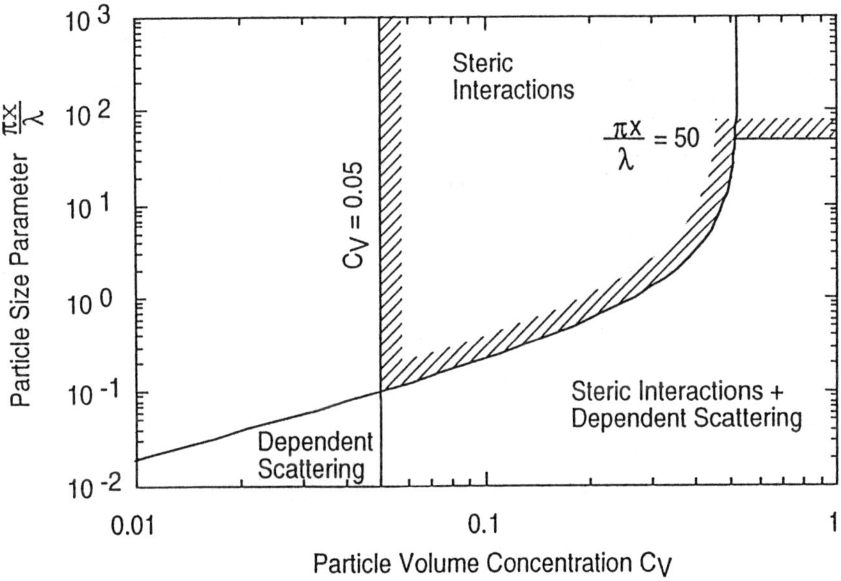

Figure 6. Regimes of particle-particle interaction.

Figure 6 shows a map with the different interaction regimes. Steric interactions were taken to be significant above 5 % by volume, where the effects are already well above the limit of detection.

Equation (9) shows that steric interactions are accounted for simply by multiplying extinction according to Bouguer-Lambert-Beer`s law with a function of volume concentration. If the extinction efficiency is evaluated from measurements of extinction spectra by using BLBL (Eq. (3) instead of Eq. (9)), steric interactions should just lead to a parallel shift in the curves $K_{ext} < \pi x / \lambda >$, while dependent scattering effects would express in a distortion of the curves.

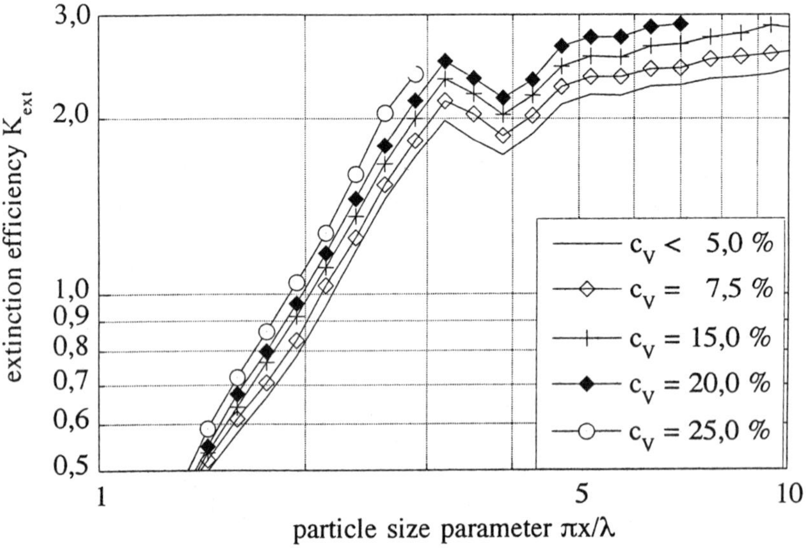

Figure 7. K_{ext} for monodisperse bronze beads (65 μm) in water/57 wt % glycerol.

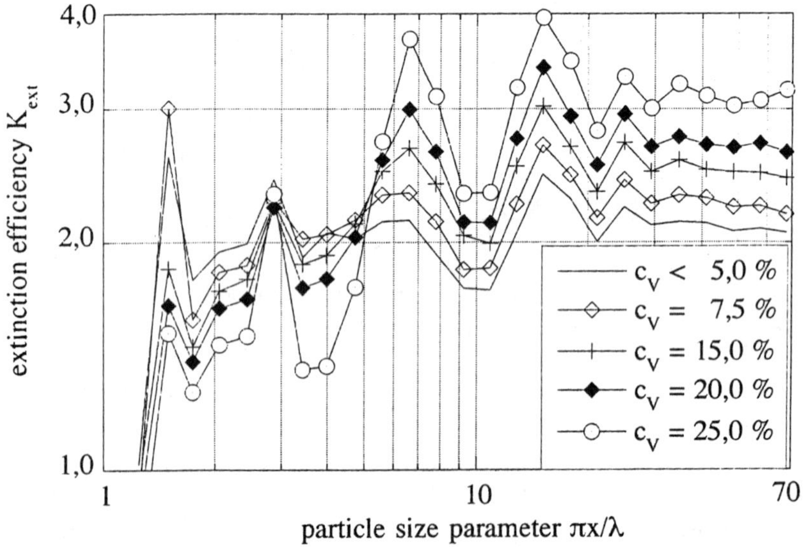

Figure 8. K_{ext} for monodisperse polystyrene beads (450 μm) in water.

Figures 7 and 8 show two out of a series of such measurements with different materials. It appears that in the short wavelength regime, steric effects are always the dominating mechanism of interaction. With intermediate wavelengths, dependent scattering may occur additionally, however only if the impedance mismatch between particles and fluid is not too strong. In the case of a very strong mismatch, such as with bronze beads in water/glycerol, the wave cannot penetrate significantly into the particles, and dependent scattering remains restricted to the long wavelengths.

Other than extinction, the phase velocity seems to be quite independent of suspension structure. This is seen from a comparison of the phase functions for burst transmission measured at three different particle concentrations (Fig. 9). There is almost no effect for $\pi x / \lambda > 1$.

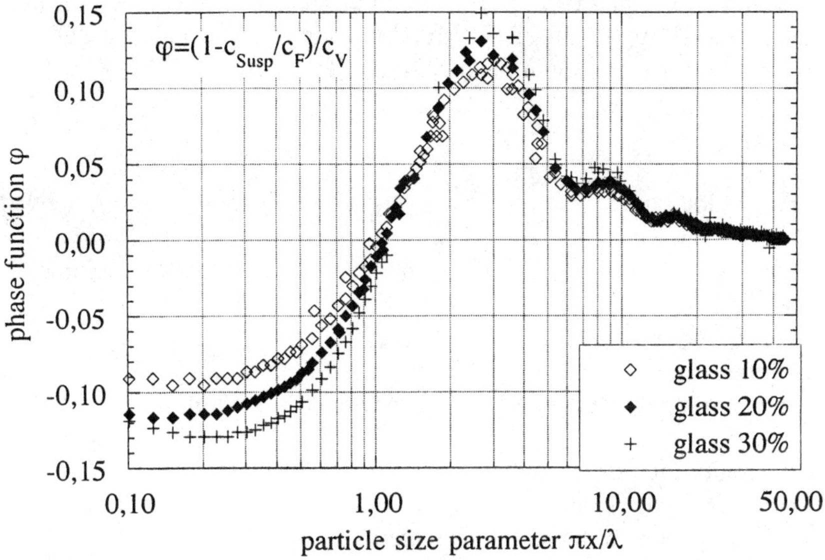

Figure 9. Phase function of glass beads in water measured at different concentrations by burst transmission.

5. Second Conclusion

The discussion of particle-particle interaction mechanisms has revealed that the common terminology, using "multiple scattering" as a general explanation for deviations from Bouguer-Lambert-Beer`s law, is quite wrong. From a physical point of view, steric interactions are the most basic effect, occurring whenever the volume concentration exceeds a few percent.

Dependent scattering superimposes on the steric effects, but occurs only with the regime of long and intermediate wavelengths.

Multiple scattering is always present as a physical phenomenon, but it is, in general, not relevant for ultrasonic measurements. Whenever multiple scattering effects are found, they are strongly dependent on the configuration of the experiment and can be reduced or even avoided by a proper design.

Acknowledgements

The overview given here resumes results from the projects Lo 142/11, Ri 533/3-1, and Ri 533/3-2 funded by Deutsche Forschungsgemeinschaft (DFG). The support given by the late Prof. Friedrich Löffler and the help of numerous students, who contributed their work and their ideas is gratefully acknowledged, as well as the additional financial support from SYMPATEC GmbH of Clausthal-Zellerfeld, Germany and the American Ceramic Society.

Nomenclature

c	phase velocity
x	particle diameter
q<x>	particle size density distribution by number (0), surface (2) or volume (3)
z	coordinate
A_{ext}	extinction cross section
An	complex efficients
C	concentration by number (N), projected area (PA) or volume (V)
E	extinction
F< >	function of Intensity received in presence of particles and without (o)
K_{ext}	extinction coefficiency
P	structure parameter for steric interactions
T	transmission
α_{ext}	extinction coefficient
λ	wavelength
Φ	phase function

References

1. U. Riebel and F. Löffler, "The Fundamentals of Particle Analysis by Means of Ultrasonic Spectrometry", *Part. Part. Syst. Charact.*, **6**, 135-143 (1989).

2. U. Kräuter, "Grundlagen zur in-situ Partikelgrößenanalyse mit Licht und Ultraschall", Doctoral thesis, Karlsruhe, 1995.

3. A.E. Hay and D.G. Mercer, "On the theory of Sound Scattering and Viscous Absorption in Aqueous Suspensions at Medium and Short Wavelengths", *J. Acoust. Soc. Am.*, **78**, 1761-1771 (1972).

4. U. Riebel and U. Kräuter, "Extinction of Radiations in Sterically Interacting Systems of Monodisperse Spheres. Part 1: Theory", *Part. Part. Syst. Charact.*, **11**, 212-221 (1994).

5. U. Kräuter and U. Riebel, "Extinction of Radiation in Sterically Interaction Systems of Monodisperse Spheres. Part 2: Experimental Results", *Part. Part. Syst. Charact.*, **12**, 132-138 (1995).

Ultrasound Spectroscopy: A Sound Approach to Sizing of Concentrated Particulates

Felix Alba
Felix ALBA Consultants, Inc., 5760 S. Ridge Creek Rd., Murray, UT 84107 USA

David Higgs, Robert Jack, and Paul Kippax
Malvern Instruments Ltd., Spring Lane South, Malvern, Worcestershire WR 14 1AT UK

Abstract

The three main issues associated with the design, implementation, and use of a particle size analyzer based on Acoustic Spectroscopy are conceptually discussed. They are *mathematical modeling*, *spectral measurement*, and *mathematical inversion*. Classical theoretical physics, used to model the interaction between sound waves and particulates, is reviewed and experimentally validated. The ubiquitous phenomenon of multiple scattering is encountered in ultrasound, and denounced as still a non-solved problem from a fundamental point of view. The principle of continuous waves (CW) is presented as the one with the best performance, when combined with an expert system which optimally associates frequency ranges with acoustical paths. Spectral deconvolution is treated through the concepts of *Spectral Information Content* and *Spectral Sensitivity to the Particle Size Distribution Concentration Pair (PSDCP)*. This paper intends to unfold –if only superficially- the necessary basic knowledge and criteria to assess the performance of a particle size analyzer based on ultrasound spectroscopy.

1. Introduction

Particle size analysis using acoustic spectroscopy consists of: a) predicting attenuation and velocity spectra for any particle size distribution (PSD) and concentration (mathematical modeling); b) measuring both spectra for the suspension, (spectral measurement); and c) inverting the mathematical model to extract the PSDCP out of the measured spectra (mathematical inversion).

The physics of ultrasound interaction with particles goes back, together with Optics, to the great scientist Lord Rayleigh more than a century ago. Phenomena occurring when light waves propagate through particulates also take place when acoustic waves interact with particles. However, due to the mechanical character of ultrasound waves, several other transport and thermodynamical phenomena occur, which render the process of mathematically modeling them more intricate, and its practical implementation more difficult. Besides, because of the subtle PSDCP information content available in the spectral measurements, and the needed high accuracy and resolution over broad frequency ranges, the engineering of gathering spectral data and deconvoluting them becomes a formidable challenge. These facts explain why, even though ultrasound –for the last thirty years- has been considered the technology of the future for industrial measurements, only very recently it has acquired the status of a proven usable technology.

2. Mathematical Modeling

2.1 Theoretical Background

A suspension/emulsion is a two-phase composite. There are two basic approaches to modeling the interaction of ultrasound with composites: *microscopic* (fundamental) and *macroscopic* (phenomenological). In the fundamental approach –the Wave Scattering Theory (WST)- the suspension is treated as what it is: a two-phase system, with the element of volume being much smaller than both the smallest particle and the smallest wavelength. This description is the most complex but the most accurate and general.

In the macroscopic approach instead, the dimension of the volume element is chosen such that it is much smaller than the minimum wavelength transmitted but still big enough to contain a large number of particles. In this fashion, the suspension appears –from point to point in space- to be homogeneous to the acoustical wave. This approach is much simpler, but it delivers not so accurate predictions and it is definitely very limited in scope. This oversimplification of the problem was originally devised -not surprisingly- by the same conspicuous physicist who initiated the WST. The reason for Lord Rayleigh to conceive such an approach was not a disbelief in his own fundamental creature, but the numerical intractability of it back in those times. The extreme exponent of this approach during this century has been Biot's theory of saturated porous solids.[1]

The above classification is not sharp: there are some intermediate approaches like the *cell-model*, *the coupled-phase theory*, and *closed-form approximations* to the fundamental one which avoid the numerical solution of the wave-equations by assuming the wavelength much longer than the size of the particles, the viscous wave penetration depth much smaller than the particle radius, the particle size much larger than the wavelength, etc.[2-7]

2.2 Engineering Background

There are three major patents which paved the way to engineering particle size analyzers based on acoustic spectroscopy: Cushman C.R. et al, in 1973;[8] Riebel U. in 1986;[9] and Alba in 1992.[10,11] The mathematical model employed by Cushman consisted of an empirical calibration between the attenuation at a single frequency and a single point in the PSD of mineral slurries. Such a simple approach has been employed in the mineral industry for decades, demonstrating the tolerance of ultrasound instruments to industrial environments.[12]

Riebel made a step further by postulating, as Flammer did in 1950,[13] and Alba et al in 1983,[14] that the attenuation at each frequency could be predicted as a linear combination of a number of size fractions representing the PSD of the particles. Riebel proposed measuring the spectrum and inverting the linear system using "per se known" techniques. He considered his extinction coefficients also as "per se known". There is no reference in his patent as to how to determine them, either from first principle physics or by any empirical calibration means.

Riebel also specified that, the minimum size in the suspension had to be larger than the minimum wavelength, and the maximum size had to be smaller than the maximum wavelength. As a result, the technique could not measure particle sizes under a few microns unless extremely and impractical high frequencies (>GHz) were transmitted.

Alba fully described a fundamental approach to modeling sound interaction with particulates by following the classical work of Epstein, Carhart, Allegra, Foldy, Waterman and Truell and others.[15-18] The predicted attenuation spectrum for a particle population having a range of particle sizes was not any more a linear function of the size fractions. By knowing a number of physical properties of both phases in the suspension, the attenuation and velocity spectra could be accurately predicted through fundamental Physics. The accurate measurement of particles a hundred times smaller than the minimum wavelength was clearly demonstrated with titanium dioxide slurries.

2.3 Our Approach: The Wave Scattering Theory

Because the term *scattering* was coined in Optics, some authors use it in Ultrasonics to mean only those phenomena analogous to light scattering, and some others with the general meaning of including all mechanical, thermodynamical and transport phenomena. As a result, the term multiple scattering has also the same ambiguous meaning. We propose to use the term *scattering* in its most general acceptation. After all, ultrasound WST describes – through its *scattering* coefficients- all phenomena analogous or not to light scattering. Finally, consistent with this meaning, we will call *multiple scattering* all phenomena occurring when the acoustical fields around the particles overlap and, as a consequence, the attenuation spectrum fails to be proportional to particle concentration. The terms *concentrated/diluted* are misleading for the purpose, because the overlapping can occur at concentrations as low as 0.1% or as high as 80%.

In 1945, Foldy published a paper on multiple scattering of electromagnetic waves. His concept of a configurational average over the ensemble of all possible spatial configurations of scatterers was extended to ultrasound waves by Waterman and Truell.[15-18] Epstein and Carhart, as well as Isakovich in 1950 independently pointed out that temperature gradients between particles and medium could provoke significant sound absorption through heat conduction.[15]

In 1970, Allegra gave proper recognition to Epstein & Carhart's theory in his thesis at Harvard University.[16] His outstanding conclusions were:

- The thermal conduction mechanism plays a decisive role in low density-contrast suspensions and all emulsions.

- For suspensions of solid particles in a liquid medium, thermal and viscous losses do not decouple, i.e., simple separate explicit expressions for the different extinction mechanisms cannot be obtained.

- For the first time, the boundary-value differential problem (wave equations) was solved numerically, with higher order contributions to the attenuation in its series solution being computed (up to about five terms).

- All the explicit expressions available in the literature by then were proved to fail for large particles/high frequencies. This was caused by the usual "long wave assumption" employed to derive those approximations.

- The absolute necessity of solving numerically the wave equations, instead of using explicit equations with limited validity regions, was established.

Epstein & Carhart, Allegra & Hawley, and Waterman & Truell papers have become classical references during this decade. The reason is that they provide a solid basis for the mathematical model needed in any particle size analyzer based on ultrasound. Even so, the phenomenological trend continues in spite of having available the necessary computational power to fully apply the WST.[19] An important human factor for this trend is that the WST does not offer an intuitive understanding of the underlying physical phenomena. We agree with Allegra: *it is not the WST itself, but the physical phenomena, which are overwhelmingly complex to allow for a fully intuitive understanding.*

The basic equations given by Waterman and Truell were:

$$[k_{sp}/k_{sm}]^2 = \{1+12C_v[f(0)]/(k_{sm}^2 D^3)\}^2 - \{12C_v[f(\pi)]/(k_{sm}^2 D^3)\}^2 \cdots [Ia]$$

$$[f(0)] = f(0,D) = (1/k_{sm}).\sum_{n=0}^{\infty}(2n+1).A_n \cdots [Ib]$$

$$[f(\pi)] = f(\pi,D) = (1/k_{sm}).\sum_{n=0}^{\infty}(-1)^n.(2n+1).A_n \cdots [Ic]$$

Where k_{sm} and k_{sp} are the wave numbers of the medium and the suspension respectively, Cv is the concentration, and [f(0)] and [f(π)] are the far acoustical fields for a population of size D. The {An} are the coefficients for the reflected wave in terms of Bessel, Hankel, and Legendre functions.[15,16]

Lloyd and Berry showed that the above formula had been derived under implicit incorrect assumptions.[20] However, their equation still does not predict the spectrum with the accuracy needed for a particle size analyzer. A basic statistical assumption for the previous model is that there is no correlation between positions of the scatterers, except that they should not interpenetrate. Some researchers have chosen different pair-correlation functions, e.g., the Percus-Yevick function.[21] Others have considered particles and medium as scatterers of different kind in an effective medium solving the equations "self consistently".[22]

The single-scattering model is obtained from the multiple scattering one by neglecting in equation Ia the term quadratic in Cv.[f(π)] which gives:

$$\mathrm{Im}(k_{sp}) = \mathrm{Im}(k_{sm}) - (12/D^3).\left\{\sum_{n=0}^{n=\infty}(2n+1).\mathrm{Re}(A_n(f,D)/k_{sm}^2)\right\}.C_v \cdots [IIa]$$

$$\mathrm{Re}(k_{sp}) = \mathrm{Re}(k_{sm}) + (12/D^3).\left\{\sum_{n=0}^{n=\infty}(2n+1).\mathrm{Im}(A_n(f,D)/k_{sm}^2)\right\}.C_v \cdots [IIb]$$

$$\alpha_{sp}(f,D,C_v) = \mathrm{Im}(k_{sm}) + \mathrm{Im}(k_{sp} - k_{sm}) = \alpha_{sm} + \alpha_{sp}^E \cdots [IIIa]$$

$$v_{sp}(f,D,C_v) = v_{sm}/\{1+v_{sm}\,\mathrm{Re}(k_{sp}-k_{sm})/2\pi f\} = v_{sm}/\{1+v_{sm}/v_{sp}^E\} \cdots [IIIb]$$

where the superscript E denotes the so-called excess attenuation. Finally, calculating the acoustical fields for the case of a polysize population with density function by volume F(D) -and using the classical Waterman and Truell assumption for scatterers of different size[18]- the attenuation and velocity for the suspension are obtained as:

$$\alpha(f)=\alpha_{sm}+\int_0^\infty \alpha_{sp}^E(f,D,C_v).F(D).dD \cdots [IVa]$$

$$v(f)=v_{sm}/\left\{1+\left[\left[\int_0^\infty \frac{v_{sm}}{v_{sp}^E}(f,D,C_v).F(D).dD\right]\right]\right\} \cdots [IVb]$$

From equations IIa, IIIa and IVa, the attenuation spectrum varies linearly with the concentration of the particles, as expected when there is no overlap between the acoustical fields around the particles. A similar approach applied to equations Ia,Ib,Ic instead of to equations IIa,IIb,IIIa,IIIb, will not obviously produce an attenuation spectrum linearly varying with the volume particle concentration.

It is worth-noting that the prediction of both spectra requires the calculation of an infinite series, each term of which requires the calculation of Bessel and Hankel functions of a complex argument, a numerical exercise far from trivial. Many authors, to avoid such a calculation to as many terms as needed for convergence, either assume again the long-wave regime for which two or three terms are enough for convergence, or simply obtain analytical approximations to the solution, or assume the short/intermediate wave regimes, with the thermal effect negligible, etc.[2,13,23] All these approaches have one intrinsic pitfall: *due to their hypotheses, they cannot predict the spectrum over several decades of frequency and particle size.*

Figure 1 shows that the explicit approximation that combines the viscous-inertial effect equation given by Urick[23] with the well-known formula for Rayleigh scattering[16] plus the Flammer's approximation[13] for Fraunhofer diffraction is substantially in error throughout the whole frequency range. Truncating the series after the fifth term, instead, gives virtually the same results up to about 30 MHz because the long-wave assumption is still valid. At higher frequencies, the five-term approximation totally deviates from the correct prediction (over 100 terms). Upon deconvolution of the spectrum, a serious error (>200%) in the PSD will result, if we use any but the accurate model.

2.4 Model Experimental Validation

We will compare experimental data with the model formulated by equations II, III, and IV. Amongst the numerous results we have collected over

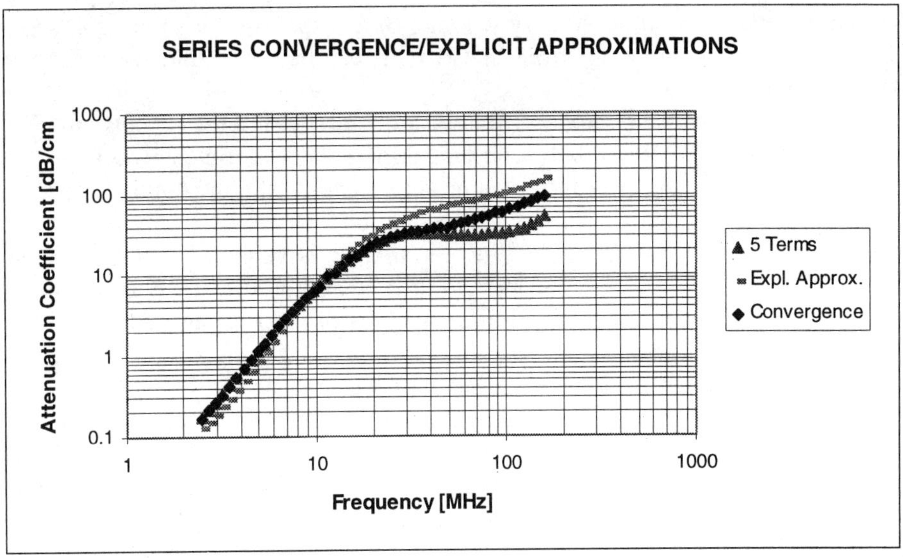

Figure 1. Glass Beads, 54 μm. Pitfalls of Explicit Approx. / Series Truncation.

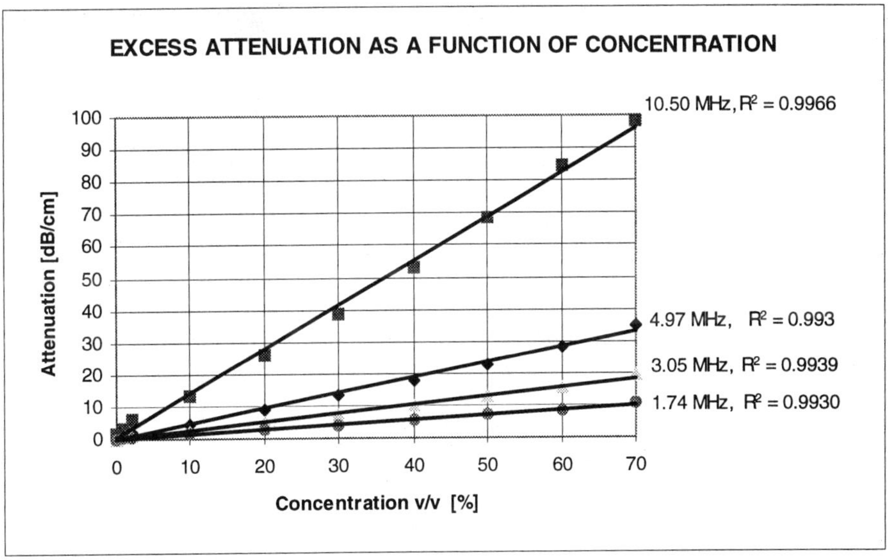

Figure 2. Bitumen Emulsion. No Evidence of Multiple Scattering. D50=10 μm.

more than 10 years of intensively testing the WST, we will show two conspicuous extreme examples: a) a bitumen emulsion for which the linearity with concentration is valid up to 70% by volume; b) a suspension of submicron solid particles for which deviations from linearity happen at concentrations as low as 4% by volume.

Figure 2 displays the excess attenuation for the bitumen emulsion. The droplet size remained essentially constant during the experiments because the product was adequately surfacted when produced and during the dilution process. Linearity of the attenuation with concentration is valid at any measured frequency. When the hypothesis of single-scattering is valid, we determined the accuracy of the WST to be better than 5% in the frequency range [1MHz,150MHz], over hundreds of different types of suspensions and emulsions at different concentrations.[10,24-26]

Figures 3 and 4 show the excess attenuation for a suspension of submicron calcium carbonate particles in water. Figure 3 corresponds to 1.5 MHz displaying significant deviations from linearity at concentrations as low as 4%. A third degree polynomial with zero intercept fits perfectly the measured spectrum. Figure 4 corresponds to a much higher frequency of about 34 MHz, and multiple scattering effects become significant only over 15% by volume. Same good polynomial fit applies. In these plots, the linear term of the polynomial fit (slope at the origin) agrees with the single-scattering prediction, and knowing the values of the empirical cubical and quadratic terms as a function of frequency, allows to calibrate the instrument for multiple scattering phenomena.

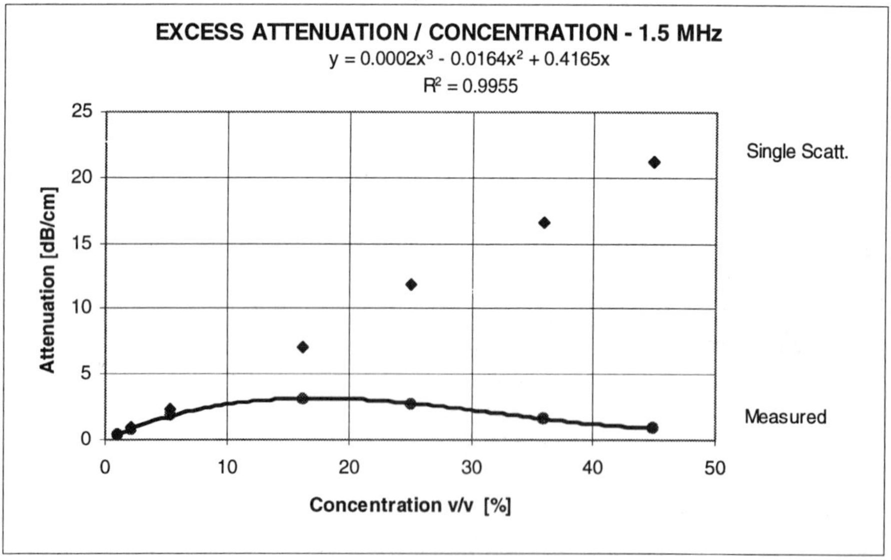

Figure 3. Calcium Carbonate. D50=0.5μm. Strong Multiple Scattering in effect.

The dependence of multiple scattering upon the acoustic frequency is easily understood, because the penetration depths of both viscous and thermal waves are inversely proportional to the square root of the frequency and, therefore, the higher the frequency the less overlap of the acoustical fields occurs between neighbor particles. Besides, because multiple scattering increases with the number of particles per unit volume, the larger the particle size, the weaker the deviations from linearity will be at a fixed concentration and frequency.

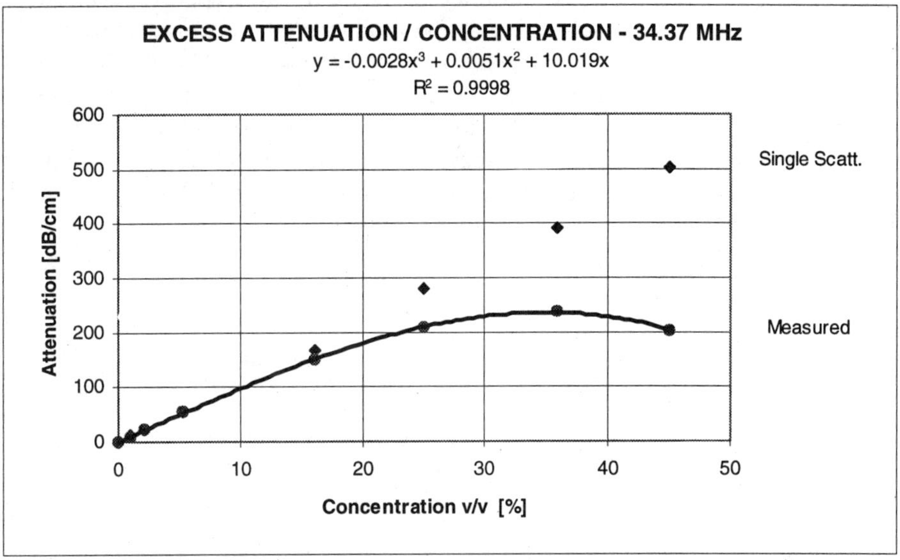

Figure 4. Calcium Carbonate. D50=0.5μm. Weaker Multiple Scattering in effect.

The strong multiple scattering shown for the submicron suspension as compared with the bitumen one can be explained as follows. First, in the former, the main attenuation mechanism is viscous-inertial while, in the latter, the dominant mechanism is thermal. At 1 MHz, the penetration depth of the viscous shear wave in the medium is around 530 nm, while the penetration depth of the thermal wave spreading from the bitumen droplet is only 213 nm. Second, being the bitumen droplets about 30 times larger than the submicron particles, a simple calculation gives a maximum concentration -below which the acoustical fields do not overlap- of about 72% v/v for the bitumen emulsion, and 3% v/v for the calcium carbonate suspension. This agrees with the shown experimental facts.

2.5 Practical Approaches to Modeling Multiple Scattering

To overcome the complexities as well as the denounced deficiency in the

WST, some researchers have proposed the cell-model, the coupled-phased theory, as well as combinations of them.[2-7,19] An extension of the cell-model approach to model multiple scattering is proposed in,[6,7] but, again, the assumption of long-wave regime is made, and adding separate different attenuation mechanisms is used as a modeling strategy.

The authors, instead, recognizing WST's lack of accuracy for some cases, as well as the complexities and time-scales for fundamentally solving the multiple scattering problem, have adopted an interim semi-empirical approach. We have developed a hybrid model comprising of the fundamental wave-scattering model plus empirical data fitted with polynomials. In this fashion, a robust model is built that relies more and more on the fundamental WST as the concentration decreases, while adjusts correspondingly the fundamental model with empirical information as the concentration increases. The usual drawbacks of empirical adjustments apply. A fundamental extension of the current WST should be a priority in this field.

3. Spectral Measurement

3.1 Theoretical Background

The most accurate model will look poor when tested against measurements with poor accuracy. A good fraction of the data published for the last 50 years did not agree with the WST simply because of poor-quality measurements.

To univocally determine the PSDCP from the spectrum, the latter has to be a signature for the first, i.e. there has to exist a one-to-one relation between spectra and PSDCPs. We then say the spectrum contains the maximum possible information. Independent pieces of information are added to the spectrum by broadening the frequency range.

The best inversion algorithm performs poorly when the information content of the spectrum is poor. Besides, when the frequency is swept over two decades or more, the wave attenuation may vary greatly from less than 0.5 dB/cm to hundreds of dB/cm. Because a typical dynamic range of a measurement electronics is little more than 100 dB, reproducibility and accuracy in the spectrum will depend upon the length of the acoustical path: *very low attenuation coefficients require long paths and very large attenuation coefficients require short paths.*

The above statements indicate that measuring the spectrum requires the best accuracy and dynamic range possible, as well as the widest possible frequency range. *The last requirement calls for not using a mathematical model based on hypotheses like long-wave, short-wave, or intermediate wavelength regimes. Inevitably, those hypotheses will be violated simply because the frequency has to be swept over a wide frequency range (to make the spectra/PSDCP relation one-to-one) or simply because the PSD being measured covers a wide range of sizes.*

3.2 Engineering Background

There are basically three technical approaches to measure ultrasound spectral information: tone burst, continuous waves (CW), and broadband pulse. The first technique consists of transmitting a few cycles of a sinusoidal wave modulated by a square pulse, conducting the measurement, and changing the frequency. In the second method, a pure sinusoidal wave is transmitted until steady-state is reached, performing the measurement, and changing the frequency. The third technique consists of transmitting a broadband pulse, numerically transforming both the transmitted and received signals to the frequency domain, and obtaining the attenuation spectrum through the ratio of both transforms.

3.3 Our Approach: CW, plus HPD, plus DFT, plus OMS

Discussions on the merits and drawbacks of each one of the above techniques would take a paper by itself.[10,27] Suffices to say that, of all them, the CW approach combined with the Heterodyne Principle of Detection (HPD), plus Discrete Fourier Transformations (DFT) to eliminate wave reverberations,[27] provide the best accuracy and resolution, the largest dynamic range, and the widest frequency range. As a disadvantage, the measurement may take longer, because the time for steady-state depends upon the length of the acoustical path.

With the best electronics, still the problem remains that the optimal acoustical path changes with the level of attenuation being measured which, in turn, depends strongly upon frequency. Besides, the optimal path is also a function of the frequency response of the transducers, which depends on the acoustical coupling between them and the sample. All these reduce the effective dynamic range of the instrument, calling for automatic adjustment of the distance.

Every unknown sample thus requires what we call an Optimal Measurement Strategy (OMS), so the measurements have maximum accuracy, reproducibility, and resolution over the widest possible frequency range. The expertise acquired over many years conducting these measurements has been coded into the computer through an expert system that scans the unknown sample, determining the OMS before the actual spectral measurements and particle size analysis are performed.

4. Mathematical Inversion

4.1 Theoretical Background

Particle size analysis is essentially an inversion process: using a model, we predict the spectrum with a given accuracy for any PSDCP. We then measure the spectrum with a given accuracy, and "look backwards" throughout the model to determine for which one of the PSDCPs, its predicted spectrum agrees with the measured one within the combined uncertainty.

The main difficulty in solving this problem resides in its inherent instability: experimental errors, model errors, even round-off errors in the measured and/or predicted spectra, can produce large variations in the solution. Rephrasing it, the relation between PSDCPs and spectra is neither one-to-one nor stable: *within the functional space of solutions (Mathematical Feasibility Space, MFS), an infinite number of them are purely mathematical without any physical meaning. Even more, within the functional space of physical solutions (Physical Feasibility Space, PFS), the relation between PSDCPs and spectra may not be one-to-one either, throughout the desired five decades of size range. We say the relation between the PSDCPs and the spectra displays singularities.*

The problem requires additional information so the non-physical solutions can be filtered. Obvious new information includes that the solution should belong to the class of non-negative functions with their integral equal to unity, and that the concentration should be in the interval [0,1]. Once the obvious non-physical solutions have been eliminated, the relation will still be multi-valued, i.e., several very different physically acceptable PSDCPs can have spectra that are experimentally indistinguishable over the measured frequency range. *The ability to expand the frequency range and –sometimes- to combine attenuation with velocity spectra are crucial to converting the relation into mono-valued and, only then, the PSDCP can be successfully determined.*

The next important two related concepts are: a) the *information content* of the spectrum, i.e., how much detail of the PSD's fine structure is really contained in the measured spectrum; and b) *sensitivity of the spectrum to the PSDCP,* i.e., how close can two different PSDCPs be and still their associated spectra being experimentally distinguishable. We would like to point out a common misconception that reads: *the information contained in the measured spectrum is very limited, ergo, very few frequency points will suffice to extract it.* The correct statement reads totally the opposite: because the information contained in the measured spectrum is quite subtle and limited, the accuracy and resolution with which we measure the spectrum are paramount.

Figure 5 displays spectra for a 2% TiO2 slurry. The spectrum with the largest curvature corresponds to a very narrow distribution, the one with the smallest curvature corresponds to a very broad distribution.[10,11] The two extremes in distribution spread show spectra not greatly different to each other: *the distribution spread information is subtly encrypted into the spectrum curvature.* Consequently, because of natural errors in the measurements, the more frequency points we measure, the more accurate the estimate of the curvature will be.

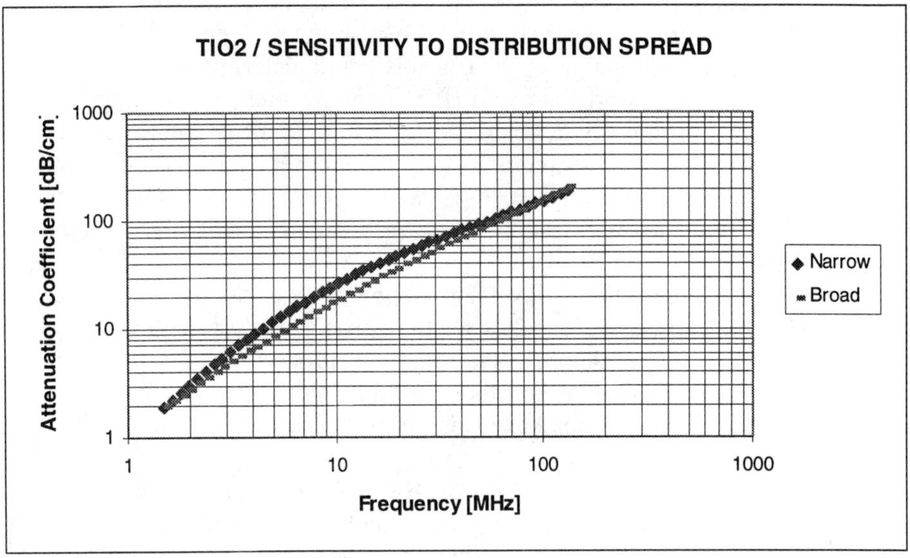

Figure 5. Spectra of TiO2 at 2% for very narrow and very broad distributions.

4.2 Engineering Background

There is a considerable amount of work and experience in dealing with inversion of integral equations, as it is a common place throughout optical particle size analyzers, radiometers, spectrometers, interferometers, seismometers, etc. However, every solution is unique to each one of those fields, because the kernel of the integral determines how difficult the inversion process is, and the kernel comes from the specific physics on which the instrument is based.[28]

Once the obvious non-physical solutions have been filtered out, the problems of converting a several-to-one relation into a one-to-one relation and stabilizing the solution become very subtle: *any further attempt to reduce the feasibility space will inevitably affect the PFS.* There are basically two approaches improperly named as "model-independent" and "model-dependent" size analyses. These misnomers led to the widely held unsupported specious notion that the first inversion technique provides "true, natural" size distributions while the second one delivers "fallacious, unnaturally forced" size distributions. We would like to unmask this folklore so the users of acoustic spectrometers base their evaluation of instrument performance on genuine more solid arguments.

The "*model*-independent" technique consists in *modeling* the density function by discretizing it into its differential form using a number of size intervals, and additionally *modeling* the solution by requesting it to belong to the class of functions with "reasonably continuous" transitions between contiguous intervals. The inversion problem so defined has as many degrees of freedom as size intervals. How reasonable is "reasonably" will determine how much the inversion process will distort the solution in order to filter out the singularities, and how many real solutions will be missed (the stronger the continuity condition, the broader the feasible distributions, and the poorer the resolution of the instrument as a whole). Clearly, the "*model independent*" inversion technique uses a *hidden model* to filter out the singularities and, inevitably, distorts the solution to an extent.

The "*model-dependent*" inversion approach consists in restricting the solution to belong to a class of functions with a "reasonable" structural form (log-normal, Rosin-Rammler, Gamma, etc.). The approach openly and a-priori assumes a functional structure for the solution, and that is why it is more vulnerable to criticism. The number of degrees of freedom, however, decreases drastically.

Again, how reasonable is "reasonable" will determine how much the solution will be distorted and how many and which real solutions will be missed. Choosing, e.g., the log-normal class, because the PFS will then include all monosize distributions (geometrical deviation=1), reference particles can be efficiently measured, <u>if the information content of the spectrum is rich enough</u>.

On the other hand, we know there are plenty of physical distributions not representable by a simple model like the log-normal or Rosin-Rammler. *The problem thus becomes one of choosing a functional structure that, on one side is flexible enough to represent any physically meaningful distribution and, on the other, the number of parameters (degrees of freedom) is a minimum, so stability of the solution is maximized, and the chances for singularities are reduced.*

In terms of *information content* of the spectrum, if the predicted spectrum for, e.g., a Rosin-Rammler distribution agrees with the measured one within experimental and modeling errors, then there is no more retrievable information in the spectrum using the current frequency range. Any fine-structure differences between the Rosin-Rammler and any other model like the log-normal, or any distribution obtained with a "model-independent" technique are indistinguishable for the instrument. However, *the fact that the "model-independent" distribution looks more irregular than the nice and clean Rosin-Rammler is misleading the user to believe the former is more "real" than the latter.* This objection to a nice and clean distribution is sustained even by experienced professionals who know that –even among particle size analyzers using the same principle- results on the same sample may differ by more than 10%.

Now, if there is no distribution within the class of, e.g., log-normal functions which predicted spectrum agrees -within experimental and modeling errors- with the measured spectrum then, clearly, the log-normal class is not rich enough to include a good approximation to the real PSD. There are more than just fine-structure differences: *the information content in the spectrum is calling for a size distribution not representable by the log-normal model.*

4.3 Our Approach: Multi-Functional Representation (MFR) of the PFS

From the above basic discussions, even though we also use the "model-independent" size analysis, we prefer -for acoustic spectroscopy- the MFR as an inversion technique more accurate, stable, and with a higher resolution. The MFR starts by assuming that the unknown density function *f(d)* can be represented by a linear combination of density functions each one of which is representable by well-known functional models like Log-Normal, Rosin-Rammler, Gamma, etc. For the last 8 years, we have used two log-normal components very successfully, and no practical need for other models or additional components has been found yet.[10,24-26]

$$f(d) = \sum_{1}^{N} p_i \cdot f_i(d) \quad ; \quad \sum_{1}^{N} p_i = 1 \quad ; \quad f_i\text{: Log-Normal } (GM_i, GD_i)\ \forall\, i$$

$$GM_i \geq 0 \qquad GD_i \geq 1 \qquad 0 \leq C_V \leq 1 \qquad \text{[V]}$$

Because the log-normal class of functions requires two parameters, namely the geometrical mean (GM) and geometrical deviation (GD), if N=2, five parameters are necessary to completely define a function within the class. Consequently, given the measured spectrum, the inversion problem consists in estimating six parameters, five for the size distribution plus the concentration Cv.

It is worth noting that the class includes all unimodal log-normal density functions (p2=0), all bimodals with two well-separated modes each one log-normal, all single size spikes (Dirac's δs, p2=0, GD1=1), all bimodal size spikes (pi>0,GDi=1 i=1,2), and –more importantly- *an infinite set of <u>non-log-normal size distributions with almost arbitrary shape</u>*. This simple representation is therefore flexible and powerful enough to cover an ample universe of practical situations.

5. Conclusions

The Wave Scattering Theory, when fully solving the wave-equations, accurately predicts spectra for emulsions at extremely high concentrations (70% v/v). For suspensions of solid submicron particles for which the thermal effect is not the dominant one, its accuracy deteriorates at much lower concentrations. Development of a fundamental theory of multiple scattering, accurate enough for its integration into a particle size analyzer, should be a high priority in this field.

Accuracy, reproducibility, resolution, and frequency range are paramount to retrieving the information content carried by the spectral data.

Both "*model* independent" and "*model* dependent" techniques for particle size analysis use *models* to filter out the singularities in the convoluted Spectra/PSDCP relation. The MFR approach extracts more efficiently the information content in the spectra, particularly when model equations are non-linear.

References

1. M.A. Biot, "Generalized Theory of Acoustic Propagation in Porous Dissipative Media", *J. Acoust. Soc. Am.*, **34** [9] 1254-1264 (1962).

2. A.E. Hay and D.G. Mercer, "On the theory of sound scattering and viscous absorption in aqueous suspensions at medium and short wavelengths". *J. Acoust. Soc. Am.*, **78** [5] 1761-1771 (1985).

3. U. Krauter and U. Riebel, "Extinction of Radiation in Sterically Interacting Systems of Monodisperse Spheres. Part 2: Experimental Results", *Part. Part. Syst. Charact.* **12**, 132-138 (1995).

4. T.A. Strout, "Attenuation of Sound in High-Concentration Suspensions: Development and Application of an Oscillatory Cell Model", Ph.D. Dissertation, University of Main (1991).

5. A.H. Harker and J.A.G. Temple, "Velocity and attenuation of ultrasound in suspensions of particles in fluids", *J. Phys.*, **21** 1576-1588 (1988).

6. A.S. Dukhin and P.J. Goetz, "Acoustic Spectroscopy for Concentrated Polydisperse Colloids with High Density Contrast", *Langmuir*, **12**, 4987-4997 (1996).

7. A.S. Dukhin and P.J. Goetz, "Acoustic Spectroscopy for Concentrated Polydisperse Colloids with Low Density Contrast", *Langmuir*, **12**, 4998-5003 (1996).

8. C.R. Cushman et al., "Particle Size and Percent Solids Monitor" U.S. Pat. No. 3,779,070, 1973.

9. U. Riebel, "Method of and an Apparatus for Ultrasonic Measuring of the Solids Concentration and Particle Size Distribution in a Suspension", U.S. Pat. No. 4,706,509, November 17, 1987, DE Pat. No. 3438798, April 24, 1986.

10. F. Alba, "Method and Apparatus for Determining Particle Size Distribution and Concentration in a Suspension using Ultrasonics", U.S. Pat. No. 5,121,629, June 16, 1992.

11. F. Alba, "Method and Apparatus for Determining Particle Size Distribution and Concentration in a Suspension using Ultrasonics", European Pat. No. EP 0605409, January 31st, 1996. Also granted in Japan.

12. F. Alba and J.A. Herbst, "Ultrasound Absorption Phenomena in Slurries makes Possible the Estimation of Complete Particle Size Distributions and Solids Concentrations", in *Particulate and Multiphase Processes*, Vol 3, Colloidal and Interfacial Phenomena. Edited by T. Ariman, T.N. Veziroglu. Hemisphere Publishing Corp. Distribution outside North America by Springer-Verlag, 1987.

13. G.H. Flammer, "The Use of Ultrasonics in the Measurement of Suspended Sediment Size Distribution and Concentration", PhD Thesis, University of Minnesota, August 1958.

14. F. Alba and J.A. Herbst, "Techniques for On-Line Determination of Multiple Points on a Particle Size Distribution using a Two Sonic Beam Attenuation Analyzer", Patent Disclosure to the University of Utah, Nov. 16, 1983.

15. P.S. Epstein and R.R. Carhart, "The Absorption of Sound in Suspensions and Emulsions. I. Water Fog in Air", *J. Acoust. Soc. Am.*, **25** [3] 553-565 (1953).

16. J.R. Allegra and S.A. Hawley, "Attenuation of Sound in Suspensions and Emulsions: Theory and Experiments", *J. Acoust. Soc. Am.*, **51**, 1545-1564 (1972).

17. L.L. Foldy, "The Multiple Scattering of Waves", *Phys. Rev.*, **67** [3-4] 107-119 (1945).

18. P.C. Waterman and R. Truell, "Multiple Scattering of Waves", *J. Math. Phys.*, **2** [4], 512-537 (1961).

19. J.M. Evans and K. Attenborough, "Coupled phase theory for sound propagation in emulsions", *J. Acoust. Soc. Am.*, **102** [1] 278-282 (1997).

20. P. Lloyd and M.V. Berry, "Wave Propagation through an Assembly of Spheres, IV, Relations between Different Multiple-Scattering Theories", *Proc. Phys. Soc.*, **91** 678-689 (1967).

21. L. Tsang, "Multiple Scattering of Acoustic Waves by Random Distribution of Discrete Scatterers with the Quasicrystalline and Percus-Yevick Approximation", *J. Acoust. Soc. Am.*, **71** [3] 552-558 (1982).

22. C.M. Sayers, "On the Propagation of Ultrasound in Highly Concentrated Mixtures and Suspensions", *J. Phys. D: Appl. Phys.*, **13**, 179-184 (1980).

23. R.J. Urick, "The Absorption of Sound in Suspensions of Irregular Particles", *J. Acoust. Soc. Am.*, **20** [3] 283-289 (1948).

24. F. Alba, et al, "On-Line High Solids Concentration Particle Size Analysis by Ultrasonic Spectroscopy", 13th Symposium on Industrial Crystallization, September 16-19 1996, Toulose, France.

25. Malvern Instruments Ltd., "Acoustic Spectroscopy, New Product Update", 1995.

27. T. Oja and F. Alba, "Acoustic Attenuation Spectroscopy for Particle Sizing of High Concentration Dispersions", in *Ultrasonic and Dielectric Characterization Techniques for Suspended Particulates*, Edited by V. A. Hackley and J. Texter. American Ceramic Society, Westerville, OH, 1998.

27. Hewlett Packard Co., "Test and Measurement Catalog", 1997.

28. S. Twomey, *Introduction to the Mathematics of Inversion in Remote Sensing and Indirect Measurements*, Elsevier, 1977.

Unified Coupled Phase Continuum Model for Acoustic Attenuation in Concentrated Dispersions

Wei Han* and Hemant P. Pendse

Chemical Engineering Department, University of Maine, Orono, Maine 04469-5737 USA

Abstract

This paper describes a Unified Coupled Phase Continuum (UCPC) model developed to analyze sound propagation through colloidal dispersions. The governing equations involve equations of state, continuity, momentum and energy for the fluid medium and suspended phase, respectively. The viscous (thermal) coupling between the two phases is accounted through matching terms in the equations of momentum (energy) formulated separately for each phase. Frequency- and concentration-dependent viscous and thermal coupling coefficients are developed using the oscillating 'sphere-in-cell' framework. An approximate similarity between the 'viscous skin drag' and 'heat transfer flux' associated with the suspended phase was used. Expressions for the coupling coefficients explicitly account for particle size, concentration, and sound frequency. A closed form expression for the effective complex wavenumber, as a function of particle size, frequency, concentration, as well as dimensionless parameters pertinent to physical and thermodynamic properties for the two constituent phases, is derived for the viscothermal couplings. The UCPC model predictions of attenuation spectra provide satisfactory agreement with reported data on several concentrated suspensions and emulsions, covering wide ranges of dispersed phase volume fractions, operating frequencies, and particle diameters. The UCPC model can provide for a theoretical framework in characterization of concentrated dispersions.

1. Introduction

Concentrated dispersions are commonly found in industry, and include suspensions and emulsions containing particles or drops suspended in a fluid medium. Measurements of ultrasonic attenuation spectra have been recently used for characterization of colloidal dispersions of practical significance in terms of particle size and volume fraction. Successful determination of dispersion properties with ultrasonic spectroscopy depends on validated theoretical models. These models must accurately account for the variation in attenuation and phase

* Current address: Los Alamos National Laboratory, MST-11, MS D429, Los Alamos, NM 87545

speed with frequency on pertinent parameters describing the dispersion. The propagation of sound through suspensions and emulsions has been the subject of several studies. Harker and Temple[1] have reviewed various analytical approaches.

Some studies are based on analyses involving series expansions of scattered wave functions associated with an isolated single sphere that yield linear dependence on suspended phase volume fraction, ϕ for the suspended phase contribution to the attenuation. This procedure has been reviewed by Allegra and Hawley[2]. Later, Davis[3], and McClements[4] applied results based on multiple scattering formulations (Waterman and Truell [5], Ma *et al* [6], Lloyd and Berry [7]) in combination with the results by Allegra and Hawley to extend the applicable volume fraction domain by using a second order term for the nonlinear ϕ dependence. Holmes and Challis[8] suggested using the results given by Lloyd and Berry[7], instead of those by Waterman and Truell[5], for the $(\phi)^2$ term.

An alternative approach in analyzing sound propagation through dispersions was taken by Temkin and Dobbins[9] for dilute suspensions of solid rigid particles in a perfect gas. This approach involves the so called "Coupled Phase Continuum (CPC)" model that treats a two-phase mixture, with a dispersed phase suspended in a second continuous fluid medium phase, as a single continuum characterized by an effective complex wave number K. Various attempts have been made to extend this approach by Gibson and Toksöz[10], Harker and Temple[1], and Atkinson and Kytömaa[11] by proposing different expressions for the interphase viscous coupling term accounting for particle concentration effects. The same analytical expression was obtained by these investigators for the effective wave number in terms of a viscous coupling coefficient, but they suggested different expressions for the dependence of the viscous coupling coefficient on ϕ. These analyses take into account the viscous coupling only, and not the thermal coupling. The results are thus limited to the case of high density contrast, $\left|(\rho_s - \rho_f)/\rho_f\right| \geq 1$, suspensions for which the viscous contribution to attenuation is much greater than the thermal contribution. Strout[12] suggested using the results from the oscillating 'sphere-in-cell' formulation to obtain the concentration and frequency dependent viscous coupling coefficient, so that the model can be applied over wide ranges of concentration and frequency. An interesting extension that takes into account both viscous and thermal couplings is recently provided by Margulies and Schwarz[13] for the case of dilute suspensions involving viscoelastic fluid media. They suggest using the frequency dependent complex viscosity to replace the shear viscosity in the expressions for the viscous coupling term.

In this paper we present a Unified Coupled Phase Continuum (UCPC) model that incorporates concentration and frequency dependent coupling coefficients, viscous as well as thermal, obtained using the 'sphere-in-cell' formulation. We have extended the coupled phase continuum model by including the viscous and thermal coupling terms that take into account various parameters such as size of the suspended particles/drops, the dispersed phase concentration, and the sound frequency. Details of the UCPC model development and evaluation are given in Han's thesis[14]. The UCPC model predictions of attenuation coefficient spectra have been compared by Han[14] with experimental measurements on a variety of colloids including aqueous suspensions and emulsions, as well as aerosols. The current UCPC model is found to successfully predict the attenuation spectra of concentrated colloids covering a wide range of relative magnitudes of viscous verses thermal attenuation contributions. Here we focus our attention to suspensions and emulsions with low density-contrast, $\left|\left(\rho_s - \rho_f\right)/\rho_f\right| \leq 1$, and with comparable thermal and viscous attenuations, at moderate to high particle volume fractions.

2. Unified Coupled Phase Continuum (UCPC) Model Formulation

The dispersed phase (suspended solid particles or liquid drops) and the continuous phase (fluid medium) mixture is represented as an effective continuum, with its complex wave number, K, leading to the eight relevant acoustic variables, $\bar{\varphi}$ $\left\{\bar{v}_s, \bar{v}_f, \bar{\rho}_s, \bar{\rho}_f, \bar{P}, \bar{\phi}, \bar{T}_s, \bar{T}_f\right\}$, in the following 'complex' form:

$$\varphi^{perturbed} = \varphi^0 + \bar{\varphi} \cdot \exp\left[j \cdot \left(\omega\, t - K\, z\right)\right] \quad (1)$$

where $\bar{\varphi}$ denotes small acoustic perturbations from the corresponding undisturbed steady state values, φ^0. The superscript ^{0} is dropped for convenience, when not necessary for clarification.

The effective complex wave number, K is

$$K \equiv \frac{\omega}{c} - j\alpha_{UCPC} \quad (2)$$

The wave numbers for the fluid medium K_f and suspended phase K_s are given by:

$$K_f \equiv \frac{\omega}{c_f} - j\alpha_f \qquad (3)$$

$$K_s \equiv \frac{\omega}{c_s} - j\alpha_s \qquad (4)$$

The attenuation coefficient, α_{UCPC} when modified by adding appropriate contributions to account for scattering losses and intrinsic absorption within the suspended phase yields a suitable estimate for the total attenuation coefficient of the two phase mixture.

The governing equations, listed in APPENDIX A, involve equations of state (A1-2), continuity (A3-4), momentum (A5-6) and energy (A7-8) corresponding to the suspended phase and the fluid medium phase, respectively. The individual phase momentum equations are related through the viscous coupling term, F^*, representing the viscous force exerted by the fluid medium onto the dispersed phase per unit total volume. Similarly, the individual phase energy equations are related through the thermal coupling term, Q^*, representing the rate of heat transfer from the dispersed phase in to the fluid medium, per unit total volume. The viscous (thermal) coupling between the two phases is accounted through matching terms in the equations of momentum (energy) formulated separately for each phase. The formulation of these governing equations has been discussed in detail elsewhere [14].

Consideration of effects of suspended particle size, dispersed phase volume fraction, ϕ and sound frequency, ω on to these coupling terms F^* and Q^* constitutes a critical step in the UCPC analysis of sound propagation through concentrated dispersions. We use the following general form to relate F^* to single particle drag force F_p , and Q^* to single particle heat loss rate Q_p:

$$F^* = \phi \cdot \left(1 - \phi\right) \cdot F_p / V_p \qquad (5)$$

$$Q^* = \phi \cdot \left(1 - \phi\right) \cdot Q_p / V_p \qquad (6)$$

Solutions for F_p and Q_p based on the oscillating 'sphere-in-cell' formulation provide relevant functions characterizing the concentration, frequency, and particle size effects in terms of the dispersed phase volume fraction, ϕ and two oscillatory viscous and thermal boundary layer parameters, s_V and s_T. The dimensionless parameters s_V and s_T are proportional to ($D\sqrt{\omega}$). Relationships between F_p and (v_s - v_f), and Q_p and (T_s - T_f) can be written as:

$$F_p = -g(\phi\ , s_V) \cdot A_p \cdot (v_s - v_f) \tag{7}$$

$$Q_p = h(\phi\ , s_T) \cdot A_p \cdot (T_s - T_f) \tag{8}$$

where functions g and h denote the viscous drag coefficient and the heat transfer coefficient, respectively. It is convenient to express the interphase transfer coefficients, g and h, in terms of corresponding dimensionless parameters, namely the Stokes' drag correction factor, Ω, and the Nusselt number, Nu, defined as follows:

$$3 \cdot \Omega(\phi\ , s_V) \equiv \frac{A_p\, g(\phi\ , s_V)}{\pi \eta_f D} \tag{9}$$

$$Nu(\phi\ , s_T) \equiv \frac{h(\phi\ , s_T) D}{k_f} \tag{10}$$

Note that g, h, Ω, and Nu are all 'complex' quantities with both real and imaginary parts, i.e. amplitudes and phase lags.

The concentration and frequency effects on the viscous drag can be adequately accounted for using a cell model approach[15, 16], as long as particle-particle contact friction and/or chain formation is not involved. Generally, the cell model approach solves for the flow fields in the fluid near a reference particle using steady Navier-Stokes and continuity equations with boundary conditions applied at the surface of the particle and an outer surrounding imaginary envelope. Happel[15] proposed a famous sphere-in-cell free surface model to represent the concentration effects for multiparticle systems. A random assemblage is considered to consist of a number of cells. Each cell contains a solid particle of radius R, moving with steady velocity, and an outer fluid spherical envelope of radius R_{cell}. The size of the cell is chosen such that the porosity of the cell entity matches that of the system as a whole, that is, $(R/R_{cell})^3 = \phi$. On the outer envelope shell, the boundary conditions insure that normal velocity and tangential stress vanish, implying that the disturbance due to the movement of the inner particle is confined within the cell of the fluid with which it is associated. The cell model approach, which is based on abstract and hypothetical assumptions and excludes the particle-particle contact and interactions, is easy to apply and has been found to provide excellent agreement with experimental values[16]. Strout[12] extended Happel's steady cell model by considering an oscillatory sphere in the cell, and derived an expression for $\Omega(\phi , s_V)$ using the oscillating 'sphere-in-cell' formulation to account for both skin and form drag contributions, on the basis of zero tangential stress condition (so called Happle's free surface condition[15, 16]) at the cell boundary. Later, Han[14] modified that expression to take into account the aspect ratio and orientations of spheroids. Using realistic particle size and aspect ratio distributions, Pendse and Han[17] reported quantitative agreements of the predicted viscous attenuation coefficients with the experimental measurements for the suspensions of plate-like clay particles in water. It was found that kaolin slurry attenuation coefficient data is very sensitive to particle shape changes at relatively high frequencies (>30 MHz) and high suspended phase volume fractions (>10%).

Knowledge of the frequency- and concentration- dependent *Nu* is the key issue to the solutions of the thermal absorption. Using $Nu = 2.0$ is valid as long as vibrational Reynolds number $s_V \to 0$ and the dispersion is very dilute [9]. For the finite or very high values of s_V or high suspended phase concentration ϕ, simply using $Nu = 2.0$ would not be applicable, as is shown in the later discussion. To account for the concentration and frequency effects on the heat transfer coefficient, Han[14] developed an expression for $Nu(\phi, s_T)$ using an approximate analogy between the 'skin drag' and the 'heat conduction flux' associated with a spherical particle, given as follows:

$$3 \cdot \Omega_{skin} = \text{a function of } (\phi , s_V), \tag{11}$$

$$Nu = \text{the same function of } (\phi , \Psi \cdot s_T), \tag{12}$$

where Ψ is the so called "shift factor". The total viscous drag on a suspended spherical particle is composed of the form drag and the skin drag, where the skin friction drag results from the tangential stress of viscous fluid at the surface of the particle. It is well known that the "form drag", unlike the skin drag, has no counterpart in "interphase heat transfer" as discussed by Bird, Stewart and Lightfoot[18]. Furthermore, the need for the shift factor can be understood since, unlike the case of steady boundary layers over a flat plate, the analogy between viscous and thermal boundary layers of oscillating submerged spheres is only approximate. Hence, an analogous relationship between (s_V) and ($\Psi \cdot s_T$) is proposed, instead of that between (s_V) and (s_T).

Strout's expressions for $\Omega(\phi, s_V)$ and Han's expression for $Nu(\phi, s_T)$

allow one to specify the viscous and thermal coupling terms (F^* and Q^*) in a unified fashion on a common basis of a 'sphere-in-cell' formulation. In the limit that both $\phi = 0$ and $D\sqrt{\omega} = 0$, $\Omega(\phi, s_V) = 1$ and $\Omega_{skin}(\phi, s_V) = \frac{2}{3}$, corresponding to the case of an isolated sphere undergoing steady translation in an infinite body of fluid. Similarly, as ϕ and ($D\sqrt{\omega}$) go to zero, $Nu(\phi, s_T) = 2$, corresponding to the case of steady heat loss from an isolated sphere immersed in an infinite body of fluid. Furthermore, in the limit of only $\phi \rightarrow 0$, $\Omega_{skin}(\phi \rightarrow 0, s_V) = (1 + s_V + j\, s_V)(2/3)$, and correspondingly, $Nu(\phi \rightarrow 0, s_T) = 2\,(1 + \Psi\, s_T + j\, \Psi\, s_T)$. The general relationships for $\Omega(\phi, s_V)$ and $Nu(\phi, s_T)$ are discussed elsewhere [14].

3. Characteristic Equations for the Effective Wave Number

By substituting the eight purturbated variables expressed in Eq. (1), into the governing equations (A.1 - A.8), and ignoring the small cross products and higher order terms, one can expand the governing equations into eight linearized equations (A9 - A16) in terms of eight perturbation variables $\left\{\bar{v}_s, \bar{v}_f, \bar{\rho}_s, \bar{\rho}_f, \bar{P}, \bar{\phi}, \bar{T}_s, \bar{T}_f\right\}$. The eight linearized governing equations for the coupled phase continuum model are listed in Appendix A. The effective complex wave number, K, can be obtained by equating the determinant of the 8 by 8 coefficient matrix to zero, yielding a-cubic equation for (K^2) of the following form:

$$C_3\,(K^2)^3 + C_2\,(K^2)^2 + C_1\,(K^2) + C_0 = 0 \tag{13}$$

where C_0, C_1, C_2, and C_3 are complicated complex valued functions given elsewhere (Han[14]). Han[14] has developed a robust iterative procedure for solving the bi-cubic characteristic equation. K obtained using the bi-cubic equation contains the general α_{UCPC}, that takes into account the contributions arising from the interphase viscothermal couplings, and the intrinsic absorption within the fluid medium. In the limit of ϕ approaching zero, the characteristic equation becomes a quadratic equation for (K^2), from which one gets the classical 'Stokes-Kirchhoff' absorption coefficient, corresponding to the case of particle-free pure fluid medium of viscosity, η_f and thermal conductivity, k_f.

Simplifying the governing equations, by dropping the terms of viscous dissipation [the last term in (A.6) and thermal conduction (the first term in the right side of (A.8)] that contribute to the viscous and thermal absorption within the fluid medium, one can obtain analytical solutions for $(K_{V+T})^2$, wavenumber due to the viscothermal coupling effects. Here, $K_{V+T} = \omega/c - j\alpha_{V+T}$, the attenuation coefficient α_{V+T} accounts for the effects arising from interphase viscous and thermal couplings only, not including the intrinsic absorption in the fluid medium. Correspondingly, the characteristic equation becomes $\boldsymbol{C_1}\ (K_{V+T})^2 + \boldsymbol{C_0} = 0$, resulting in the analytical solution for $(K_{V+T})^2$:

$$(K_{V+T})^2 = \frac{\omega^2 \cdot \langle\gamma \cdot \beta\rangle \cdot [\phi \cdot \rho_s + (1-\phi) \cdot \rho_f \cdot X]}{[\phi + (1-\phi) \cdot X] - j \cdot [\phi \cdot Y_s + (1-\phi) \cdot Y_f \cdot X]}. \tag{14}$$

where $\langle\gamma \cdot \beta\rangle = [\phi \cdot (\gamma_s \beta_s) + (1-\phi) \cdot (\gamma_f \beta_f)]$. $X = \bar{v}_f / \bar{v}_s$, is the ratio of the fluid medium velocity perturbation to suspended phase perturbation, a function f_0 of dimensionless parameters ϕ, s_V, Ω, and density ratio (ρ_S / ρ_f), given by:

$$X = f_0\left(\phi, s_V, \Omega, (\frac{\rho_s}{\rho_f})\right) \tag{15}$$

Y_f and Y_s are functions relating ratios of the temperature perturbations to the corresponding velocity perturbations, for fluid and particle phases. They are complex valued complicated functions of dimensionless parameters ϕ, s_V, s_T, viscous and thermal coupling coefficients $\Omega(\phi, s_V)$ and $Nu(\phi, s_T)$, suspended particle geometry parameter $\Phi = D\ A_p/(8V_p)$, density ratio $\zeta = \rho_f / \rho_S$, density - thermal capacity ratio $\Pi = {}^{C_{pf}\,\rho_f}\!/\!_{(C_{ps}\,\rho_s)}$, thermal expansion coefficient ratio $E = \frac{\varepsilon_f}{\varepsilon_s}$, and heat capacity ratio factors γ_f and γ_S. The functions of Y_f and Y_S can be written as:

$$Y_f = f_1(\phi, s_V, s_T, \Phi, \Omega, Nu, \zeta, \Pi, E, \gamma_f, \gamma_s) \tag{16}$$

$$Y_s = f_2(\phi, s_V, s_T, \Phi, \Omega, Nu, \zeta, \Pi, E, \gamma_f, \gamma_s) \tag{17}$$

The full expressions of X, Y_S, Y_f, the viscous drag coupling coefficient, $\Omega(\phi, s_V)$, and the thermal coupling coefficient, $Nu(\phi, s_T)$, are presented in Appendix B.

Therefore, attenuation coefficient $\alpha_{V+T} = -\text{Imag}\ (K_{V+T})$ resulting from the interphase viscous and thermal couplings, depends on $\Omega(\phi, s_V)$ and $Nu(\phi, s_T)$ in a complicated way. In addition, it is also affected by various dimensionless parameters relating to the physical and thermodynamic properties of the two constituent phases, as described above.

As the thermal coupling term Q^* approaches zero, $Y_s \rightarrow j(\gamma_s - 1)$ and $Y_f \rightarrow j(\gamma_f - 1)$. As the viscous coupling term, F^* approaches zero, $X \rightarrow (\rho_s / \rho_f)$. Thus for the case of no interphase couplings K is a real value with zero attenuation, as given by:

$$\left(K_{no-coupling}\right)^2 = \frac{\omega^2 \cdot \langle \gamma.\beta \rangle}{\left\langle \frac{1}{\rho} \right\rangle + \left\langle \frac{\gamma - 1}{\rho} \right\rangle}. \tag{18}$$

Note that <(-)> denotes the volume average of a property, (-), for the two phase mixture. With further simplifications using $\rho_s = \rho_f$ and $\gamma_s = \gamma_f$, one gets: $K^2 = \langle \rho \rangle \langle \beta \rangle \cdot \omega^2$. The above expression yields the well-known Wood's sound speed,

$$c_{Wood} = \sqrt{\frac{1}{\langle \rho \rangle \langle \beta \rangle}}. \tag{19}$$

From Eq (14), one can obtain simpler analytical expressions for α_V considering the interphase viscous coupling alone (by letting Y_s = Y_f = 0), and for α_T considering the interphase thermal coupling alone (by letting X = 1), respectively. The expression for α_V in terms of the viscous coupling coefficient, $\Omega(\phi\ , s_V)$, thus obtained is identical to the expression obtained by Gibson and ToksÖz[10], Harker and Temple[1], and Atkinson and KytÖmaa[11] based on their coupled phase analyses. Various analytical expressions for α_{V+T}, α_V, and α_T are given by Han[14]. These results combined with Strout's expression for $\Omega(\phi, s_V)$ and Han's expressions for $Nu(\phi, s_T)$ allow one to predict appropriate attenuation contributions arising from the interphase couplings. For the case of non-spherical particles, one would use Han's modification of Strout's $\Omega(\phi, s_V)$ that takes into account size dependent aspect ratios and particle orientations relative the direction of sound propagation. The scattering effect due to the particles is not included in the UCPC formulation. The scattering loss can be estimated using the expressions of Allegra and Hawley[2] and added separately to obtain the total attenuation arising from the suspended phase. Scattering contribution $\alpha_{scat}\ \lambda/\phi$ is primarily determined by the acoustic size parameter $K_0 D = \frac{\omega D}{c_f}$, to the power of three, as well as compressibilities and densities of both phases. In the long wavelength limit K_0D <<2, viscothermal contribution dominates the total attenuation and scattering effect can be safely neglected. As $K_0D \sim 2$, scattering attenuation becomes dominant.

4. Concentration Effects on Viscothermal Attenuation

The UCPC model yields α_{V+T} resulting from simultaneous accounting of both viscous and thermal couplings. It is instructive to compare this result to the sum of α_V and α_T which are the estimates obtained by calculating the viscous and thermal couplings once at a time, neglecting mutual interactions. The viscous attenuation coefficient α_V is obtained by ignoring thermal coupling with Q^* set equal to zero. On the other hand, the thermal attenuation coefficient, α_T is obtained by ignoring viscous coupling with F^* set equal to zero.

The ratio of thermal to viscous losses can be a strong function of frequency and concentration as illustrated in Figure 1. For low density-contrast systems, the thermal losses dominate over the viscous losses. Although, the separate α_V and α_T estimates are useful their sum does not in general provide an accurate estimate of the combined effects of viscous and thermal couplings, as $\alpha_{V+T} \neq \alpha_V + \alpha_T$, and the viscous and thermal absorption can not be decoupled from each other, as noted by Allegra and Hawley[2]. In this context, it should be noted that the result of Temkin and Dobbin[9], $\alpha_{V+T} = \alpha_V + \alpha_T$, can be expected to be valid only when their assumptions of the viscous and thermal coupling coefficients being independent of frequency and concentration are appropriate, that is as ϕ and $(D\sqrt{\omega})$ go to zero. However, errors caused by using $(\alpha_V+\alpha_T)$ instead of α_{V+T} , are within +/- 5% for viscous loss dominant systems such as clay-in-water suspension or thermal loss dominant systems such as polystyrene-in-water suspensions. Thus, for systems where one of the two contributions is much larger than the other by several orders of magnitude, the sum of separate viscous and thermal contributions , though not rigorous, would still give acceptable estimates of total viscothermal losses. However, for the systems such as hexadecane-water emulsions where both viscous and thermal losses are important and have comparable values, simple addition of the separate contribution could cause much greater error, even at low volume fractions of the dispersed phase when the ratio of two contributions α_T/α_V is between 0.1 to 10.

Figure 2 contrasts the relative errors $(\alpha_{V+T} - \alpha_V - \alpha_T) / (\alpha_V + \alpha_T)$ arising from using $(\alpha_V+\alpha_T)$ instead of α_{V+T} for hexadecane-in-water emulsions to that for polystyrene-in-water suspensions at dispersed phase volume fractions ranging from 1% to 48%. This clearly indicates the need for caution in using a simple sum of separate viscous and thermal loss contributions to obtain the estimates of total viscothermal losses. In the following discussion, results based on the UCPC model predictions of α_{V+T} are presented.

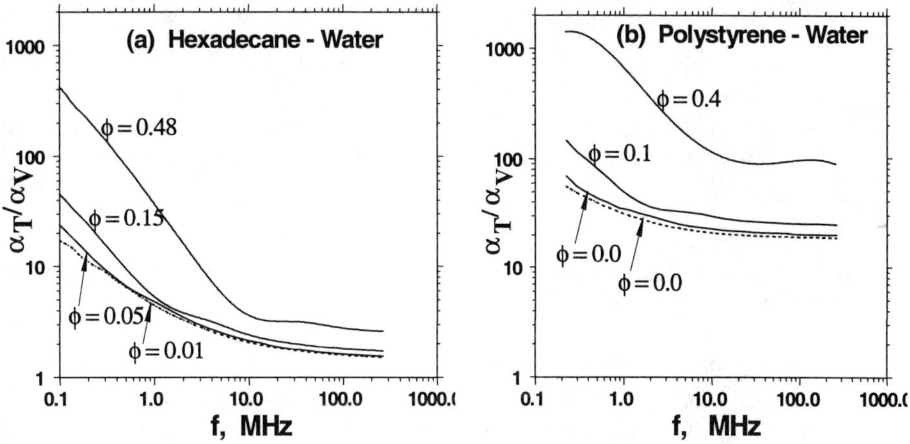

Figure 1. Estimated ratio of thermal to viscous attenuation contributions for two dispersions of D=3.0 μm. (a) hexadecane-in-water emulsion. (b) polystyrene-in-water suspension.

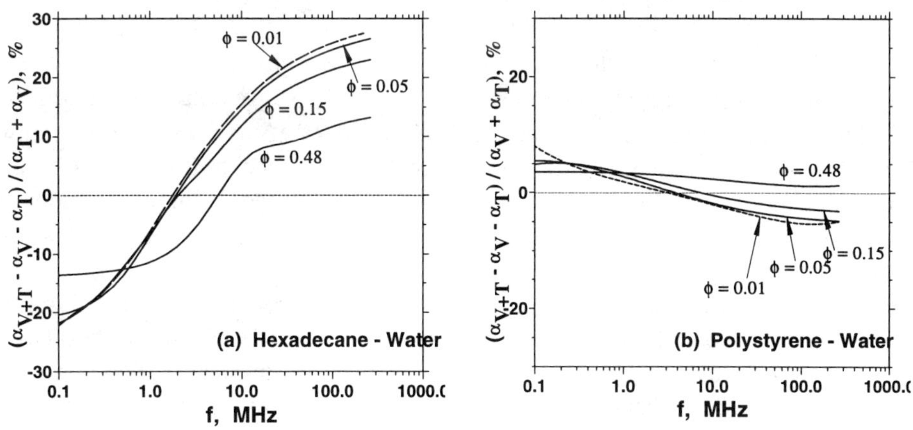

Figure 2. Typical relative error arising from using the sum of separate thermal and viscous attenuation contributions to obtain total viscothermal attenuation for D=3.0 μm. (a) hexadecane-in-water emulsion. (b) polystyrene-in-water suspension.

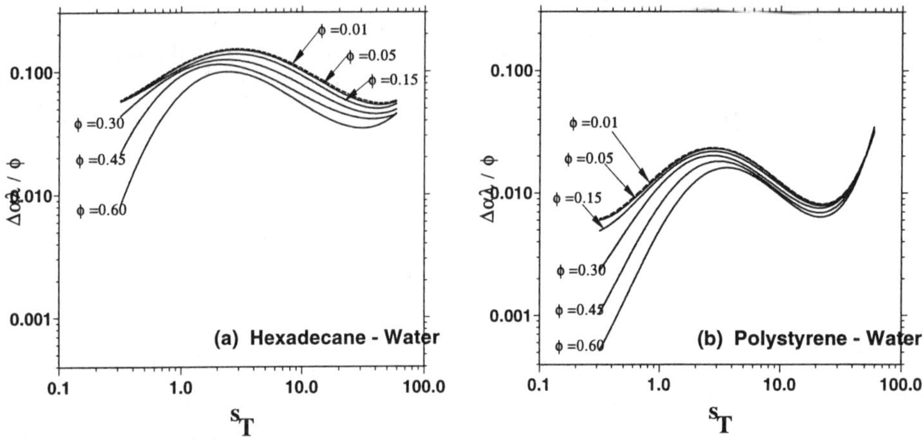

Figure 3. Predicted non-linear volume fraction dependence of excess attenuation spectra for the two dispersions with thermal losses dominating over viscous losses.

Predictions of excess attenuation spectra at various dispersed phase volume fractions for two low density-contrast dispersed systems dominated by the thermal loss contributions relative to the viscous losses are shown in Figure 3. Note that the vertical coordinates denote the excess attenuation per wavelength divided by the volume fraction and the horizontal coordinate is s_T, which is proportional to $(D\sqrt{\omega})$. The shifts in curves corresponding to different concentrations indicate significant nonlinear concentration effects. Compared to viscous loss dominated systems, for this case of thermal loss dominated systems shifts in the peak locations are much less pronounced, $(2 < [s_T \text{ at peak}] < 4)$.

As is seen in Figure 4 for the thermal loss dominant hexadecane-water emulsions, $\Delta\alpha\lambda/\phi$ sharply declines as $\phi > 0.15$, suggesting the significant nonlinear concentration effects for $\phi > 0.15$. As $\phi < 0.15$, $\Delta\alpha\lambda/\phi$ vs. ϕ is almost constant for fixed values of s_T, indicating that linear concentration effects on the excess attenuation in the low concentrations. To show the detailed variations in attenuation over wide concentration ranges, ϕ in the x- axis is plotted in log scale. In Figure 4, the measured data of McClements[4], for $s_T = 7.1$ and 15.1 at $\phi = 0.126$, 0.357, and 0.564, are also plotted, showing good agreement with the corresponding predictions. Unlike hexadecane-in-water emulsions, the results for polystyrene-in-water suspensions in Figure 3(b) show that the nonlinear concentration effects become less significant when $s_T > 3$. However, this result shows that much caution needs to be exercised when neglecting the non-linearity of $\Delta\alpha$ vs. ϕ relationship for a given colloid dispersion of a certain particle size at any frequency.

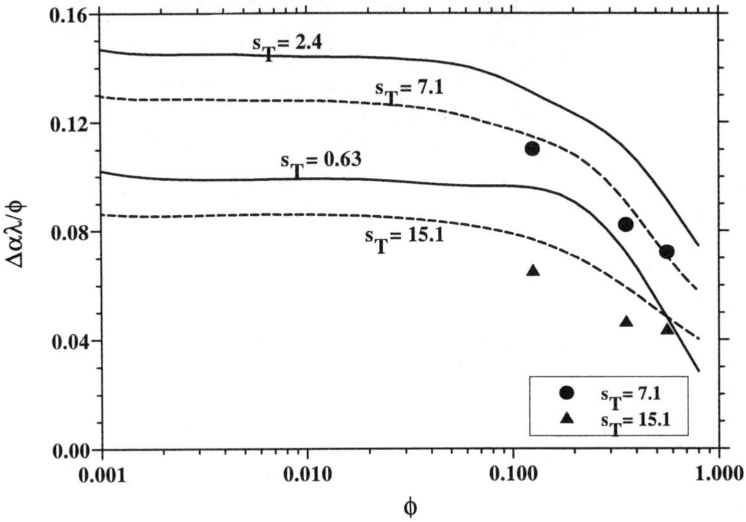

Figure 4. Predicted $\Delta\alpha\,\lambda/\phi$ vs. ϕ for emulsions of hexadecane in water, showing the strong nonlinear concentration effects on normalized attenuation at $\phi > 0.10$ and the linear dependence of concentration at $\phi < 0.10$. Experimental data (McClements[4]), (●) $s_T = 7.1$, (▲) $s_T = 15.1$, are plotted for comparison.

5. Comparison of Predictions with Experimental Data

A number of investigators have reported data on attenuation spectra for different colloidal dispersions. Han has recently reported a systematic comparison of UCPC prediction with experimental data on two suspensions, two emulsions, and two aerosols [Temkin and Dobbins[19], Zink and Delsasso[20], Allegra[21], Hawley[22], McClements[4], Han[14]]. Here we present some results dealing with concentrated dispersions and emulsions with low density-contrast. These predictions of attenuation spectra are based on the following set of 14 input parameters $\left\{D,\phi,T,k_f,\eta_f,\rho_f,c_f,C_{pf},\varepsilon_f,\rho_s,c_s,C_{ps},\varepsilon_s,\Psi\right\}$. Except for the "shift factor", Ψ, all other input parameters were prescribed based on the data provided by the experimenters. Han's comparison of predictions from the present UCPC model with available data for widely different colloidal systems suggests that the shift factor ranges from 2 to 5. Effects of using $Nu(\phi,s_T)$ instead of Nu = 2, as well as the effects of adjusting the shift factor, Ψ are discussed in detail by Han. So far as available data justifies, a reasonable tentative value for the shift factor can be given as 3. A better specification of the shift factor would require additional reliable experimental measurements under well-defined conditions over wide ranges of particle size, concentration and frequency.

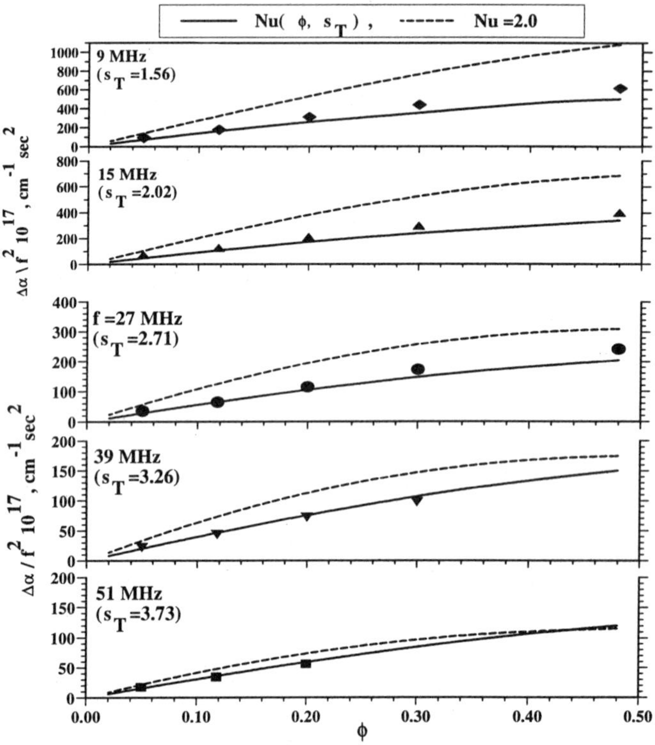

Figure 5. Predictions for Hawley's polystyrene-in-water suspension attenuation data [D=0.22 μm] showing improvements resulting from the use of ϕ and ω dependent thermal coupling coefficient *Nu*.

The concentration and frequency dependence of attenuation coefficients is illustrated in Figure 5 in which Hawley's data (Hawley[22]) for polystyrene-in-water suspensions covering 5% < ϕ < 48% are presented. Improvements in the predictive capabilities of the UCPC model resulting from the use of a general thermal coupling coefficient, $Nu(\phi, s_T)$, with $\Psi = 3$, is quite clear. Han's analysis of Allegra's data (Allegra[21]) on a toluene-in-water emulsion [ϕ = 20% with Ψ = 3] and McClements' data (McClements[4]) on three hexadecane-in-water emulsions [ϕ =12.6%, 35.7%, 56.4% with Ψ = 5] also shows significant improvements resulting from the use of the general thermal coupling coefficient, $Nu(\phi, s_T)$ instead of using $Nu = 2$.

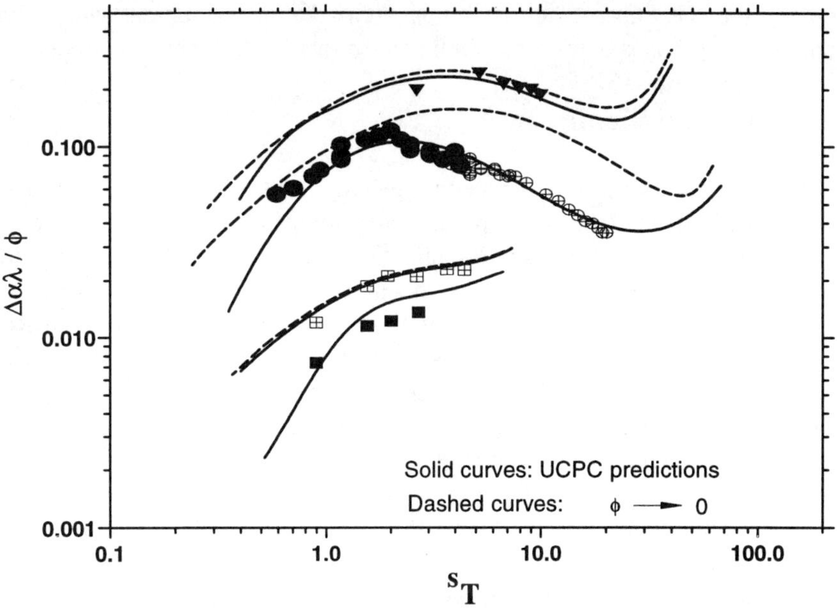

Figure 6. Excess attenuation coefficient per wavelength normalized by volume fraction as a function of the thermal boundary layer parameter, s_T, for five different dispersions: 1. Toluene-in-water emulsion, $\phi = 20\%$, $D = 0.6$ μm (▼) [Allegra[21]]; 2. Hexadecane-in-water emulsion, $\phi = 56.4\%$, $D = 1.0$ μm (●) [McClements[4]]; 3. Hexadecane-in-water emulsion, $\phi = 56.4\%$, $D = 3.6$ μm (⊕) [McClements[4]]; 4. Polystyrene-in-water suspension, $\phi = 5\%$, $D = 0.22$ μm (⊠) [Hawley[22]]; 5. Polystyrene-in-water suspension, $\phi = 48\%$, $D = 0.22$ μm (■) [Hawley[22]]

Comparison of UCPC model predictions over a wide range of s_T, $0.5 < s_T < 20$ is shown in Figure 6 along with experimental data for several colloid systems. The reported values of the mean particle sizes are used in the theoretical predictions. Unlike the inorganic pigment suspensions, for all these systems the thermal contributions to sound attenuation are significantly higher than the viscous contributions. In comparison with the UCPC predictions (solid curves) for appropriate dispersed phase concentrations, the corresponding dilute-limit predictions (dashed curves) are also shown. Shifts from the dilute-limit curves indicate nonlinear concentration effects. The spectra for two hexadecane emulsions with two different droplet sizes clearly show that the effects of variations in particle diameter, D and frequency, ω can be expressed through changes in the oscillatory boundary layer parameter, such as s_T, which is proportional to ($D\sqrt{\omega}$). While the dilute limit predictions are seen to consistently overestimate, the UCPC predictions show remarkable agreement with the experimental attenuation spectra for several concentrated colloidal dispersions.

6. Conclusions and Significance

Ultrasonic spectroscopy has received increased attention, in recent years, in connection with its application to particle size monitoring in both laboratory and on-line sensor settings. Successful applications of ultrasonic spectroscopy in characterization of colloidal dispersions of industrial significance depend on appropriate use of validated predictive theoretical models. In this work a unified coupled phase continuum (UCPC) model is presented, and is shown to be capable of accounting for the effects of dispersed phase volume fractions, ϕ as high 60%, and particle size and frequency combinations covering wide ranges of relevant viscous and thermal boundary layer parameters.

The UCPC model incorporates results for the viscous and thermal coupling coefficients, based on the 'sphere-in-cell' model formulation applied to periodic temporal variations of particle velocities and temperatures covering wide ranges of (ϕ, s_V) and (ϕ, s_T). Both viscous and thermal couplings are taken into account simultaneously. Careful analysis of predictions of viscothermal attenuation based on such simultaneous accounting clearly shows that conventional superposition of separate estimates can be in substantial error, especially when the viscous and thermal contributions are of the same order. Only in limited cases, when one of the two contributions is orders of magnitude larger the other, one may add separate viscous and thermal contribution estimates to obtain the total viscothermal attenuation.

One of the significant features of the UCPC model is that the nonlinear effects of particle volume fraction on the excess attenuation can be properly predicted using this model. These effects can be significant when the dispersed phase volume fraction is more than 15%.

Evaluation of the predictions through comparison with experimental data on widely different colloidal dispersions indicates that the UCPC model is surprisingly effective and capable of predicting on quantitative basis some intricate details of measured attenuation spectra for both dilute and concentrated systems.

Acknowledgement

The authors gratefully acknowledge the partial support of Pen Kem and the Office of Industrial Technologies at the U. S. Department of Energy.

References

1. A.H. Harker and J.A.G. Temple, "Velocity and Attenuation of Ultrasound in Suspensions of Particles in Fluids," *J. Phys. D: Appl. Phys.*, **21**, 1576-1588 (1988).

2. J.R. Allegra and S.A. Hawley, "Attenuation of Sound in Suspensions and Emulsions: Theory and Experiments," *J. Acoust. Soc. Am.*, **51**, 1545-1564 (1972).

3. M.C. Davis, "Attenuation of Sound in Highly Concentrated Suspensions and Emulsions," *J. Acoust. Soc. Am.*, **36**, 387-390 (1979).

4. D.J. McClements, "Comparison of Multiple Scattering Theories with Experiments", *J. Acoust. Soc. Am.*, **91**, 849-853 (1992).

5. P. C. Waterman and R. Truell, "Multiple Scattering of Waves," *J. Math. Phys.*, **2**, 512-537 (1962).

6. Y. Ma, V.K. Varadan, and V.V. Varadan, "Comments on Ultrasonic Propagation in Suspensions," *J. Acoust. Soc. Am.*, **87**, 2779-2782 (1990).

7. P. Lloyd and M.V. Berry, "Wave Propagation through an Assembly of Spheres," *Proc. Phys. Soc. London*, **91**, 678-688 (1967).

8. A.K. Holmes and R.E. Challis, "Ultrasonic Scattering in Concentrated Colloidal Suspensions," *Colloid Surf. A: Phys.. Engineer Asp.*, **77**, 65-74 (1993)

9. S. Temkin and R.A. Dobbins, "Attenuation and Dispersion of Sound by Particulate-Relaxation Processes," *J. Acoust. Soc. Am.*, **40**, 317-324 (1966).

10. R.L. Gibson, Jr. and M.N. Toksöz, "Viscous Attenuation of Acoustic Waves in Suspensions," *J. Acoust. Soc. Am.*, **85**, 1925-1934 (1989).

11. C.M. Atkinson and H.K. Kytômaa, "Acoustic Wave Speed and Attenuation in Suspensions," *Int. J. Multiphase Flow*, **18**, 577-592 (1992).

12. T.A. Strout, "Attenuation of Sound in High-Concentration Suspensions: Development of an Oscillatory Cell Model," Ph.D. Thesis, Dept. of Chemical Engineering, University of Maine, Orono, Maine, May (1991).

13. T.S. Margulies and W.H. Schwarz, "A Multiphase Continuum Theory for Sound Wave Propagation through Dilute Suspensions of Particles," *J. Acoust. Soc. Am.*, **96**, 319-331 (1994).

14. W. Han, "Viscothermal Coupling Effects on Sound Attenuation in Concentrated Colloidal Dispersions," Ph.D. Thesis, Dept. of Chemical Engineering, University of Maine, Orono, Maine, December (1995).

15. J. Happel, "Viscous Flow in Multiparticle Systems: Slow Motion of Fluid Relative to Beds of Spherical Particles," *AIChE J.*, **5**, 197-201 (1959).

16. J. Happel and H. Brenner, *Low Reynolds Number Hydrodynamics,* Martinus Nijhoff Publishers, Dordrecht, The Netherlands, 1973.

17. H.P. Pendse and W. Han, "Ultrasound Characterization of Concentrated Colloidal Suspensions Containing Nonspherical Particles," pp. 142-147 in *Proceedings of the First International Particle Technology Forum*, Vol. 1, 1994.

18. R.B. Bird, W.E. Stewart and E.N. Lightfoot, *Transport Phenomena*, p. 409, John Wiley & Sons, 1960.

19. S. Temkin and R.A. Dobbins, "Measurements of Attenuation and Dispersion of Sound by an Aerosol," *J. Acoust. Soc. Am.*, **40**, 1016-1024 (1966).

20. J.W. Zink and L.P. Delsasso, "Attenuation and Dispersion of Sound by Solid Particles Suspended in a Gas," *J. Acoust. Soc. Am.*, **30**, 765-771 (1958).

21. J.R. Allegra, "Theoretical and Experimental Investigation of the Attenuation of Sound in Suspensions and Emulsions," Ph.D. Thesis, Physics Dept., Harvard University (1970).

22. S.A. Hawley, "Ultrasonic Absorption in Aqueous Suspensions of Small Elastic Particles," Ph.D. Thesis, Dept. Biophysics, University of Illinois, Urbana (1967).

LIST OF SYMBOLS

A_p	single particle surface area
C	sound speed, longitudinal wave speed
$\tilde{C}_A$	complex function , Eq.(B.11)
$\tilde{C}_B$	complex function , Eq.(B.12)
C_p	specific heat at constant pressure
C_v	specific heat at constant volume
D	equivalent sphere diameter
E	$= \varepsilon_f / \varepsilon_s$, ratio of thermal expansion coefficient for fluid and suspended phase, Eq.(B.7)
f	sound frequency, cycles/time
F^*	viscous coupling term, viscous force exerted by the fluid medium on to the dispersed phase, per unit total volume
F_p	single particle viscous drag force
$g(\phi, s_V)$	interfacial viscous drag coefficient, complex quantity, Eq. (7)
$h(\phi, s_T)$	interfacial heat transfer coefficient, complex quantity, Eq. (8)
j	imaginary complex number, $\sqrt{-1}$
k_f	thermal conductivity of fluid medium
K	complex wave number
K_0	$= \omega / c_f$
$Nu(\phi, s_T)$	Generalized complex Nusselt number, hD/k_f
P	Pressure
P_0	pressure at undisturbed steady state
Q^*	thermal coupling term, rate of heat transfer from the dispersed phase into the fluid medium, per unit total volume
Q_p	single particle heat transfer loss rate
R	sphere radius
R_{cell}	sphere radius, in a sphere-in-cell model
R_s	equivalent sphere radius, $D/2$

S	$= (1 + j)\, s_V$, Eq. (B.14)
s_T	Dimensionless oscillating thermal boundary layer, $(D/2)[\rho_f C_{Pf} \omega / (2k_f)]^{1/2}$
s_V	Dimensionless oscillating viscous boundary layer parameter, $(D/2)[\rho_f \omega / (2\eta_f]^{1/2}$
t	Time
T	Temperature
T_0	Temperature at undisturbed steady state
V_p	single particle volume
v_f	fluid medium phase velocity
v_s	suspended phase velocity
X	ratio of fluctuation velocity $\overline{v}_f / \overline{v}_s$
Y_f	function of ratio of temperature perturbation to velocity perturbation for fluid medium
Y_s	function of ratio of temperature perturbation to velocity perturbation for suspended phase
z	distance in the direction of sound propagation
ω	$2\pi f$, angular frequency, radians/time
α	amplitude attenuation coefficient, (Neper/length)
α_f	absorption coefficient in pure fluid, (Neper/length)
α_s	absorption coefficient in pure suspended phase, (Neper/length)
α_T	attenuation coefficient due to thermal coupling effect only, (Neper/length)
α_V	attenuation coefficient due to viscous coupling effect only, (Neper/length)
α_{V+T}	attenuation coefficient due to viscothermal coupling effects, (Neper/length)
α_{UCPC}	attenuation coefficient due to viscothermal coupling effects, and the intrinsic absorption in the fluid medium, (Neper/length)
$\Delta\alpha$	excess attenuation coefficient, $[\alpha_{dispersion} - \alpha_f]$, (Neper/length)
β	adiabatic compressibility, $= 1/(\rho\, c^2)$
λ	Wavelength of sound in the fluid medium
ϕ	volume fraction of suspended phase
Φ	$= D\, A_p/(8V_p)$, suspended particle geometry parameter
Γ	complex function, Eq. (B.13)

Π	$C_{pf}\rho_f/(C_{ps}\rho_s)$, ratio of density-thermal capacity for fluid and suspended phase
Ψ	shift factor for the viscothermal analogy
$\Omega(\phi, s_V)$	Generalized complex Stokes' drag correction factor
Ω_{skin}	skin drag correction factor
η_f	shear viscosity of fluid medium
ε	thermal expansion coefficient
γ	ratio of specific heat, $=C_p/C_V=1+\varepsilon^2 c^2 T_0/C_P$
ζ	$=\rho_f/\rho_s$, ratio of density for fluid and suspended phase
ρ	density (mass/volume)
Subscript	
s	suspended particle phase
f	fluid medium phase

APPENDIX A. General Equations for the UCPC Model Formulation

The general equations of states, continuity, momentum, and energy for the suspended phase and suspending fluid medium phase are written as:

Equations of State:

$$P-P_0=\frac{1}{\gamma_s\rho_s^0\beta_s}(\rho_s-\rho_s^0)+\frac{\varepsilon_s}{\gamma_s\beta_s}(T_s-T_0). \tag{A.1}$$

$$P-P_0=\frac{1}{\gamma_f\rho_f^0\beta_f}(\rho_f-\rho_f^0)+\frac{\varepsilon_f}{\gamma_f\beta_f}(T_f-T_0). \tag{A.2}$$

Continuity:

$$\frac{\partial(\phi\rho_s)}{\partial t}+\frac{\partial(\phi\rho_s v_s)}{\partial z}=0 \tag{A.3}$$

$$\frac{\partial[(1-\phi)\rho_f)]}{\partial t}+\frac{\partial[(1-\phi)\rho_f v_f]}{\partial z}=0. \tag{A.4}$$

Momentum:

$$\frac{\partial(\phi\rho_s v_s)}{\partial t}+v_s\frac{\partial(\phi\rho_s v_s)}{\partial z}+\phi\frac{\partial P}{\partial z}=F^*. \tag{A.5}$$

$$\frac{\partial[(1-\phi)\rho_f v_f]}{\partial t}+v_f\frac{\partial[(1-\phi)\rho_f v_f]}{\partial z}+(1-\phi)\frac{\partial P}{\partial z}=-F^*+(1-\phi)\frac{4}{3}\frac{\partial}{\partial z}(\eta_f\frac{\partial v_f}{\partial z})$$

(A.6)

Energy: $$\phi\rho_s c_{vs}\left(\frac{\partial T_s}{\partial t}+v_s\frac{\partial T_s}{\partial z}\right)=-\phi T_s\frac{\varepsilon_s}{\gamma_s\beta_s}\frac{\partial v_s}{\partial z}-Q^* \qquad (A.7)$$

$$(1-\phi)\rho_f c_{vf}\left(\frac{\partial T_f}{\partial t}+v_f\frac{\partial T_f}{\partial z}\right)=(1-\phi)k_{tf}\frac{\partial^2 T_f}{\partial z^2}-(1-\phi)T_f\frac{\varepsilon_f}{\gamma_f\beta_f}\frac{\partial v_f}{\partial z}+Q^*$$
$$+F^*(v_s-v_f)+(1-\phi)\frac{4}{3}\eta_f\left(\frac{\partial v_f}{\partial z}\right)^2 \qquad (A.8)$$

The linearized equations for Eqs. (A.1-8) are given as:

$$\overline{\rho}_s=\gamma_s\rho_s^0\beta_s\overline{P}-\rho_s^0\varepsilon_s\overline{T}_s. \qquad (A.9)$$

$$\overline{\rho}_f=\gamma_f\rho_f^0\beta_f\overline{P}-\rho_f^0\varepsilon_f\overline{T}_f. \qquad (A.10)$$

$$(j\omega)(\phi^0\overline{\rho}_s+\rho_s^0\overline{\phi})+(-jK)\rho_s^0\phi^0\overline{v}_s=0 \qquad (A.11)$$

$$(j\omega)[(1-\phi^0)\overline{\rho}_f-\rho_f^0\overline{\phi})]+(-jK)\rho_f^0(1-\phi^0)\overline{v}_f=0. \qquad (A.12)$$

$$(j\omega)\phi^0\rho_s^0\overline{v}_s+(-jK)\phi^0\overline{P}=(j\omega)\phi^0\rho_f^0\;[j(1-\phi)\frac{9}{4s_v^2}\;\Omega(\phi,s_v)]\;(\overline{v}_s-\overline{v}_f)$$

(A.13)

$$(j\omega)(1-\phi^0)\rho_f^0\overline{v}_f+(-jK)(1-\phi^0)\overline{P}=-(j\omega)\phi^0\rho_f^0[j(1-\phi^0)\frac{9}{4s_v^2}\;\Omega(\phi,s_v)]\;(\overline{v}_s-\overline{v}_f)$$
$$+(1-\phi^0)\frac{4}{3}\eta_f(-jK)^2\overline{v}_f$$

(A.14)

$$(j\omega)c_{vs}\phi^0\rho_s^0\overline{T}_s=-\frac{\varepsilon_s}{\gamma_s\beta_s}\phi^0T_0(-jK)\overline{v}_s-(A_p\frac{k_f}{DV_p})(1-\phi^0)\phi^0Nu(\phi,s_T)\;(\overline{T}_s-\overline{T}_f)$$

(A.15)

$$(j\omega)c_{vf}(1-\phi^0)\rho_f^0\overline{T}_f = k_f(-jK)^2(1-\phi^0)\overline{T}_f - \frac{\varepsilon_f}{\gamma_f\beta_f}(-jK)(1-\phi^0)T_0\overline{v}_f$$

$$+(A_p\frac{k_f}{DV_r})(1-\phi^0)\phi^0\, Nu(\phi, s_T)\,(\overline{T}_s-\overline{T}_f) \tag{A.16}$$

APPENDIX B. Detailed Expressions of X, Y_s, Y_f, Ω, and Nu in the UCPC Model Solutions

The ratio of fluid medium perturbation velocity to suspended phase perturbation velocity in Eq. (15) is expressed as

$$X = f_0\left(\phi, s_V, \Omega, (\frac{\rho_s}{\rho_f})\right) = \frac{(1-\phi)(\rho_s/\rho_f) - j\frac{9}{4s_V^2}\,\Omega(\phi, s_V)}{1-\phi-j\frac{9}{4\,s_V^2}\,\Omega(\phi, s_V)} \tag{B.1}$$

Complex valued functions Y_f and Y_s in Eqs.(16-17) can be expressed as

$$Y_f = \frac{-A_1 - j\,(\gamma_f - 1)}{j\,B_1 - 1} \tag{B.2}$$

$$Y_s = \frac{-A_1\,X\,\mathrm{E} - j\,(\gamma_s - 1)}{j\,B_1 - 1} \tag{B.3}$$

where intermediate parameters A_1 and B_1 are given as

$$A_1 = \frac{Nu}{s_T^2}\frac{RA_p}{4V_p}\left[(1-\phi)\gamma_s(\gamma_f - 1)\,\Pi + \phi\,\gamma_f(\gamma_s - 1)\,\frac{\mathrm{E}}{X}\right] \tag{B.4}$$

$$B_1 = \frac{Nu}{s_T^2}\frac{RA_p}{4V_p}\left[(1-\phi)\gamma_s\,\Pi + \phi\,\gamma_f\right] \tag{B.5}$$

The dimensionless parameters E and Π are related to thermodynamic properties thermal expansion, specific heat, and density of two phases, given by

$$\Pi = \frac{C_{pf}\,\rho_f}{C_{pf}\,\rho_s} \tag{B.6}$$

$$\mathrm{E} = \frac{\varepsilon_f}{\varepsilon_s} \tag{B.7}$$

The viscous and the thermal coupling coefficients viscous and thermal coupling coefficients $\Omega(\phi, s_V)$ and $Nu(\phi, s_T)$ in the UCPC model solution are given as follows:

$$\Omega(\phi, s_V) = \frac{4{s_V}^2}{9}\left[-j(1-\frac{3}{2}\tilde{C}_A)\right] \tag{B.8}$$

$$Nu = 3\,\Omega_{skin}(\phi,\ s_V = \Psi\, s_T) \tag{B.9}$$

where Ψ is the so-called shift factor and the function $\Omega_{skin}(\phi, s_V)$ is given as

$$\Omega_{skin}(\phi, s_V) = \frac{4\,{s_V}^2}{9}\left[-j\,(1-\tilde{C}_A-\tilde{C}_B)\right] \tag{B.10}$$

$\tilde{C}_A$ and $\tilde{C}_B$ are complex complicated functions of ϕ and s_V, given from the oscillating "sphere-in-cell" model solutions by Strout[12]:

$$\tilde{C}_A = \frac{S^5+3(1-\phi^{1/3})S^4+3(2\phi^{2/3}-3\phi^{1/3}+1)S^3+3(-2\phi+6\phi^{2/3}-3\phi^{1/3})S^2+18(\phi^{2/3}-\phi)S-18\phi}{S^2\,\Gamma} + e^{2(1-\phi^{-1/3})S}\,\frac{\left[\begin{array}{l}S^5-3(1-\phi^{1/3})S^4+3(2\phi^{2/3}-3\phi^{1/3}+1)S^3\\-3(-2\phi+6\phi^{2/3}-3\phi^{1/3})S^2+18(\phi^{2/3}-\phi)S+18\phi\end{array}\right]}{S^2\,\Gamma}$$

(B.11)

$$\tilde{C}_B = \frac{-\phi\left[S^3+3(1-\phi^{1/3})S^2+3(1-3\phi^{1/3})S-9\phi^{1/3}+12S\,e^{1-\phi^{-1/3}}\right]}{\Gamma} - \phi\,\frac{\left[S^3-3(1-\phi^{1/3})S^2+3(1-3\phi^{1/3})S+9\phi^{1/3}\right]}{\Gamma}\,e^{2(1-\phi^{-1/3})S}$$

(B.12)

Where

$$\Gamma = \left[(1-\phi)S^3 + 3(\phi^{4/3} - \phi - \phi^{1/3})S^2 + (9\phi^{4/3} - 3\phi + 6\phi^{2/3})S + (9\phi^{4/3} - 6\phi)\right]$$
$$+ e^{2(1-\phi^{-1/3})S}\left[(1-\phi)S^3 - 3(\phi^{4/3} - \phi - \phi^{1/3})S^2 + (9\phi^{4/3} - 3\phi + 6\phi^{2/3})S - (9\phi^{4/3} - 6\phi)\right]$$
$$- 24\, e^{(1-\phi^{-1/3})S}\, \phi\, S$$

(B.13)

$$S = (1 + j)s_V. \quad \text{(B.14)}$$

Applications Development and Industrial Process Monitoring

Industrial Applications of In-Line Ultrasonic Spectroscopy

David M. Scott
DuPont Central Research & Development, Particle Science & Technology, (PARSAT), Experimental Station, Wilmington, Delaware 19880-0357 USA

Abstract

The use of ultrasonic attenuation spectroscopy to measure particle size distribution (PSD) in colloidal dispersions is well known, and it is the basis for several commercial instruments. With few exceptions, to date the most successful applications of this technology have been confined to off-line (laboratory-based) instrumentation. The possibility of using ultrasound for on-line PSD measurements in concentrated slurries and emulsions continues to attract the interest of industry, where dynamic measurements are needed for process control. Although ultrasonic techniques hold much promise for on-line and in-line applications, there are several issues that must be addressed in a practical instrument. These issues include proper flow cell design, extraction of the PSD from ultrasonic attenuation spectra, and difficulties associated with high concentrations and particle morphology. Characteristics of a robust ultrasonic system are presented here, along with a brief description of an in-line ultrasonic spectrometer which has proven to be reliable under industrial plant conditions. Two examples of in-line size measurement with this instrument demonstrate its application in typical industrial processes.

1. Introduction

In-line and on-line methods of characterizing particulate systems continue to attract the attention of industry for two reasons: Firstly, the physical characteristics of the particles moving through a chemical plant are known to impact the performance of both the process and the product. Process control addresses this need to control certain characteristics (such as size distribution), but it requires dynamic measurements of the quantity to be controlled. Secondly, characterization of the dynamics of a chemical process is necessary in order to build up an understanding of the process. This knowledge eventually leads to improvements in process. Thus, in-line characterization contributes to process control and process improvements, both of which lead to improvement of the product.

For most particulate systems, size (including size distribution) is a key parameter. In the past, characterization of industrial slurries and emulsions typically relied upon off-line measurements of the particle size distribution (PSD). Due to the difficulty of obtaining representative samples and the relatively slow feedback from the laboratory, this approach is not suitable for process control applications. Consequently, industrial process engineers have sought measurement techniques which are suitable for on-line applications. Since ultrasonic techniques have been used in the mineral processing industry for several decades,[1] it seems likely that rugged in-line ultrasonic sensors can be developed to measure the PSD in concentrated slurries.

A great deal has already been written about the use of ultrasound to determine the PSD in slurries and suspensions,[2,3] including many contributions to the present volume. The basic concept is to measure the frequency-dependent attenuation or velocity of the ultrasound as it passes through the sample. This data is related to the PSD and concentration of the sample by empirical observation or by theoretical calculations. Once this connection is established, one can in principle estimate the PSD from the ultrasonic data. This approach forms the basis for several commercial instruments, including both off-line and on-line configurations. As a rule, the off-line ultrasonic instruments perform adequately, but the on-line instruments have not withstood the rigors of industrial on-line applications. The main challenges facing industrial applications of ultrasonic sensing, as well as the salient features of suitably robust system designs, are enumerated below.

2. Design Considerations

Ultrasonic PSD instruments face a number of challenges which are particularly difficult in on-line and in-line applications. One class of difficulties is associated with acquiring the ultrasonic spectral data. These difficulties include

- the presence of gas bubbles in the process stream,
- incomplete filling of the process line,
- potential plugging of the flow cell, process line, or side stream,
- harsh environment (heat, pressure, vibration, abrasion, corrosion), and
- extreme attenuation of the ultrasonic signal.

At these frequencies, ultrasound does not propagate through gas; therefore the presence of air bubbles (or a partially filled process line) prevents the ultrasonic signal from being detected. The potential for plugging in the instrument flow cell, process line, or side stream cannot be overstated; plugging within the instrument or its associated piping has been the failure mechanism of every commercial on-line ultrasonic instrument tested at this company to date. The harsh environment found in industrial plants can cause failure of the ultrasonic transducers and the associated electronic systems; in addition, temperature excursions affect the transducers as well as the interaction between the ultrasound and the process stream. Finally, some process streams (such as highly concentrated slurries of dense sub-micron particles) so effectively attenuate the ultrasound that measurements on the received signal become difficult.

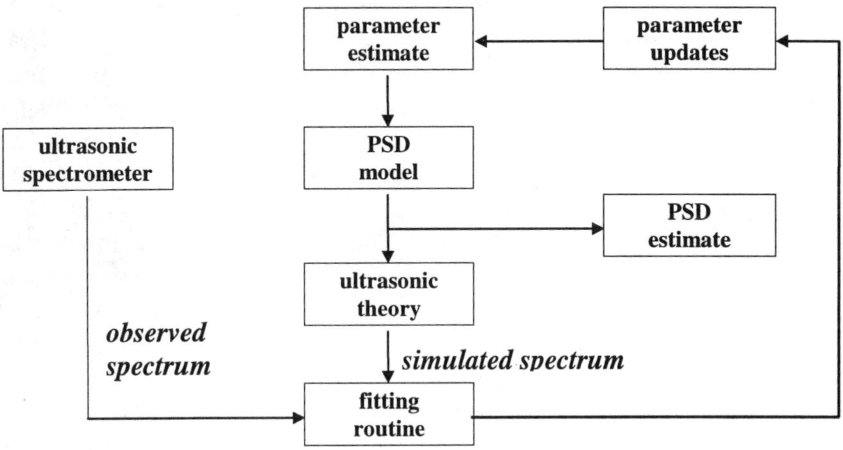

Figure 1. Algorithm for determining PSD from ultrasonic spectra.

Other challenges are associated with extracting the PSD from the measured ultrasonic spectrum. Due to the ill-posed nature of this problem, the spectral data cannot be inverted directly. Instead, most workers have resorted to an iterative fitting scheme similar to that depicted in Figure 1. Typically the spectra are very smooth, so that very little information appears to be contained in the data. In order to extract the PSD from the attenuation data, a model function (such as a log-normal or bi-log-normal distribution) is assumed, effectively reducing the number of free parameters to be fit to the data. Accounts have also been given[4] of model independent inversion of the data, wherein a larger number of basis functions are used to describe the PSD; in order to obtain a reasonable PSD it is necessary to require that the predicted spectrum be sufficiently smooth. A theoretical description of the interaction between the ultrasound and the particulate system is used to calculate the predicted attenuation spectrum from the estimated PSD; in the present work the Allegra-Hawley model is used.[5] The estimated PSD is modified until agreement is reached (to a specified accuracy) between the observed and predicted spectra. This iterative process has to be sufficiently rugged to produce useful results even in the case of limited, missing, or contradictory spectral data. In the case of the model dependent (i.e., log-normal) PSD, it must be noted that the choice of PSD function affects the results obtained. It should also be noted that changes in particle morphology can affect the ultrasonic signal, but at present none of the ultrasonic theories take morphology into account.

A further difficulty is encountered in the case of highly concentrated slurries of sub-micron particles. There are currently no general theories for predicting ultrasonic attenuation when the wavelength of sound is much greater than both the interparticle spacing and the particle diameter. Linear theories (such as the Allegra-Hawley model[5]) and multiple-scattering theories (such as that of Waterman and Truell[6]) do not describe the attenuation spectra observed at high concentrations, where the interparticle spacing is smaller than the particle diameter. Riebel[7] has found that in the short wavelength limit, a modified Beer-Lambert law can be used to describe the spectrum even at high concentrations, but due to practical limitations this theory cannot be used for sub-micron particles (which would require frequencies over 1 GHz). Thus, empirical calibration curves must be used to determine PSD from the observed ultrasonic spectra in the case of concentrated slurries of sub-micron particles.

In view of the obstacles just listed, the design of a robust in-line ultrasonic instrument for measuring PSD in industrial applications should incorporate the following guidelines:

- The flow cell must have a large bore to prevent plugging, and the gap between transducers should be as large as feasible. The use of valves should be avoided.
- Flow through the instrument must be streamlined so that particle segregation does not occur.
- There should few (if any) moving parts.
- Materials of construction must be suitable to the application, as many industrial process streams are highly corrosive.
- In the case of crystallization applications, the flow cell must not act as a heat sink; otherwise, the instrument will become encrusted.
- The placement of the flow cell in the process stream must take into

account such factors as complete filling of the pipe, possible entrainment of gas bubbles in the process stream at that point, and the expected temperature and pressure to be encountered.

- In-line placement is to be preferred over side-streams in most applications for two reasons: adequate sampling is assured, and the potential for plugging is reduced (although the consequences of plugging are more severe).

- The data inversion algorithm must be sufficiently robust to extract PSD from noisy or limited ultrasonic data. The software should produce a goodness of fit parameter that indicates how well the estimated PSD accounts for the observed spectrum; ideally, the PSD estimates should include an estimated error in the quoted values to indicate the reliability of the measurement.

- The instrument should require no operator intervention and minimal calibration.

- The simplicity of the overall design should lead to ease of maintenance.

These principles have been incorporated into the design of an in-line ultrasonic instrument for measuring PSD in concentrated slurries, as described below.

3. In-Line PSD Instruments

Several in-line ultrasonic devices have been built in this laboratory according to the guidelines listed above. The new instruments determine PSD from ultrasonic attenuation spectra, and they are designed for reliable operation in concentrated slurries and emulsions. A typical instrument schematic is shown in Figure 2. The attenuation measurements are made with the use of wide-band pulses: a short pulse (which contains a wide range of frequency components) is sent to the transmitting transducer, thereby launching an ultrasonic wave through the slurry; the attenuated wave is detected by the receiving transducer, amplified, digitized, and Fourier-transformed into its frequency components. This approach is easily implemented with standard, economical components, and the Fourier transform contains phase velocity as well as frequency-dependent attenuation data. Single-sided measurements can be made by using the pulse echo technique and a suitable ultrasonic reflector. Although these instruments were developed independently, other pulse-based systems have been described in the literature.[8,9] Results obtained with these new instruments have been described elsewhere.[10,11]

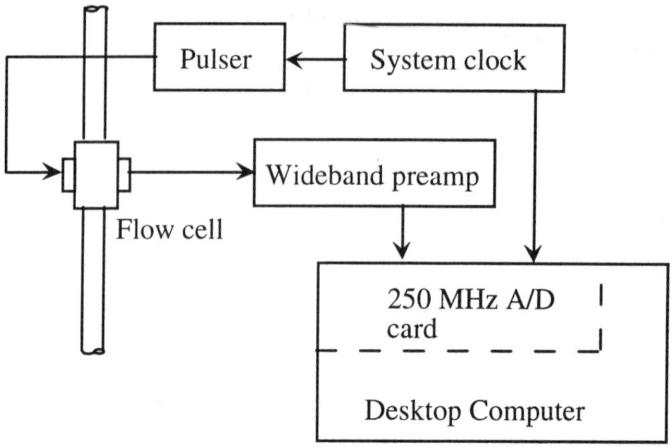

Figure 2. Schematic of a generic pulsed system.

Once the attenuation spectrum is measured, the PSD is estimated by a fitting procedure based on the Allegra-Hawley model as described above. A distinguishing aspect of the inversion software is that the validity of the resulting size information is quantified. In particular, the standard deviation of the estimated size is determined from the standard deviations of the spectral data. This ability to estimate the error in the measured size is generally absent in commercial instruments.

The use of a pulsed signal provides a very high data capture rate: up to 1000 spectra per second, at which rate transient processes may be studied. A more typical rate achieved in normal operation, including signal averaging and spectral inversion, is about 4 PSD per minute. The dynamic range of the spectrometer is over 66 dB, and the frequency range (which is determined in part by the transducers) typically lies between 1-70 MHz. These instruments have been used for on-line and in-line industrial applications without dilution of the sample stream.

4. Applications

There are many potential applications of ultrasonic instruments within the chemical industry. Most of these applications involve concentrated slurries of sub-micron particles, for which no other suitable measurement techniques have been identified. Among the more prominent applications are:

- monitoring and control of milling circuits,

- droplet size characterization in emulsions,
- process stream monitoring,
- characterization of dispersion, and
- stability (shelf life) studies of dispersions.

Two of these applications, monitoring of milling circuits and emulsion characterization, will be illustrated below.

4.1 Monitoring and Control of Milling

Previously, on-line measurements of PSD in milling applications (via laser diffraction) have relied on isokinetic sampling and dilution/dispersion.[12] The in-line ultrasonic instrument described above has been placed in a milling loop in order to study the comminution of sub-micron particles of pigment (both organic and inorganic). During startup, the mill generated large quantities of bubbles, making ultrasonic measurements difficult. The high data capture rate of the pulsed system provided enough redundant information to determine the slurry attenuation in spite of the noisy data.

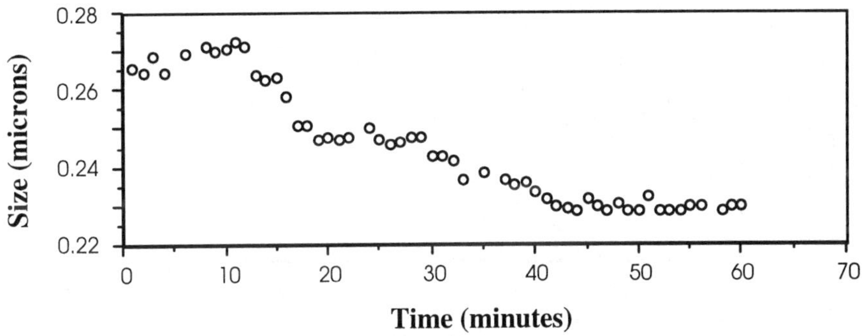

Figure 3. Particle size of sub-micron pigment (at 75 wt % concentration) measured during milling.

Figure 3 shows the decrease in median particle size during the milling of sub-micron inorganic pigment in a semiworks installation. The residence time of the mill accounts for the delay between the start of milling and the point in time when the size started to decrease. The concentration of the slurry was 75% by weight. Previous experience showed that the milling operation increases the available surface area to the point where the dispersion becomes unstable. In order to prevent floccing (which would be observed in the trend line as an increase in particle size), the slurry was buffered to a pH of 10. Off-line measurements of samples taken during the run corroborated the size estimates based on the ultrasonic data. It is evident from the ultrasonic size data that the milling was complete at 40 minutes. The size of organic pigments has also been monitored during milling with this ultrasonic instrument.

4.2 Droplet Size Measurements in Emulsions

This instrument has also been used to study the droplet size distribution (DSD) in emulsions of chlorobenzene with water. Figure 4 shows the results of measuring median droplet size in an emulsification experiment. A 50% emulsion was produced in a beaker, and the DSD was monitored by circulating the emulsion through the ultrasonic flow cell. The shear of the pump reduced the median droplet size from 0.6 to 0.4 microns, which proved to be the equilibrium size under those operating conditions. A few minutes later, a high-speed emulsifier was used to decrease the droplet size to about 50 nm. A soon as the emulsifier was turned off, the droplet size increased until it reached the equilibrium value.

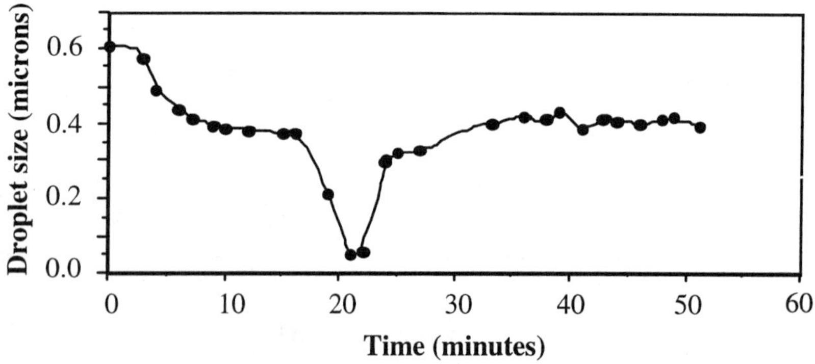

Figure 4. Dynamic droplet size measurement in an emulsion of chlorobenzene and water; a line has been added to guide the eye. A temporary change in droplet size was effected by the use of a high speed/high shear mixer; droplet coalescence commenced immediately after the mixer was stopped (at 21 minutes).

In tests at a plant site, different solvent/water emulsion samples were prepared. An experiment was arranged so that droplet size measurements could be made simultaneously with two instruments: the ultrasonic sensor and a commercial laser-backscatter device. Several emulsions were prepared with various droplet sizes, and the size of the droplets were estimated with both instruments. The results are compared in Figure 5, in which a linear fit to the data has been used to determine the correlation (R^2 = 0.991). The strong correlation between these techniques validates the use of ultrasound to measure droplet size. The systematic discrepancy between the two measurements is attributed in part to inaccuracies in the physical parameters of the solvent; these parameters can be amended, but for process control applications the absolute droplet size is unimportant.

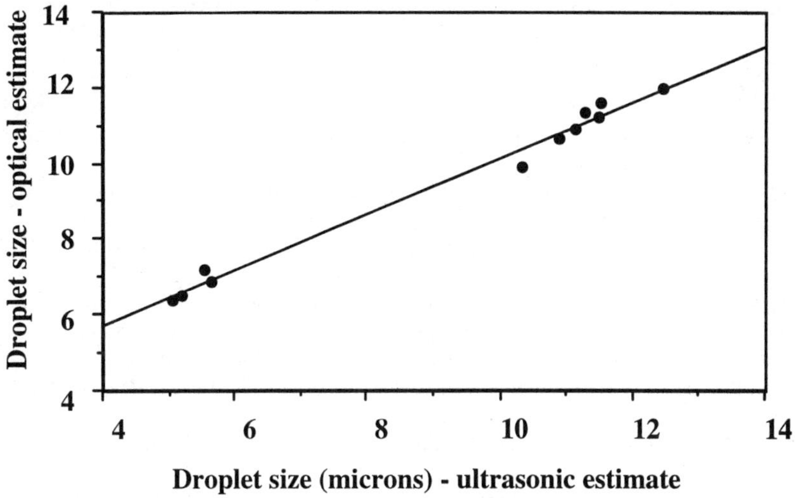

Figure 5. A comparison between the droplet size in a solvent/water emulsion as estimated by simultaneous ultrasonic and laser back-scatter measurements. The correlation between the two data sets is depicted as a line, which is a linear fit to the data (R^2 = 0.991).

5. Conclusions

Process instrumentation needs are very different from those of scientific (i.e., laboratory) instrumentation. Process instruments must be rugged and easy to maintain, and the absolute accuracy of the measurement is generally unimportant as long as it is reproducible. Scientific instruments, on the other hand, typically have a greater complexity in order to provide absolute accuracy. Given these two opposing requirements, it is clear that one cannot create a process instrument simply by running a process stream through a laboratory instrument. Unfortunately, many of the early on-line ultrasonic instruments have essentially taken that approach.

This paper has enumerated many issues that must be faced in bringing ultrasonic measurements to the factory floor. Some of them have been addressed by the in-line ultrasonic attenuation spectrometer described here. The foregoing examples indicate that, with care, such ultrasonic instruments can be used to measure PSD on industrial processes, even at high concentrations. However, work is still needed to improve the theoretical basis of these instruments, such as a better description of the effects of particle morphology and steric interactions (in the long wavelength regime). It is clear that further progress in applying ultrasound to in-process size measurements will require the close collaboration of theoreticians, particle scientists, sensor developers, and process engineers.

References

1. V. Anderson, C. Cushman, and J. Hale, "Particle Size and Percent Solids Monitor", U.S. Pat. 3,779,070 (1973).
2. U. Riebel and F. Löffler, "The Fundamentals of Particle Size Analysis by Means of Ultrasonic Spectrometry", *Part. Part. Syst. Charact.*, **6**, 135-143 (1989).
3. H.P. Pendse and A. Sharma, "Particle Size Distribution Analysis of Industrial Colloidal Slurries Using Ultrasonic Spectroscopy", *Part. Part. Syst. Charact.*, **10**, 229-233 (1993).
4. U. Riebel, "Ultrasonic Extinction and Velocity in Dense Suspensions", in *Handbook on Ultrasonic and Dielectric Characterization Techniques for Suspended Particulates*. Edited by V.A. Hackley and J. Texter. American Ceramic Society, Westerville, OH, 1998.
5. J.R. Allegra and S.A. Hawley, "Attenuation of Sound in Suspensions and Emulsions: Theory and Experiments", *J. Acoust. Soc. Am.*, **51**, 1545-1564 (1972).
6. P.C. Waterman and R. Truell, "Multiple Scattering of Waves", *J. Math. Phys.*, **2**, 512-540 (1961).
7. U. Riebel, "Ultrasonic Spectrometry: On-Line Particle Size Analysis at Extremely High Particle Concentrations", in *Particle Size Analysis*. Edited by N. Stanley-Wood and R.W. Lines. Royal Society of Chemistry, 1992.
8. A.K. Holmes, R.E. Challis, and D.J. Wedlock, "A Wide Bandwidth Study of Ultrasound Velocity and Attenuation in Suspensions: Comparison of Theory with Experimental Measurements", *J. Colloid Interface Sci.*, **156**, 261-268 (1993).
9. D.J. McClements and M.J.W. Povey, "Ultrasonic Velocity as a Probe of Emulsions and Suspensions", *Adv. Colloid and Interface Sci.*, **27**, 285-316 (1987).
10. D.M. Scott, A. Boxman, and C.E. Jochen, "Ultrasonic Measurement of Sub-Micron Particles", *Part. Part. Syst. Charact.*, **12**, 269-273 (1995).
11. D.M. Scott, A. Boxman, and C.E. Jochen, "In-Line Particle Characterization", *Part. Part. Syst. Charact.*, **15**, in press (1998).
12. R. Polke and N. Scholz, "On-Line Particle Measurement Technology for Controlling a Mill Classifier Circuit", pp. 461-471 in *Proceedings of PARTEC*, Nürnberg, Particle Characterization, 1995.

Particle Shape Effects and Active Ultrasonic Spectroscopy

Hemant P. Pendse, Timothy C. Bliss[*] and Wei Han[**]
Industrial Process Control Sensor Systems (IPCoSS) Program, Chemical Engineering Department, University of Maine, Orono, ME 04469 USA

Abstract

In active ultrasonic spectroscopy, we propose to use a continuously-changing-gap mode to take advantage of the evolving orientation field of suspended non-spherical particles under the influence of squeezing flows when monitoring slurry response. Predictions of line-of-sight averages of orientation angles can be useful in sensing applications that use squeezing flows. A procedure (Pendse-Bliss Algorithm) was developed to simulate evolution of particle orientation fields for a simple squeezing flow geometry, using the Transversely Isotropic Fluid (TIF) model. Highly non-spherical particles spend considerable time in parallel orientation, but nearly-spherical particles tend to spend considerable time undergoing repeated rotations of decreasing periods. Attenuation spectra under random orientations versus parallel orientations are expected to be significantly different from each other. Two methodologies to infer shape characteristics have been developed based on (1) attenuation spectra obtained with a cross-flow using typical on-axis transducer configuration, and (2) attenuation spectra obtained in a stop-flow mode using off-axis transducer configuration while sensor gap decreases continuously. Only the former methodology has been tested experimentally. Experimental attenuation spectra for 11 kaolin slurries were obtained with the recirculating cross-flow, (i.e. in a flow-through mode). The data were analyzed to assess sensitivity to shape factors. It is shown that predictions based on realistic particle size and shape distribution (PSSD) functions allow one to interpret measured ultrasonic spectra for kaolin slurries.

1. Introduction

Clay pigment particles in general have a platelet shape. For platelet or disk-like particles the shape factor, R, can be defined as the thickness to face diameter ratio. We obtained attenuation coefficient spectra over a wide frequency

[*] Presently at Foxboro Company, Foxboro, MA 02035
[**] Presently at Los Alamos National Laboratory, MS D249, Los Alamos, NM 87545

range using slurries of different particle concentrations. These slurries were prepared by careful dilution of a predispersed kaolin slurry, formulated at 35 wt % solids. The present model[1,2] uses the coupled-phase framework with the viscous coupling coefficient related to the drag force experienced by an oscillating sphere in cell. In an earlier study[3,4] this model was tested using well-dispersed titanium dioxide suspensions. For clay suspensions studied here thermal contributions are much less than the viscous attenuation. The scattering contributions can be safely neglected when one restricts the frequency range of interest to 3 to 45 MHz. This model has now been modified to take into account the effects of non-sphericity of suspended particles. The model that deals with a concentrated suspension of spheroids allows us to quantitatively account for a particle size and shape distribution (PSSD) appropriate for kaolin clay. For theoretical predictions, clay particles are modeled as spheroids. For a given clay particle, an equivalent spheroid is defined as the one with its aspect ratio, A, (i.e. polar to equatorial axis ratio) set equal to the shape factor, R, and its volume set equal to that of the clay particle.

2. Particle Shape Analysis on Test Suspensions

Here we present illustrative analysis of the attenuation spectra on kaolin slurries (KWW Standard supplied by Engelhard Corporation) obtained with a flow-though chamber using typical on-axis transducer configuration. In this conventional mode, total loss of sound energy is measured for a sequence of multiple gaps at each operating frequency. Measured ultrasound attenuation spectra of concentrated clay suspensions are interpreted in terms of suitable PSSDs. Figure 1 shows attenuation coefficient data for three concentrations (2.9, 7.61 and 16.9 vol % solids) as functions of sound frequency, and similar data for two operating frequencies (6.85 and 35.7 MHz) as functions of particle concentration. The excess attenuation coefficients range from 2 to 200 dB/cm increasing with frequency and particle concentration.

Figure 1 also shows model predictions, which are based on assumed test PSSDs shown in Figure 2. A given sedimentation-based particle size distribution (PSD) corresponds to five different PSDs when expressed in terms of volume equivalent sphere diameters, corresponding to five test shape factor distributions (R vs. D) shown in Figure 2. Values of R larger than 1 correspond to prolate spheroids, while R less than 1 correspond to oblate spheroids. Random orientation was assumed to apply to sedimentation as well as acoustophoresis. The test distribution PSSD1, with constant aspect ratio R of 1, corresponds to the assumption of spherical shape where the volume equivalent diameters, D and the sedimentation equivalent diameters, D_{sed}, become identical. The test distributions

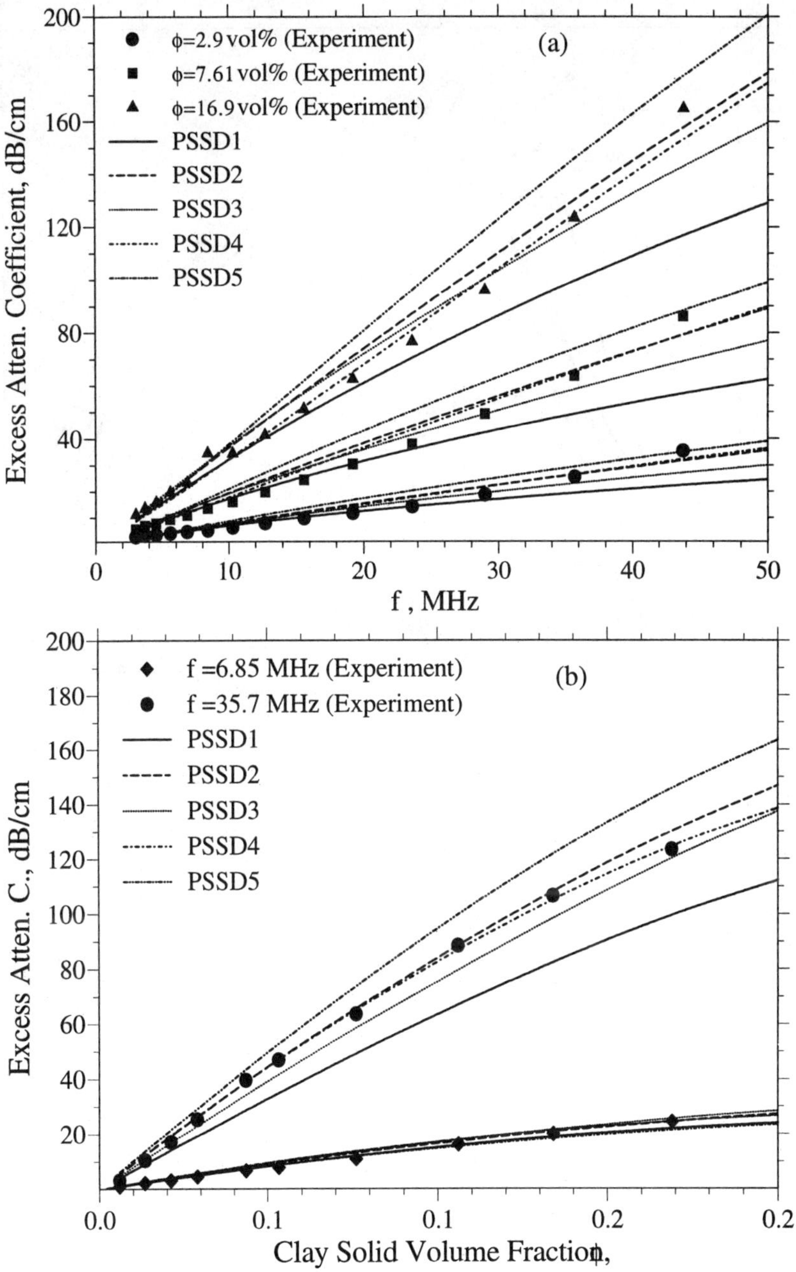

Figure 1. Attenuation coefficient as a function of (a) sound frequency and (b) particle concentration: Comparison of predictions and the experimental data, for five test PSSDs.

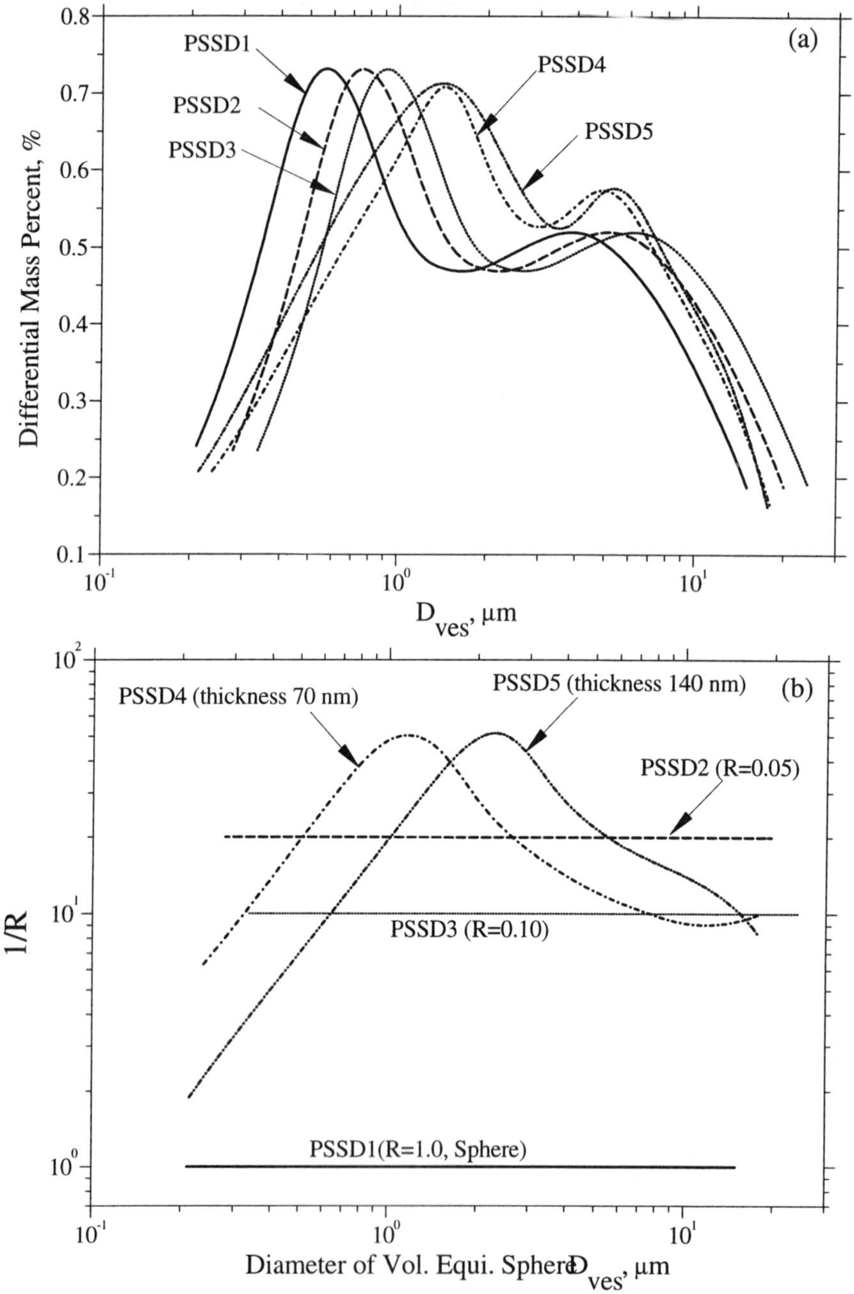

Figure 2. Test particle size and shape distributions (PSSDs): (a) differential size distributions and, (b) corresponding size dependent shape factor distributions.

PSSD2 and PSSD3 assume constant R of 0.05 and 0.10, respectively. The assumption of size-independent aspect ratio is obviously an over simplification for a natural pigment like kaolin. Very small kaolin particles can be fragments with shape factors in the range of 0.2 to 1. The coarse kaolin particles can be 'booklets' or aggregates with shape factors in the range of 0.1 to 0.4. In the intermediate size range, one encounters natural platelets of shape factors ranging from 0.1 to 0.01. Based on these considerations, we have constructed two test distributions PSSD4 and PSSD5 with size dependent shape factors corresponding to basic plate thickness of 70 nm and 140 nm, respectively. Size dependence of the shape factor of kaolin particles has been discussed by Slepetys and Cleland.[5]

Predictions based on the assumption of spherical shape are in agreement with the experimental data for dilute slurries at low frequencies. It is clear that effects of non-sphericity become more prominent as the sound frequencies (or particle concentrations) increase. The overall agreement between theoretical predictions and experimental measurements is remarkably well when shape factors of non-spherical particles are taken into account. In particular, predictions corresponding to PSSD4 appear to match experimental data over wide ranges of frequency and particle concentration. This shape distribution corresponds to 70 nm thick platelets. The preliminary analysis indicates that attenuation spectra for clay slurries with sufficiently high solids volume fractions can allow one to discriminate among several realistic shape factor distributions, especially in the high frequency region. Complete details have been documented in Chapter 5 of Han's dissertation.[6]

3. Orientation Field Evolution and Active Spectroscopy

In active spectroscopy, one takes advantage of an evolving orientation field of suspended particles under the squeezing flows associated with the continuously-changing-gap mode of operation. One needs to measure total loss of sound energy while the transducer gap, and hence the path length and the pulse transit time, changes continuously. One thus loses the simplicity of having the frequency and the gap fixed during measurement of the total loss. Additional efforts associated with the on-the-fly total loss measurements allow one to use resulting particle orientation fields in the analysis.

Recently, Phan-Thien and Graham[7] have successfully applied the Transversely Isotropic Fluid (TIF) model to describe the orientation evolution of particles in suspension subjected to squeezing flows. Analytical solutions for the TIF model are in agreement with classical solutions for the Jeffrey's orbits for spheres in a simple shear flow. For squeezing flows, the governing differential equations for the TIF model are transformed into finite difference equations using the upwind difference scheme. The unit vector field evolution is quantified in terms of the angle made by the particle's axis of revolution with respect to the approaching piston's face. We have analyzed the effects of spheroid aspect ratio on the evolution of orientation fields in a squeezing flow encountered in a sensing zone when the continuously-changing-gap mode is employed. The line-of-sight average of the orientation angles provides quantitative measure of the orientation evolution. These can be useful in sensor applications that use squeezing flows, where the sensor response is dependent on particle orientations along the line-of-sight. This forms the basis of active ultrasonic spectroscopy for particle shape analysis since attenuation is dependent on particle orientation relative to the direction of sound propagation.

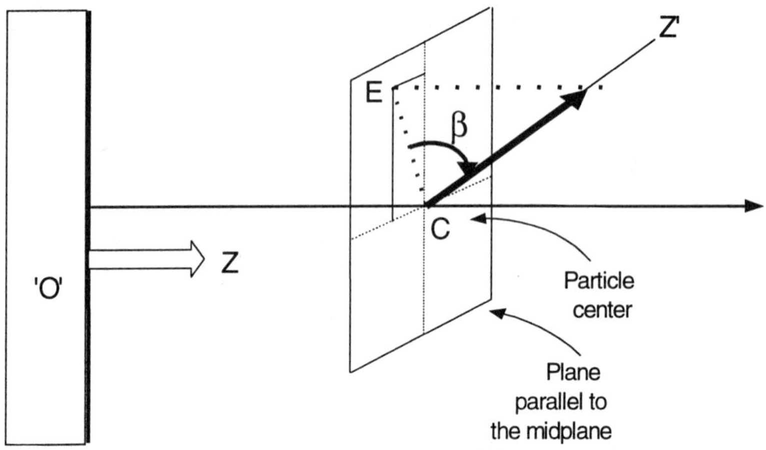

Figure 3. Particle orientation angle, β, where CZ' is the particle's axis of revolution.

Figure 3 shows the fixed reference frame and a particle orientation vector describing how a particle orientation is related to this fixed coordinate system. A schematic of the sensing zone is shown in Figure 4 for reference. For the purposes of the current work, the only angle of consequence is that describing a particle's rotation from the radial axis, β. Because axial symmetry is assumed, only a single two-dimensional plane is considered. The particle center, C, is the origin of the particle orientation vector. The angle between the axis of rotation, Z', and its projection on a plane parallel to the midplane, $\vec{CE}$, is β. The time-dependent particle orientation field differs considerably from the fluid velocity field, illustrated in Figure 5. Table 1 defines the orientation terminology.

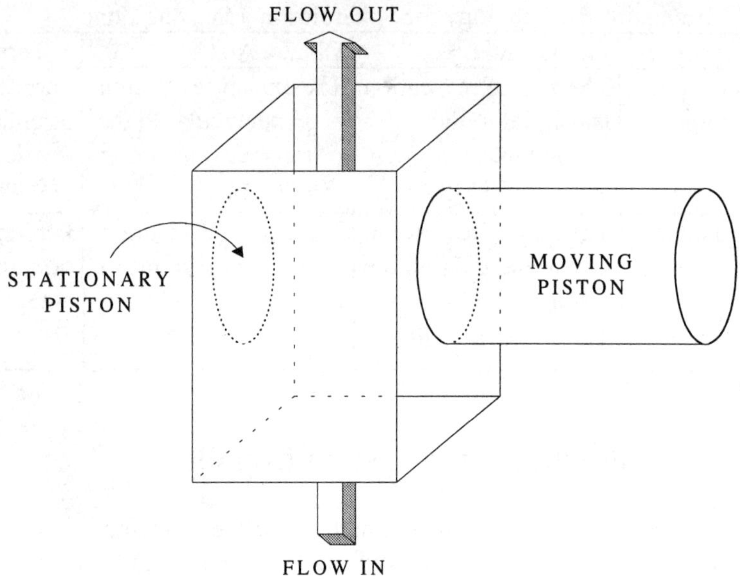

Figure 4. Simple schematic of sensing zone where squeezing flow would be encountered in a continuously decreasing gap mode.

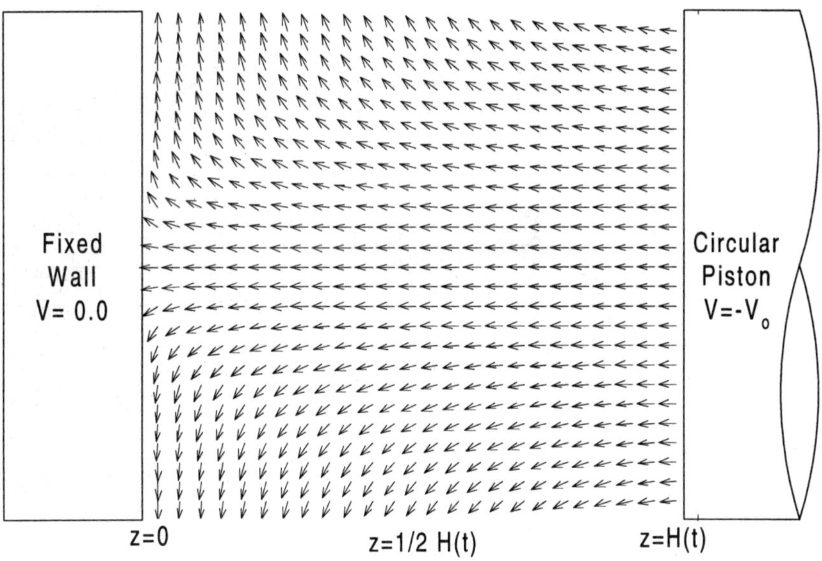

Figure 5. Fluid velocity vector orientation profile between moving piston and wall (shown using unit vectors).

Table I. Orientation Terminology for Spheroids in a Sensing Zone.

Aspect ratio, A	Prolate, A>1	Oblate, A<1	Sphere, A=1
Parallel Orientation	The axis of revolution is parallel to the piston face $\beta = 0^o$ or $\beta = 180^o$	The axis of revolution is perpendicular to the piston face $\beta = 90^o$ or $\beta = 270^o$	Parallel orientation is not defined $\beta =$ any angle
Perpendicular Orientation	The axis of revolution is perpendicular to the piston face $\beta = 90^o$ or $\beta = 270^o$	The axis of revolution is parallel to the piston face $\beta = 0^o$ or $\beta = 180^o$	Perpendicular orientation is not defined $\beta =$ any angle

4. Illustrative Results for Orientation Evolution

In general, it was found that the particles at the off-center locations would approach close to parallel orientations, which they would hold for a short duration before undergoing repeated rotations. Highly non-spherical particles (aspect ratio, A >10 or A<0.1) tend to spend considerable time in parallel orientation, but nearly-spherical particles (0.1 <A<10) tend to spend considerable time undergoing repeated rotations of decreasing periods. Figure 6 shows results for four different

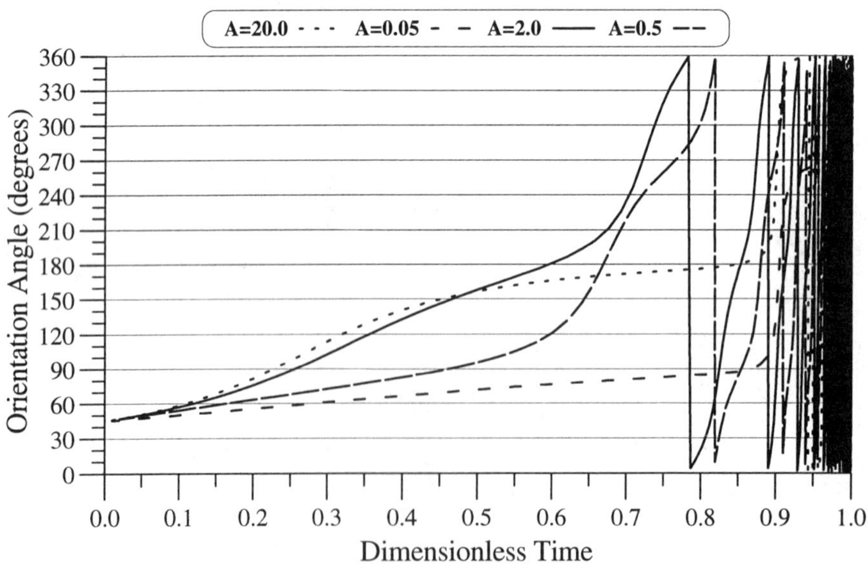

Figure 6. Comparison of oblate and prolate orientation evolution.

types of particles corresponding to the case of an initial gap of $H_o/R_o=0.25$ for $r/R_o = 0.85$ and $z/H(t) = 0.75$. The particles with aspect ratios of 20 and 0.05 appear to attain parallel orientation for the dimensionless times ranging from 0.5 to 0.85. As these particles approach parallel orientation, their rates of rotation become very small. There is much less difference between oblate and prolate particles as their aspect ratios approach 1.0. For particles with aspect ratios of 2.0 or 0.5, changes of the orientation angle with time appear to be very similar. After approximately 0.8 normalized time units, they begin to undergo rapid rotations. It is clear that the highly non-spherical particles with aspect ratios much different from 1 approach close to parallel orientation and remain in parallel orientation for a considerable duration, before undergoing repeated rotations like the so-called Jeffrey's orbits. Highly non-spherical particles (aspect ratio, $A > 5$ or $A < 0.2$) tend to spend considerable time in parallel orientation, but nearly-spherical particles ($0.5 < A < 2$) tend to spend considerable time undergoing repeated rotations of decreasing periods. Imposition of parallel orientation for suspended spheroids resulting from the squeezing flow, associated with a piston approaches a fixed wall, provides an opportunity to probe orientation sensitive phenomena.

5. Summary and Conclusions

Conventional attenuation spectra obtained with a cross-flow of slurry through the sensing zone can be used for shape analysis with success. Attenuation spectra of kaolin slurries with high solids volume fractions are shown to contain sufficient information to discriminate among several plausible PSSDs.

Furthermore, it should be noted that the attenuation coefficients of suspensions of spheroids are strongly influenced by particle orientations, as shown by Han. Figures 5.20 and 5.21 from Han's dissertation[6] are reproduced here for easy reference, as Figures 7 and 8. The ratio of excess attenuation coefficients for random versus parallel orientation is shown as functions of frequency, f, or equivalently ($kR_{ves} = 2\pi.f.R_{ves}/c_o$). R_{ves} denotes the average volume equivalent radius of the particles. It is clear that the frequency dependence of this attenuation ratio (random versus parallel) is sensitive to size dependent shape distributions.

A simple methodology has been developed to quantify orientation fields of suspended spheroids subjected squeezing flows. Information on orientation field evolution associated with squeezing flows can be used to infer shape factors from attenuation spectra obtained in the continuously-changing-gap mode, as compared to those in the multiple-gap-sequence mode. However, current challenges lie in making attenuation measurements while the gap between transducers changes continuously.

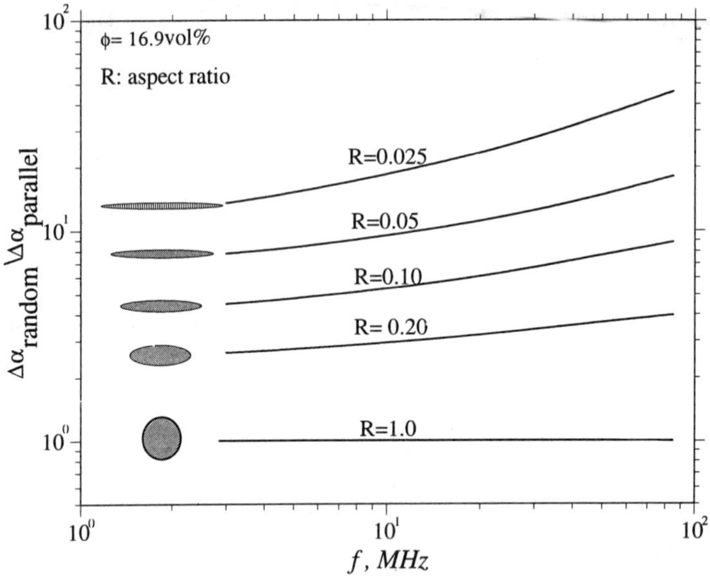

Figure 7. Model predictions of effects of orientation of clay particles on excess attenuation spectra under random and parallel orientations, based on particle size distributions with constant *R*.

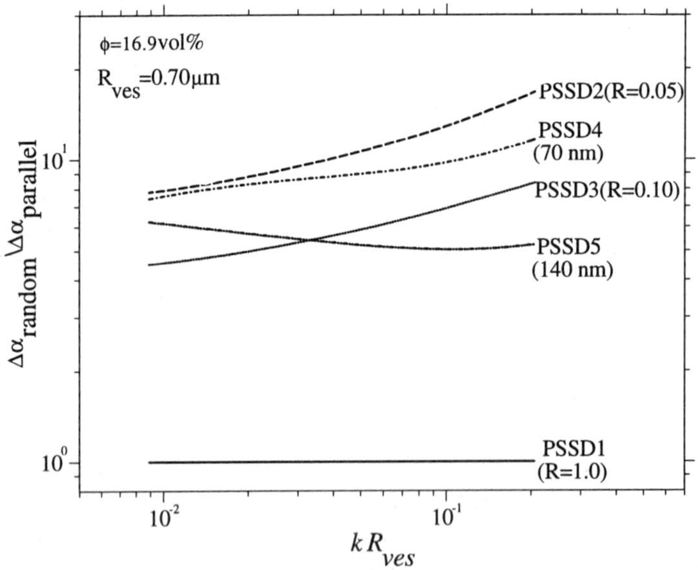

Figure 8. Model predictions of effects of orientation of clay particles on excess attenuation spectra under random and parallel orientations, based on five test PSSDs.

References

1. H.P. Pendse and T.A. Strout, *Proc. Sensor Expo West,* 205B (1990).

2. T.A. Strout, "Attenuation of Sound in High-Concentration Suspensions," Ph.D. Thesis, University of Maine, Orono, ME, 1991.

3. H.P. Pendse, T.A. Strout, and A. Sharma, "Theoretical Considerations in Acoustophoresis Analysis of Concentrated Colloids," pp. 23-39 in *Electroacoustics For Characterization of Particulates and Suspensions*, NIST Special Publication No. 856. Edited by S.G. Malghan. National Institute of Standards and Technology, Gaithersburg, MD, 1993.

4. A. Sharma, "Ultrasonic Spectroscopy of Solid Liquid Suspensions," Ph.D. Thesis, University of Maine, Orono, ME (1995).

5. R.A. Slepetys and A.J. Cleland, "Determination of Shape of Kaolin Pigment Particles", *Clay Minerals*, **28**, 495-508 (1993).

6. W. Han, "Effects of Particle Shape on Viscous Attenuation in Concentrated Clay Suspensions", Chapter 5, in *Viscothermal Coupling Effects on Sound Attenuation in Concentrated Colloidal Dispersions*, Ph.D. Thesis, University of Maine, Orono, ME, 1995.

7. N. Phan-Thien and A.L. Graham, "The squeezing Flow of Model Suspension Fluid", *Rheologica Acta*, **29**, 433-441 (1990).

Ultrasonic Spectroscopy Measurement System for On-Line Sensing of Concentrated Particulate Suspensions

Arvind Sharma* and Hemant P. Pendse
Industrial Process Control Sensor Systems (IPCoSS) Program, Department of Chemical Engineering, University of Maine, Orono, Maine 04469-5737 USA

Abstract

During field tests at a pigment manufacturing plant, the on-line sensor prototype of the automated ultrasonic spectrometer has provided attenuation spectra, and corresponding particle size distribution (PSD) estimates, under processing conditions showing excellent reproducibility and robust operation. The tests of on-line sensor prototype operation show that ultrasonic spectroscopy of concentrated colloids is a promising tool for on-line characterization of industrial slurries, and offers the possibility of its use in quality control and process control applications. Performance of the automated ultrasonic spectrometer system used in the on-line prototype has been validated using Dow Corning 710 silicone fluid as a reference liquid. Ultrasonic spectroscopy measurements of several colloidal dispersions have been obtained over 1 to 100 MHz using through-transmission loss data corresponding to several path lengths at each operating frequency. The automated ultrasonic spectrometer is found to provide reliable values of attenuation coefficients ranging from 0.1 to 2500 dB/cm. Quantitative comparison of experimental data on attenuation coefficients of well-characterized test suspensions with corresponding predictions based on theoretical modeling that accounts for attenuation dependence on PSD and particle volume fraction, over the wide range of operating frequencies is presented. In particular, complicated dependence of the excess attenuation coefficient on the volume fraction of suspended particles at low and high frequencies is revealed. Such dependence cannot be explained using simple quadratic relationships.

1. Introduction

In recent years several advances have been made in ultrasonic spectroscopy of suspensions. However, in the literature, measurement techniques and ultrasonic data are not available over wide ranges of frequencies for suspended particle concentrations commonly encountered in chemical process

---

* presently at the Los Alamos National Laboratory, Group EES-4, MS D443, Los Alamos, New Mexico 87545-1443, USA

industry. To date, there is not much work reported on highly attenuating solid-liquid suspensions. The focus of this work was on an automated ultrasonic spectroscopy measurement system for a wide frequency range, capable of handling commonly encountered solid-liquid suspensions at high particle concentrations, which are highly attenuating. Here, the method involving tone-burst signal with amplitude measurements is used for obtaining through-transmission loss data. The details of the ultrasonic measurement system are given in a thesis by one of the authors[1]. The test setup used in the study employed flow-though slurry characterization chambers developed by Techna Machine. Ultrasonic measurement electronics and software was developed to provide automated measurement setup. Transmission losses were obtained at 21 different path lengths, for each operating frequency, using the computerized automated system. These losses were then selectively used to determine the attenuation spectra. The details of attenuation coefficient calculations from the loss versus path length data were recently published elsewhere[2]. The performance of this setup was validated using Dow Corning 710 silicone fluid as a reference liquid. Measured attenuation coefficients of this silicone fluid at 22 °C range from 0.7 to 1600 dB/cm over the frequency range of 1 to 100 MHz. These spectra consisting of 18 frequencies could be obtained in less than 10 minutes in the automated mode. The results were in good agreement with other reported measurements.

In the following sections, results of ultrasonic spectroscopy of well-characterized concentrated suspensions are discussed. The focus is on the effects of concentration over a wide frequency range, when PSD is held constant.

2. Experimental Results on Pigment Suspensions

For this study, test suspensions were prepared by varying solids concentration from 42.3 to 3.5 vol% (75 to 13 wt%) through ten successive dilutions of a well-dispersed high solids titanium dioxide slurry (TI-PURE R-940). Independent PSD analyses of these slurries using SediGraph 5100 confirmed that all of these slurries are characterized by a single particle distribution curve, independent of their particle concentrations. The SediGraph PSD corresponds to median diameter of 0.35 μm, modal diameter of 0.31 μm, and geometric standard deviation (*gsd*) of 1.6.

Attenuation measurements were repeated at least six times for every test suspension. The values of the mean attenuation coefficients have been tabulated in Table I. The standard deviations are found to be less than 2% of the corresponding mean values. This comprehensive data set can be used to evaluate various theoretical prediction schemes since all the test suspensions of different particle concentrations correspond to the same PSD.

Table I. Ultrasonic Spectra for Titanium Dioxide Suspensions (mean temperature: 27 °C).

Frequency (in MHz)	*Attenuation coefficients (in dB/cm) at volume percent of:* 42.3	32.2	25.6	21.2	19.4	13.3	11.2	8.4	6.5	5.9	3.5
3.00	8.0	20.8	**26.1**	23.2	19.3	17.5	16.0	13.6	12.0	9.9	7.6
3.69	10.9	26.0	**31.4**	28.3	25.9	22.7	20.3	17.8	14.9	12.8	9.7
4.53	15.5	35.6	**40.6**	37.2	33.6	30.2	27.9	24.1	20.6	16.8	12.5
5.57	21.7	46.5	**51.4**	47.3	42.9	38.9	36.0	31.0	26.6	21.7	16.1
6.85	30.9	61.3	**67.3**	62.3	56.6	52.3	47.7	40.7	35.0	28.3	20.6
8.41	43.5	84.3	**88.8**	82.0	75.8	69.3	65.0	53.4	45.4	37.3	27.8
10.34	61.1	107.2	**110.1**	102.1	92.1	89.1	81.4	69.0	57.4	45.7	33.3
12.71	84.5	138.2	**139.2**	128.0	118.6	110.5	102.7	86.3	70.5	57.1	41.9
15.62	116.7	**179.6**	179.2	164.1	150.9	141.7	128.6	108.2	88.2	70.9	51.6
19.20	159.4	**234.2**	228.1	207.9	190.5	177.1	160.3	134.3	107.8	86.6	63.0
23.60	216.8	**304.3**	291.0	263.7	240.1	221.3	196.3	163.6	130.8	104.1	76.3
29.01	298.1	**397.1**	369.6	333.4	298.6	274.0	238.1	197.3	156.4	124.4	90.9
35.65	412.7	**510.4**	467.7	417.0	368.7	336.0	291.9	237.1	187.4	147.6	107.6
43.82	555.8	**654.7**	586.0	516.1	457.4	411.3	349.7	284.2	221.2	175.0	128.0
53.86	744.1	**816.2**	726.8	642.7	558.9	490.2	414.4	336.3	260.0	206.1	151.0
66.20	925.8	**960.8**	870.6	761.1	655.0	578.0	477.0	399.1	301.2	241.7	175.2
81.36	**1457.7**	1416.8	1194.4	1060.9	765.8	747.6	557.5	456.0	355.4	284.0	205.2
100.00	**2011.9**	1840.7	1517.9	1345.3	962.2	863.7	695.6	547.3	476.3	314.2	241.7

Bold entries show peaks in concentration responses. Note that the concentrations at the peak locations increase with increasing frequencies.

Here, a theoretical predictive model[3-5] was used to compute viscous attenuation coefficients as a function of frequency (f) over wide ranges of particle concentrations and diameters. The excess attenuation due to suspended particles is computed on the basis of a model for viscous dissipation associated with acoustophoresis. When a sound wave travels through a suspension, the suspended particles as well as the suspending medium undergo oscillatory motion. The motion of particles relative to the medium under the influence of applied acoustic field is termed acoustophoresis. Acoustophoresis takes place when the density of particle phase is different from that of the medium.

The theoretical model used in this work uses the coupled phase continuum approach. The viscous coupling coefficients explicitly account for particle size, concentration and sound frequency using the oscillating 'sphere-in-cell' framework. This model and its extension for visco-thermal coupling is discussed in detail by Han and Pendse elsewhere in this volume.

For the case of submicrometer high density-contrast particles considered here the viscous attenuation is found to be adequate. These calculations require values of solids volume fraction (ϕ), PSD given in terms of geometric mean (*gm*) and *gsd*, in addition to the properties of the medium and particulate phases, viz., density and sound speed, and the viscosity of the suspending medium. The assumed log-normal PSD is given by the following equation:

$$p(x) = \frac{1}{\sqrt{2\pi}\log_{10} gsd}\exp\left[-\frac{1}{2}\left(\frac{x-\log_{10} gm}{\log_{10} gsd}\right)^2\right] \tag{1}$$

$$\ldots\text{where} \quad x = \log_{10}(d)$$

Here, $p(x)\,dx$ is the fractional mass of particles with diameters between x and $x+dx$, and d is the particle diameter in microns. On the basis of the considerations of the rate of viscous energy dissipation in a suspension of spherical solid particles subjected to a sound wave, one obtains an expression for the attenuation coefficient contribution due to monodisperse particles, $\alpha_{\text{p, mono}}(f, \phi, x)$. For the case of a suspension of polydisperse particles, the attenuation coefficient becomes:

$$\alpha(f,\phi) = \alpha_{\text{medium}}(f) + \alpha_p(f,\phi) \tag{2}$$

where the following form is assumed for the excess attenuation:

$$\alpha_{\mathrm{p}}(f,\phi)=\int_{x_{\min}}^{x_{\max}}\alpha_{\mathrm{p,mono}}(f,\phi,x)p(x)dx \tag{3}$$

3. Comparison of Experimental Data with Theoretical Predictions

Here, we explore results of theoretical modeling that accounts for attenuation dependence on PSD and particle volume fraction, over the wide range of operating frequencies. Quantitative comparison of experimental data on attenuation coefficients at 18 frequencies and 11 concentrations with corresponding predictions is presented. The theoretical model is able to predict the experimentally observed complex functionality of frequency and concentration dependence of attenuation coefficients fairly well. The attenuation coefficient spectra are found to be quite sensitive to the PSD. This analysis provides a basis for understanding why deconvolution of measured spectra works and why one is able to extract PSD function from the measured ultrasound spectra.

Comparison of experimental measurements and corresponding predictions of attenuation coefficient spectra is shown in Figure 1. Experimental data is shown for attenuation coefficients covering a wide range of particle concentrations for selected sound frequencies. The seven concentration response curves correspond to seven operating frequencies of 3.00, 5.57, 10.34, 19.20, 35.65, 66.20, and 100.00 MHz. The curves shift up with increasing frequency. The attenuation coefficients range from 2 to 2000 dB/cm increasing monotonically with increasing frequency. Note that the vertical axis is in logarithmic scale.

The model predictions reported here were based on an assumed log-normal PSD characterized by *gm* of 0.33 μm and *gsd* of 1.65. No attempt is made here to fine-tune the log-normal distribution function parameters to fit the data. As already seen, the overall agreement between predictions and experimental measurements is remarkably well, with the overall root-mean-square relative deviation being 19% over such wide ranges of volume fraction (3 to 43 vol%) and frequency (3 to 100 MHz). From Figure 1, it is clear that the attenuation coefficient versus volume fraction relationship is non-monotonic, with the peak location shifting to higher volume fractions with increasing frequency. This comparison of measurements done on a set of well-characterized, well-dispersed slurries and corresponding predictions based on a single realistic PSD shows that the theory predicts appropriate dependence of attenuation coefficient on particle volume fraction and operating sound frequency over very wide ranges. The quantitative agreement of course will depend on adequacy of the assumed size distribution function, which was log-normal in this case.

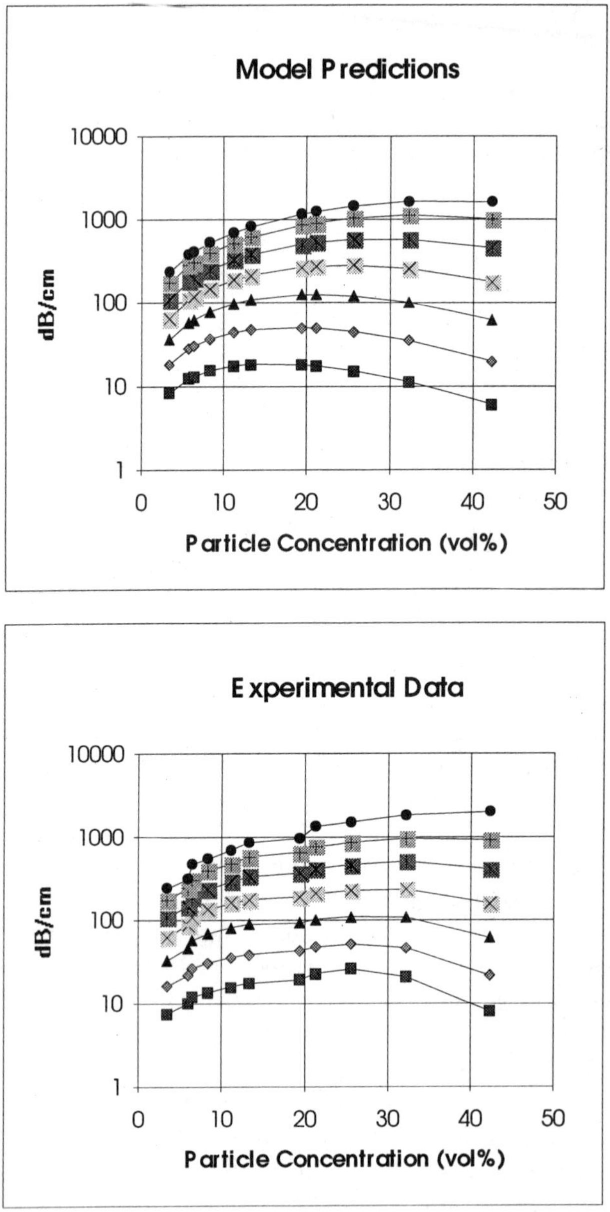

Figure 1. Comparison of measured attenuation spectra to model predictions for titanium dioxide slurry. The curves shift up with increasing frequency.

4. Ultrasound Response on a Complex Plane

Both, the sound speed and the attenuation spectra, can be presented together on a complex plane. The quantity, complex wave number, is:

$$\tilde{k} = \frac{\omega}{c} + i \cdot \alpha \tag{4}$$

Let

$$\hat{k} = \frac{\tilde{k} c_0}{\omega} \tag{5}$$

i.e.,

$$\hat{k} = \frac{c_0}{c} + i \cdot \frac{\alpha c_0}{\omega} \tag{6}$$

Here, c and c_0 are the speeds of sound in the slurry and the media respectively, and ω ($=2\pi f$) is the radian frequency. Thus, $\hat{k}$, which combines the two ultrasound measurements of sound speed and attenuation coefficient, coupled with frequency, is given in Figures 2 and 3. In these figures, both the ultrasound measurements are compared at the same time to the predictions on a complex plane, where a given curve corresponds to a fixed concentration, and labeled points correspond to different frequencies. These features are easily apparent in Figure 3 for the predictions where the curves corresponding to each solids concentration go through a peak. Due to the limited frequency range applicable to the sound speed measurements, the experimental data only spans the right-hand falling portions of the overall response curves.

5. Analyzing Normalized Excess Attenuation Coefficient

Excess attenuation coefficient $(\Delta\alpha)$ can be obtained by subtracting the attenuation coefficient of the suspending liquid, at corresponding temperature, from that of the solid-liquid suspension. Here, the normalized excess attenuation coefficient, given by $\frac{\Delta\alpha \cdot \lambda}{2\pi\phi}$ would be studied (λ is the wavelength for the suspending medium). First, the dilute-limit attenuation[4] is compared to the measurements in Figure 4, and it shows good agreement. For the case of the measurements, the zero concentration limit of the normalized excess attenuation coefficient was obtained by extending the low concentration regression lines as shown in Figures 5 and 6.

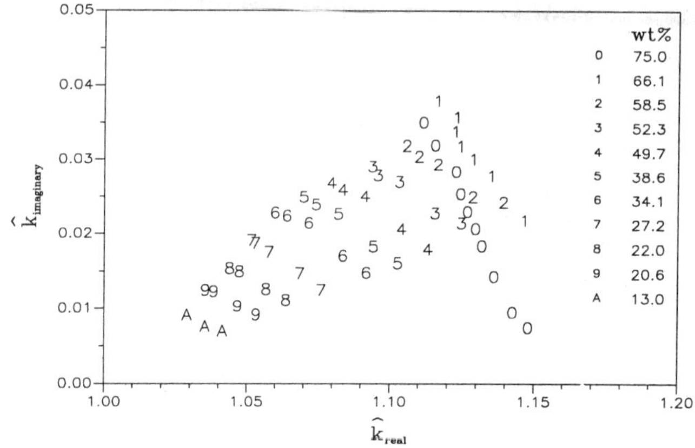

Figure 2. Measurements of titanium dioxide slurry presented on a complex plane.

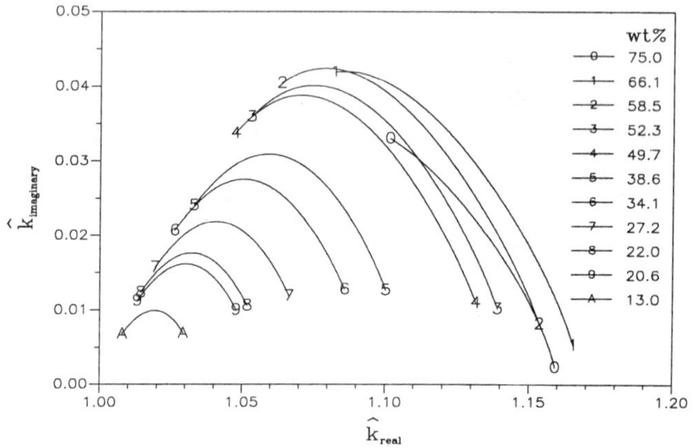

Figure 3. Model predictions of titanium dioxide slurry presented on a complex plane.

Now, as is obvious from these two figures, the quantity $\frac{\Delta\alpha \cdot \lambda}{2\pi\phi}$ is not always constant over different values of solids concentration, as the dilute limit extrapolation based on the assumption of direct proportionality between the excess attenuation and the volume fraction would indicate. For the case of 100 MHz, the low volume fraction data is consistent with the dilute limit extrapolation, but this does not apply to the 3 MHz data. Furthermore, one cannot represent the experimental data by a single straight line of a negative slope over the entire concentration range, as assumed quadratic dependence of excess attenuation on volume fraction would indicate. The response is concave upward at low frequencies, and concave downward at high frequencies. The general predictions are consistent with the observed response.

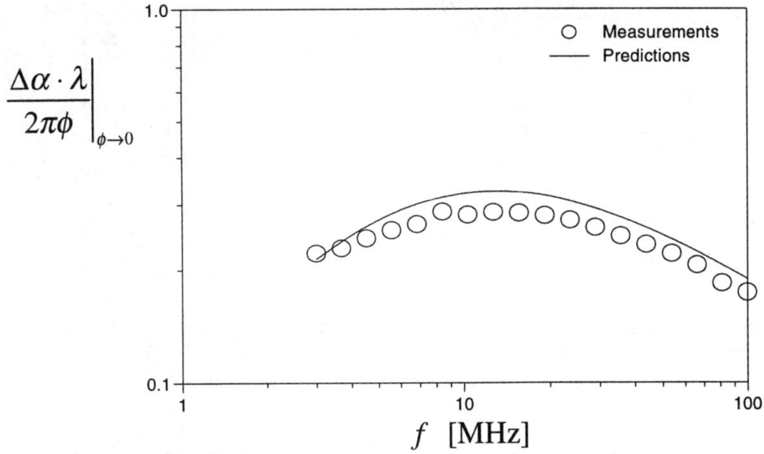

Figure 4. Comparison between the measurement data and the predictions at zero concentration limit for titanium dioxide slurry.

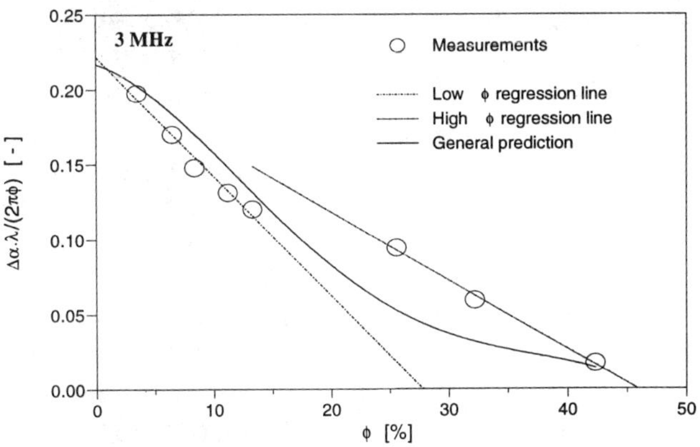

Figure 5. Comparison between the measurement data and the predictions at 3 MHz frequency for titanium dioxide slurry.

Thus, it can be seen that the predictive model can not only explain the experimental data in the dilute region (Figure 4) but it can also account for the different shapes of the curves obtained in the concentrated region at a wide range of frequency (Figures 5 and 6). It may be noted that in case this wide range of frequency and solids concentration was not covered in the experiments, then these important features given above would have been missed.

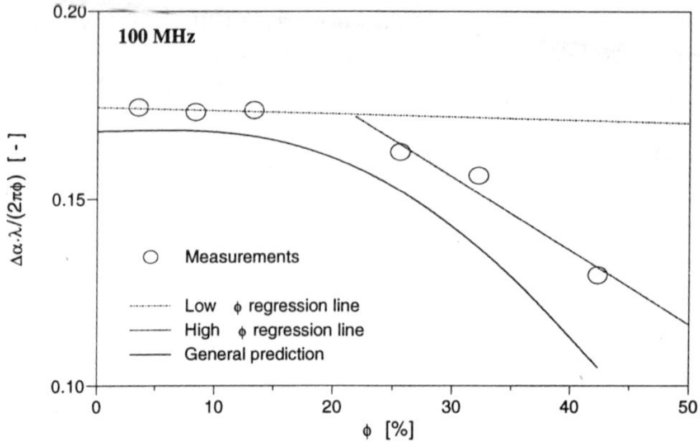

Figure 6. Comparison between the measurement data and the predictions at 100 MHz frequency for titanium dioxide slurry.

6. PSD Effects on Peak Frequencies of Attenuation Spectra

Figure 7 shows α / f versus f response of five different assumed log-normal PSDs, along with the measurements for 6.5 vol% slurry. From the figure, it can be clearly seen that the peak frequency of a normalized spectrum is dependent upon the mean particle diameter for fixed particle concentration. Higher the *gm*, lower is the peak frequency. Also, as the *gsd* increases, the attenuation spectra broadens but the peak frequency is unaffected for fixed particle concentration.

Although the experimental normalized spectrum, too, goes through a peak, it does not fit any of the five predicted spectra over the entire frequency range. Moreover, the peak resolution is not good. However, at lower frequencies (< 15 MHz), the measurements are close to the predictions based on *gm* = 0.27 μm and *gsd* = 1.65, and at higher frequencies (> 15 MHz), the measurements agree with the curve corresponding to *gm* = 0.33 μm and *gsd* = 1.65.

This points out the limitations arising from the assumption of a particular functional form for the PSD. Further fine-tuning of *gm* and *gsd* is only of limited value.

On further analysis of the peak frequencies, it was observed that at any given solids concentration,

$$f_{\text{peak}} \propto \frac{1}{ead^2} \tag{7}$$

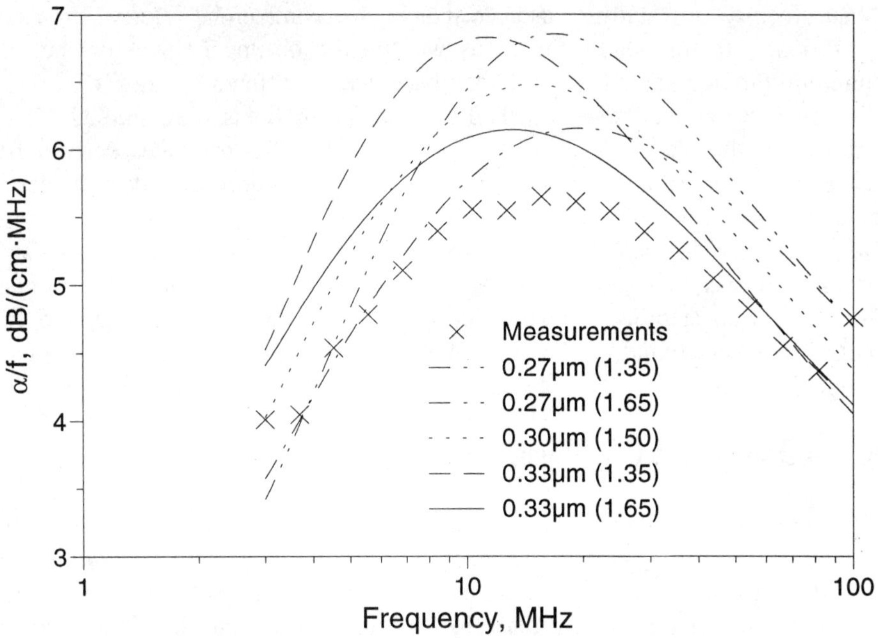

Figure 7. Comparison of measured attenuation spectrum with predictions for 6.5 vol% titanium dioxide slurry.

where we have defined a mean diameter named *equivalent acoustophoretic diameter* (ead) as:

$$\ln ead = \ln gm - \tfrac{3}{8}\ln^2 gsd \tag{8}$$

This ead can be a convenient way of representing acoustically measured PSD by means of a single parameter, which is associated with the two parameters (gm and gsd) of the log-normal distribution.

7. On-Line Experience for Prototype Sensor System

A prototype of an on-line sensor system has been developed for monitoring PSD of concentrated colloids using automated ultrasonic spectroscopy[6]. During field tests at a manufacturing plant, the on-line sensor has provided PSD data under processing conditions showing excellent reproducibility and robust operation. Earlier, the on-line sensor prototype was evaluated at a pilot plant facility to study sensor response to commercial slurry processing conditions[7]. In this study, the focus was on the on-line field experience at a titanium dioxide manufacturing facility. Several different grades of titanium dioxide pigments and some finished commercial products were studied. Careful analysis of the ultrasonic response as well as the PSD estimates derived from measured attenuation spectra of industrial titanium dioxide slurries at concentrations as high as 75 wt% has been carried out.

During the field tests, the prototype sensor has operated reliably as an on-line process monitoring tool during several protracted runs of continuous operation. Repeatability of attenuation spectra and PSD estimates derived from them has been found to be excellent in the light of normal variations in temperature.

8. Summary and Conclusions

In the current work, studies were carried out to deal with several issues with regard to making measurements of attenuation spectra and sound velocity on solid-liquid suspensions at different concentrations, over a wide frequency range. Measurements of ultrasonic attenuation spectra of titanium dioxide suspensions are given over the frequency range of 3 to 100 MHz. Solids concentration was varied from 42.3 to 3.5 vol% (i.e., 75 to 13 wt%) through ten successive dilutions of a well-dispersed high solids slurry. This comprehensive data set of ultrasonic attenuation spectra of well-characterized test suspensions of various concentrations, all at the same PSD can be used to evaluate various prediction schemes.

The results of theoretical modeling that accounts for attenuation dependence on PSD and particle volume fraction, over the wide range of operating frequencies were explored. Detailed quantitative comparison of experimental data on attenuation coefficients at 18 frequencies and 11 concentrations with corresponding predictions was carried out. The theoretical model is able to correctly predict the experimentally observed complex functionality of concentration dependence of attenuation coefficients at various frequencies.

Availability of a predictive, theoretical model that accounts for hydrodynamic interactions underlying acoustophoresis in multi-particle systems with moderate and high particle volume fractions, over a wide range of frequencies allows one to interpret measured attenuation spectra in terms of particle size distributions. This model is validated in the light of experimental data on attenuation spectra obtained using a set of well-dispersed suspensions of different solids concentrations. Ultrasound spectroscopy may find interesting applications in characterization and quality control of a wide variety of industrial colloidal dispersions including suspensions and emulsions.

Finally, on-line measurements for the prediction of PSD using ultrasound were summarized. During field tests at a manufacturing plant, the on-line sensor has provided PSD data under processing conditions showing excellent reproducibility and robust operation during several protracted runs of continuous operation. The measurements were automated and were fast, taking about ten minutes per PSD output. Also, the on-line sensor prototype has been evaluated at a pilot plant facility to study sensor response to commercial slurry processing conditions. Thus, the ultrasonic spectroscopy of concentrated colloids appears to be a promising tool for on-line characterization of industrial slurries.

Acknowledgments

We would like to acknowledge financial support from the U. S. Department of Energy's Office of Industrial Technologies, and DuPont Chemicals. We also thank Techna Machine for providing the Slurry Characterization Chambers and Pen Kem for providing the early prototypes of measurement electronics and software.

References

1. A. Sharma., "Ultrasonic Spectroscopy of Solid-Liquid Suspensions," PhD Thesis, University of Maine, Orono, 1995.

2. A. Sharma and H.P. Pendse, "Ultrasonic Attenuation Spectroscopy of Industrial Slurries," pp. 196-200 in *Proceedings of the 1997 IEEE International Frequency Control Symposium*, IEEE Ultrasonics, Ferroelectrics & Frequency Control Society, 1997.

3. H.P. Pendse and T.A. Strout, "Ultrasound-Based Sensors for Characterization of Colloidal Slurries," *Proceedings of Sensors Expo West*, 205B (1990).

4. T.A. Strout, "Attenuation of Sound in High-Concentration Suspensions: Development and Application of an Oscillatory Cell Model," PhD Thesis, University of Maine, Orono, 1991.

5. H.P. Pendse, T.A. Strout, and A. Sharma, "Theoretical Considerations in Acoustophoretic Analysis of Concentrated Colloids," pp. 23-39 in *Electroacoustics for Characterization of Particulates and Suspensions*, NIST Special Publication No. 856. Edited by S.G. Malghan, US Department of Commerce, Technology Administration, Gaithersburg, MD, 1993.

6. H.P. Pendse and A. Sharma, "On-Line Instrumentation for Particle Size Distribution Analysis of Industrial Colloidal Slurries using Ultrasound," pp. 136-141 in *Proceedings of the First International Particle Technology Forum*, American Institute of Chemical Engineers, Part I, 1994.

7. H.P. Pendse and A. Sharma, "Particle Size Distribution Analysis of Industrial Colloidal Slurries using Ultrasound Spectroscopy," *Part. Part. Syst. Charact.*, **10**, 229-233 (1993).

Electroacoustic Analysis in Processing of Advanced Ceramics

Vincent A. Hackley
National Institute of Standards and Technology, Materials Science and Engineering Laboratory, Gaithersburg, MD USA

Ungyu Paik
Changwon National University, Department of Ceramic Engineering, Chang Won, South Korea

Abstract

The application of single-frequency electrokinetic sonic amplitude (ESA) measurements in the characterization and processing of advanced ceramic materials is reviewed. Examples from a variety of systems are used to illustrate advantages and limitations of this method.

1. Introduction

An increasing demand exists for high performance, high reliability ceramic components in applications across a broad spectrum of technological fields. Examples include silicon nitride in gas turbine engines for power generation, and barium titanate in multilayer ceramic capacitors for the electronics industry. Such components must meet specific and stringent requirements in order to acquire the desired properties and make the expected impact. The reliability and performance of ceramic components are limited by the microstructure, which in turn depends on the method of processing of green bodies.[1,2] In recent years the utilization of colloidal processing methods has gained in popularity because these systems offer significant advantages over conventional powder processing methods. More specifically, colloidal processing permits greater control over microstructure homogeneity and flaw size in the green body, which is of paramount importance for the elimination of critical defects that may limit ultimate performance in the fired ceramic.[3,4]

Different shape-forming methods require different suspension properties to produce a green body with a homogeneous microstructure. Suspension properties in colloidal systems are predominantly attributed to the interactions between the constituent microscopic particles. These interparticle interactions, which result from both attractive and repulsive contributions, control the structure of particles in suspension and hence the suspension properties. Since the nature of colloids allows a wide range of properties and casting behaviors, characterization of the colloidal state *in situ* is critical for obtaining the desired characteristics in the green body.[5,6]

In order to monitor and optimize suspension properties and ensure good dispersion, the measurement of zeta potential and particle size distribution are crucial. Zeta potential is an important parameter for characterizing the stability of colloidal systems, since it reflects the variation in surface potential for a specific material in a fluid medium and is sensitive to the adsorption of ionic species at the particle-solution interface. Particle dispersion, zeta potential, and rheology are inter-related suspension properties; zeta potential directly impacts the state of dispersion and indirectly affects flow properties. Typically, zeta potential is determined from measurements of the particle electrophoretic mobility.[7]

The electrokinetic sonic amplitude (ESA) technique is a rapid and convenient analytical tool for measuring the dynamic electrophoretic mobility of particles in concentrated colloidal systems.[8] ESA can accurately determine the pH at which the electrokinetic potential changes polarity (i.e., the isoelectric point or pH_{iep}), and is particularly useful in the study of ionic surfactant interactions with the dispersed phase. This technique has a wider range of applicability compared to conventional methods such as laser-based micro-electrophoresis. Advantages include analysis of opaque or colored suspensions, measurement in flowing or stirred systems, solid loadings up to roughly 50 %, and virtually no particle size restrictions. Additionally, data acquisition can be combined with automatic titrators to provide electrokinetic data over a wide range of solution conditions. These characteristics also make ESA a good candidate for on-line monitoring of process streams. The ESA technique has been tested and evaluated with many colloid systems and found to work exceedingly well, although certain limitations do apply and interpretation of results often requires some familiarity with electroacoustic theory.[9]

Therefore, the objective of this paper is to show how the (single-frequency) ESA technique is typically applied in ceramic processing research and development. Examples are given from oxide and non-oxide systems, suspended in both aqueous and non-aqueous media. Aspects related to sample preparation, analysis, and data interpretation are discussed, as are issues such as solids concentration, background electrolyte contributions, and the influence of particle size on the inertial response. Measurements in mixed particulate-phase systems are discussed as well. An evaluation is given of the benefits and limitations of the electroacoustic technique for characterization of ceramic powder suspensions and process control applications.

Although we focus here on measurements using a single-frequency analyzer, the same information may be applied more generally to measurements using a multiple-frequency electroacoustic spectrometer.

2. Experimental

2.1 Materials

Ceramic powders used in these studies include: α-silicon nitride (SN E-10, Ube Industries, Japan)[§], barium titanate (TICON HPB, TAM Ceramic Inc., Niagara Falls, NY), silica microspheres (Geltech Inc., Alachua, FL), yttrium oxide (Rhone-Poulenc, Shelton, CT) and silicon (HQ Silgrain, Elkem Materials Inc., Pittsburg, PA). The particle size distribution of well-dispersed powders was determined by x-ray gravitational sedimentation (Sedigraph 5100, Micromeritics, Norcross, GA), the specific surface area was measured by BET nitrogen-gas adsorption (Autosorb-1, Quantachrome, Syosset, NY), and the density was measured by He pycnometry (AutoPycnometer 1320, Micromeritics). The physical characteristics of the powders are listed in Table I.

Table I. Physical properties of powders.

Powder	Median Size (μm)	Surface Area (m^2/g)	Density (g/cm^3)
Silicon nitride	0.40 ± 0.01	10.6 ± 0.2	3.18[a]
Barium titanate	1.05 ± 0.10	3.15 ± 0.10	5.9 ± 0.1
Silica	1.21 ± 0.10	---	2.2[a]
Yttria	2.88 ± 0.20	4.30 ± 0.15	4.85 ± 0.01
Silicon	3.66 ± 0.20	3.29 ± 0.10	2.31 ± 0.01

[a] Value provided by manufacturer

[§] Certain trade names and company products are mentioned in the text or identified in illustrations in order to adequately specify the experimental procedure and equipment used. In no case does identification imply recommendation or endorsement by National Institute of Standards and Technology, nor does it imply that the products are necessarily the best available for the purpose.

Well-dispersed aqueous suspensions were prepared in deionized water using high-intensity ultrasonic treatment. The pH was adjusted with reagent grade HNO_3, NH_4OH, or NaOH, and analytical grade NH_4NO_3 or $NaNO_3$ were used as supporting electrolytes where necessary. Organic additives used in these studies include, poly(acrylic acid) (denoted as PAA, M_r = 5000, Polysciences, Warrington, PA), dimethyamine-epichlorohydrin copolymer (denoted as a DMAE1, Betz 1190, M_r = 10,000 - 12,000, Betz, Trevose, PA), poly (vinyl alcohol) (denoted as PVA, M_r = 25,000, 98 mole % hydrolyzed, Polysciences, Warrington, PA), and phosphate ester (Emphos PS-21A, Witco Co., Houston, TX). Non-aqueous suspensions were dispersed in Fisher HPLC grade solvents used without further purification. Solids concentrations are reported as % volume fraction, whereas additive concentrations are reported as % mass fraction relative to the solid phase.

2.2 Methods

The electrokinetic behavior of ceramic suspensions was characterized using a single-frequency electroacoustic analyzer operating in ESA mode at a nominal frequency of 1 MHz (ESA-8000, Matec Applied Sciences, Hopkinton, MA). The instrument was calibrated using a Ludox™ silica suspension of known mobility provided by the manufacturer. Based on an analysis of previous measurement data,[8] an estimate of the single-measurement standard uncertainty yields ± 0.10 mPa·m/V (aqueous) and ± 5 μPa·m/V (non-aqueous) for ESA, and ± 0.05 (pH units) for pH measurements. Ceramic suspensions were analyzed at 2%-5% solids, well within the expected linear response region. Samples were stirred during analysis, and titrated with standard acid or base solutions (1 molar HNO_3, NH_4OH, or NaOH). Measurements were performed at 22 ± 0.2 °C.

3. Results and Discussion

3.1 Interpreting ESA Data

The ESA response of a suspension depends on several factors as shown in Equation (1):[10]

$$ESA = c\,\Delta\rho\,\phi f_g \cdot \left[\frac{\epsilon\zeta}{\eta} G(\omega a)\right] \tag{1}$$

where c is the sound speed, $\Delta\rho$ is the density contrast between solid and medium, ϕ is the particle volume fraction, f_g is an instrument factor, ϵ is the dielectric constant of the medium, ζ is the zeta potential, η is the medium viscosity, and $G(\omega a)$ is the frequency and size dependent inertial term. For a correct interpretation of ESA data, the impact of each of these parameters must be examined.

For a given suspension, the first four terms in Eq. (1) are constants. The factors c and f_g are known and calibrated quantities, respectively, while $\Delta\rho$ and ϕ are material-related parameters. High-$\Delta\rho$ systems, such as ceramic particles dispersed in water ($\Delta\rho \geq 1$), generate relatively high ESA signals, and are easily detected; whereas materials such as latex ($\Delta\rho \sim 0.05$) generate very low signals and are more difficult to analyze using electroacoustics. The influence of particle concentration is more complex. Eq. (1) predicts that the ESA signal will increase linearly with ϕ; however, this relationship only holds up to roughly $\phi = 0.1$ (i.e., the linear regime), above which the ESA response becomes increasingly insensitive to changes in ϕ (i.e., the non-linear regime). Non-linear response is a result of concentration effects that are not fully understood; empirical corrections have been used to account for high-ϕ effects with some success.[8]

The terms grouped together within parentheses in Eq. (1) represent the dynamic mobility, μ_D, which, for a given material, is a function of ζ and particle size (through the inertial term G). In order to quantitatively determine ζ from an ESA measurement, the average particle size must be known *a priori*. Figure 1 shows how the magnitude of G varies with particle size and frequency for a material with $\Delta\rho$=1.

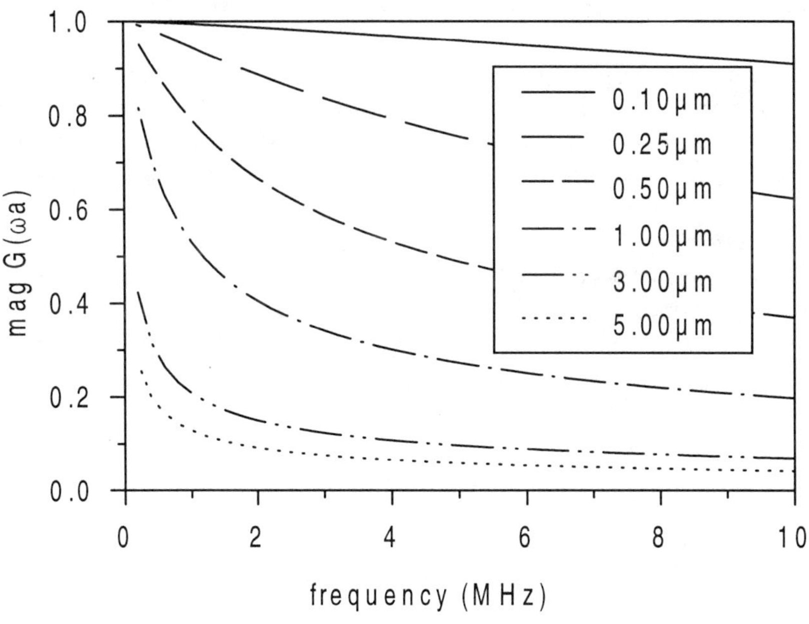

Figure 1. Inertial factor, G, calculated according to O'Brien's theory, for silica (ρ=2.2 g/cm^{-3}) as a function of particle size and frequency.

For ceramic particles smaller than about 0.25 μm, the inertial effect is insignificant at 1 MHz, and μ_D is then equivalent to the d.c. electrophoretic mobility. Obviously, any aggregation process which occurs during the course of measurements, or as a result of sample preparation, will have an impact on ESA data by increasing the inertial term and thereby lowering μ_D. The pH_{iep}, however, is not affected by the inertial response.

Another consideration when interpreting ESA data is the potential contribution of dissolved ionic species. The ESA response of ions is normally quite small relative to the more massive particulate phase, but may become significant at low solids concentrations, high electrolyte concentrations, or when the particle mobility is low (e.g., near the pH_{iep}). Under these circumstances it may be necessary to subtract the ionic signal vectorially (i.e., taking into consideration the phase and magnitude of the signal).[11] The presence of a significant background signal is often indicated by the appearance of a discontinuity near the pH_{iep} in an otherwise smooth mobility titration curve; this effect occurs because the background signal is relatively constant over the normal pH range of a titration, while the particle contribution is changing phase and magnitude. An experimental comparison of a number of ionic species has indicated that $NaNO_3$ is the preferred indifferent electrolyte for electroacoustic applications, due to its relatively low background signal.[9]

3.2 Monitoring of Processing Additives

A most promising application for ESA is in monitoring the effects of additives on the properties of ceramic suspensions.[12,13] These additives play a significant role during processing through their affect on the flow behavior requisite for shape forming, and in preventing the formation of undesirable agglomerates. Using ESA, a process engineer can evaluate the effectiveness of several polyelectrolyte dispersing agents, for example, and quickly determine the optimum dose needed to obtain the desired properties, without the necessity of time-consuming adsorption isotherm and viscosity measurements. There are essentially two methods for evaluating the behavior of dispersion additives using ESA: (1) as a function of additive concentration at constant pH, or (2) as a function of pH at constant additive concentration.

Figure 2 shows the results of titrations performed on 2 % silicon nitride suspensions with standard PAA solutions at constant pH. The electrokinetic behavior of the suspensions provides some insight into the nature of particle-polymer interactions in this system. These interactions are dominated by the electrostatic behavior of each species; the pH_{iep} of silicon nitride is near 6.2, while PAA assumes a negative charge above pH 3 due to the dissociation of carboxylic acid functional groups.[13] At pH 3, silicon nitride is positively charged and PAA is essentially neutral (undissociated); the interaction is moderate as indicated by the gradual decrease of ESA in Figure 2. At pH 9, silicon nitride is negatively charged, as is the fully dissociated PAA; the interaction is weak due to mutual electrostatic repulsion.

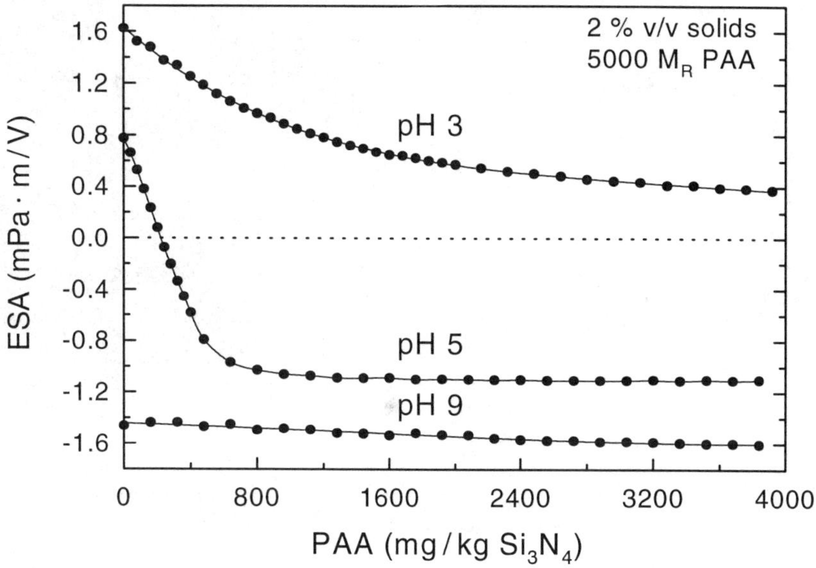

Figure 2. ESA response of 2 % silicon nitride suspension as a function of PAA concentration at constant pH. (From ref. 13)

However, at pH 5, silicon nitride carries a moderate positive charge, while PAA is roughly 80 % dissociated; the interaction is very strong as indicated by the charge reversal.

Figure 3 illustrates an experimental procedure to determine the optimum concentration of a dispersing agent, DMAE1, in a 5 % silicon powder dispersion in 0.02 mol/L NH_3NO_3. DMAE1 is a polyelectrolyte containing quaternary amine functional groups that interact with deprotonated silanol groups on the oxidized silicon surface. In Figure 3(a), acid-base titrations are performed on each suspension containing a fixed amount of dispersant. In Figure 3(b), the corresponding pH_{iep} is plotted versus additive concentration. From this data it is determined that surface saturation of silicon occurs at a DMAE1 concentration of roughly 0.25 % (relative to silicon). At this concentration the maximum positive electrokinetic potential imparted by DMAE1 becomes more or less independent of pH, and no further benefit would be derived from additional dispersant in this system.

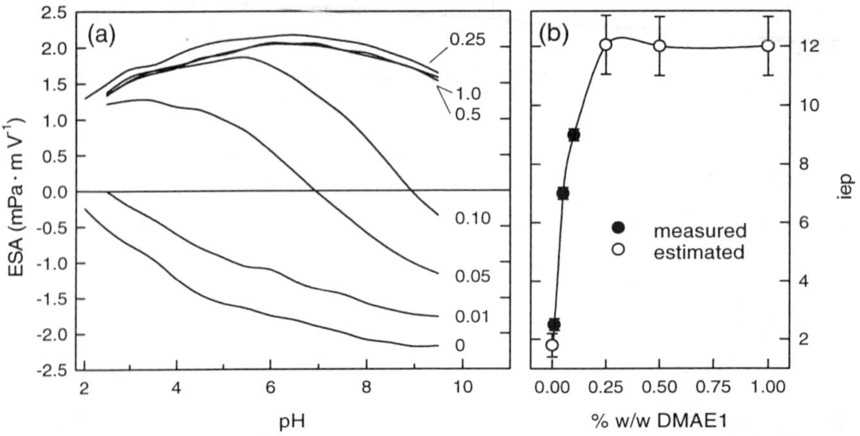

Figure 3. ESA-pH curves for 5 % silicon suspensions in the presence of different additives. (From ref. 12)

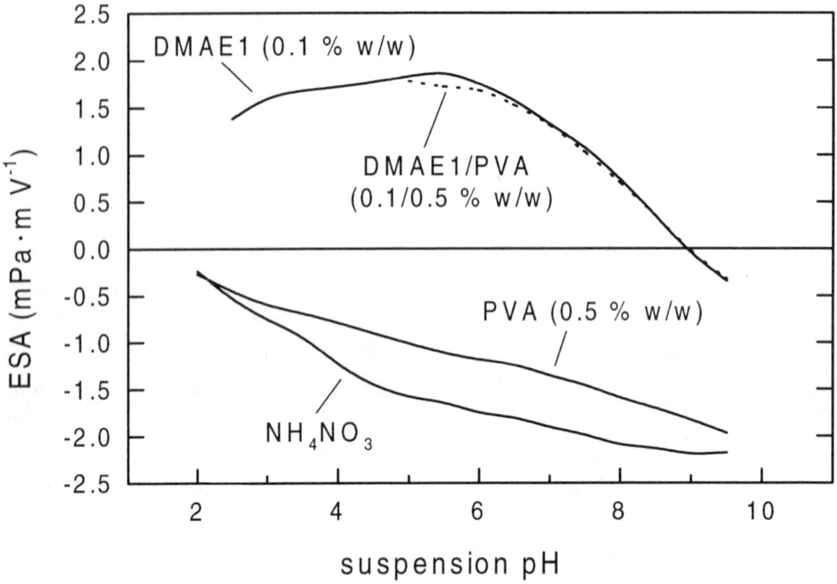

Figure 4. Dynamic mobility versus pH for mixed silicon-yttria suspension.

On the other hand, ESA measurements are not very useful for examining the interaction of neutral species, such as PVA binder. Figure 4 compares the effects of DMAE1 dispersant and PVA on the electrokinetic properties of silicon. In the presence of DMAE1, PVA has no influence on the electrokinetic behavior, as expected, and the pH_{iep} is unchanged compared to DMAE1 alone. Similarly, in the absence of DMAE1, PVA does not shift the pH_{iep}. Whereas, the modest reduction of ESA values above the pH_{iep} in the presence of PVA alone, may be interpreted as a physical phenomenon, in which PVA chains adsorbed at the particle-solution interface cause the plane of shear to expand outward from the surface, resulting in a reduced electrokinetic potential.

3.3 Mixed Particulate Systems

Although it is not possible to directly differentiate component contributions to the ESA signal in a suspension containing multiple particulate phases, nevertheless it is still possible to analyze data from these systems and extract useful information under some circumstances. The individual contributions to the measured ESA signal (magnitude and phase) from various suspension components may be considered as additive. Therefore, one can measure the ESA contribution of each component individually under the same solution conditions, and then subtract these contributions from the mixed-suspension ESA signal. In practice, however, this method is cumbersome, of limited utility, and rarely works due to the difficult problem of obtaining identical solution conditions when one or more of the phases is partially soluble or complex surface interactions exist.

A simpler approach is possible if one is primarily interested in the effects of a secondary powder on the electrokinetic properties of the suspension *as a whole*. An example of this type of simple analysis is given in Figure 5, which shows the effect of 20 % yttria (Y_2O_3) powder (relative to the primary powder) in a 5 % aqueous suspension of silicon. Since yttria is positively charged below pH 10, and silicon is negatively charged above pH 3, we expect heterocoagulation to occur between yttria and silicon, thereby reducing the dynamic mobility of the more concentrated primary phase. Although the measurement data supports this expectation, in reality there is no way to distinguish if the reduction in dynamic mobility is the result of charge screening by "adsorbed" yttria on the surface of silicon, or if it is due to the cumulative ESA signals from kinetically independent, but oppositely charged, particles. Figure 5 also demonstrates the effect of yttria solubility in acidic media on the electrokinetic behavior of silicon. The first curve (open symbols) was obtained during acid titration starting from pH 10; an arrow indicates the onset of acid dissolution of the yttria phase. The subsequent base titration (closed symbols) results in a drastically modified electrokinetic curve due to heterogeneous precipitation of yttrium hydroxide and/or carbonate onto the surface of the silicon powder.

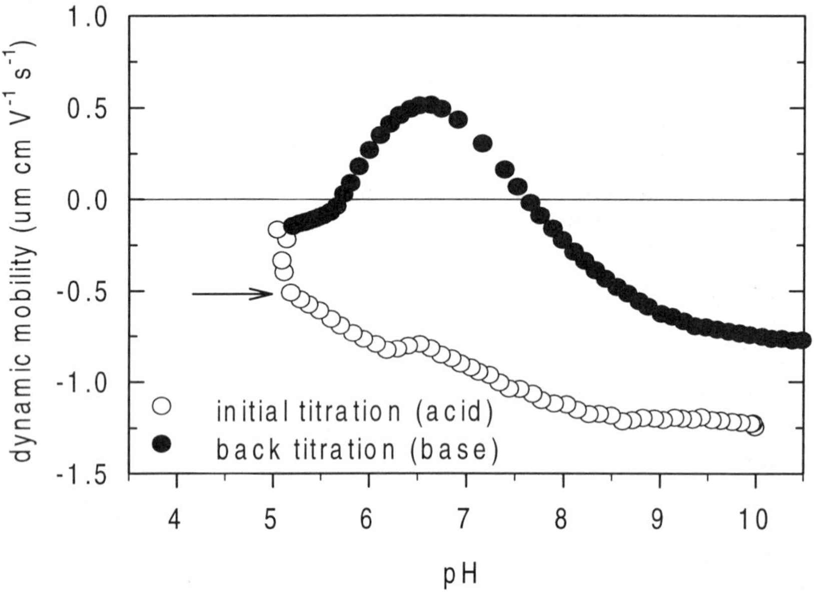

Figure 5. Dynamic mobility versus pH for mixed silicon-yttria suspension.

3.4 Measurements in Non-Aqueous Media

The primary difficulty associated with electroacoustic measurements in non-aqueous media is the characteristically weak signal resulting from the low dielectric permittivity of organic solvents. In order to perform measurements in these systems, it is necessary to increase the applied voltage, reduce the electrode gap, and increase the electrode surface area. The ESA-8000 analyzer comes equipped with a non-aqueous electrode specifically designed for such measurements. However, other issues must be considered when interpreting ESA data from non-aqueous suspensions. Because of the low dielectric constant (resulting in low concentrations of ions), the diffuse electrical double-layer surrounding the particles is greatly expanded. This means that the particle radius will often be quite small relative to the characteristic double-layer thickness, and therefore the Hückel approximation should be used for calculating zeta potential.[7] Highly expanded double-layers also mean that overlap will occur at much lower particle concentrations then would typically be found in aqueous systems. As a result, concentration effects are more likely to be encountered in non-aqueous suspensions. Keeping these issues in mind, it is possible to gain much insight into the mechanisms of charging and particle stability in non-aqueous systems using the ESA technique.

There is no single, complete theory to explain stabilization mechanisms in non-aqueous media, although steric stabilization has been generally assumed to dominate in these systems.[14,15] Significantly less attention has been paid to the influence of electrostatic mechanisms. However, a few studies[16-18] have shown that electrostatic repulsive forces in non-aqueous systems can be appreciably high, and the stability of particles suspended in non-aqueous media may depend to a large degree on the amount of surface charge developed at the solid-liquid interface. Figure 6 shows the electrokinetic behavior of 2 % $BaTiO_3$ suspensions prepared in three different organic media (acetone, toluene, and a 50/50 volume mixture of acetone and toluene), as a function of phosphate ester concentration. While acetone supports a moderate charge development, toluene does not. A comparison between the ESA data presented in Figure 6 and corresponding viscosity data (not shown), indicates that charge development, in relation to solvent polarity, may play a significant role in the stability of acetone and mixed acetone-toluene systems.

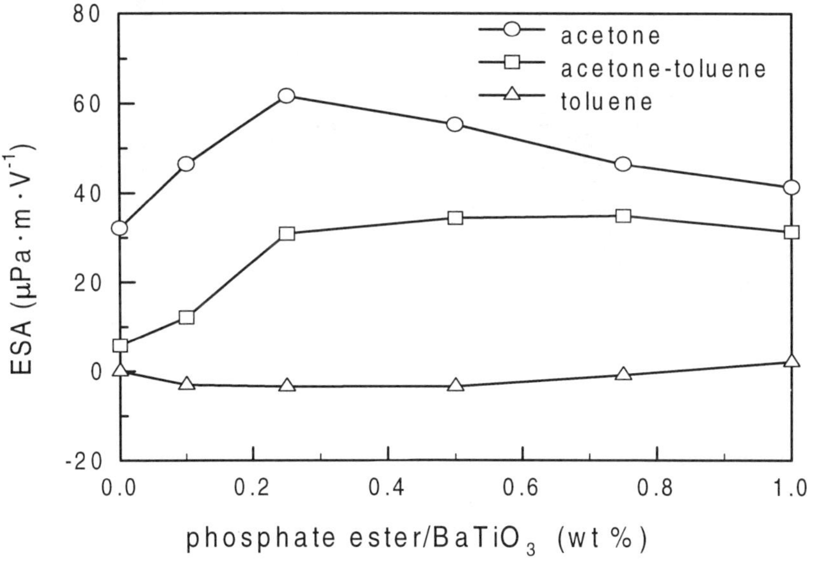

Figure 6. ESA as a function of phosphate ester concentration for 2 % barium titanate suspensions in three different organic media. (From ref. 18)

3.5 Electroacoustic Response of Semiconductors

The basic theory underlying our current treatment of electroacoustic effects is based upon the assumption of electrically insulating particles.[10] This condition applies satisfactorily to the vast majority of advanced ceramics used in structural or electronic packaging applications. However, in many cases, ceramics may be classified as dielectrics, semiconductors, or even conductors. Very little has been published on electroacoustic measurements of non-insulating particulates, so caution must be exercised when interpreting data in these systems. A clear example of such a situation may be found in the electroacoustic behavior of silicon powders used in the fabrication of reaction-bonded silicon nitride.[12]

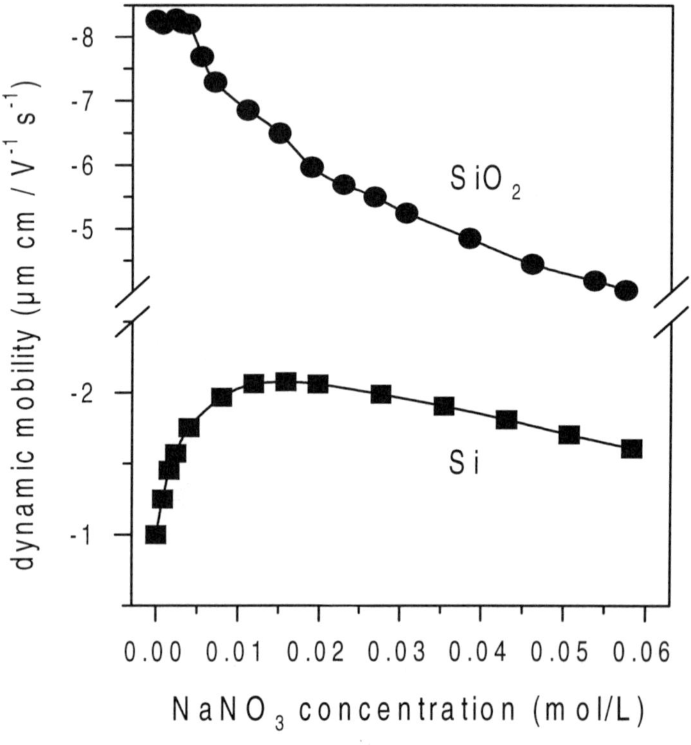

Figure 7. Comparison of electroacoustic response for silica and silicon as a function of electrolyte concentration.

Figure 7 compares the measured dynamic mobility of two suspensions as a function of electrolyte concentration, one containing silica and the other silicon. These materials have similar surface chemistries in an aqueous environment, despite differences in their bulk properties. The electrokinetic behavior exhibited by silica is typical for a negatively charged insulating sphere; the mobility *decreases* monotonically with increasing salt concentration as a result of electrostatic screening. The silicon powder, in comparison, behaves atypically; the mobility first *increases* and then decreases in magnitude, exhibiting a pronounced maximum. After further consideration and extended tests, particle size and shape effects, solubility problems, surface conductivity, and contamination were ruled out as origins for this phenomenon. An explanation was subsequently provided by O'Brien,[19] and is related to the development, under the applied field, of an internal electrical double-layer as a result of high charge carrier mobility in the semiconductor particle. This behavior leads to an enhanced dielectric permittivity, which reduces the apparent mobility. This effect can be screened by adding sufficient electrolyte, explaining why the mobility of silicon in Figure 7 initially increases at low electrolyte additions. The subsequent decrease at higher concentrations is due to simple electrostatic screening, such as is observed with silica. The enhanced dielectric affect is also dependent on particle size, which makes it an even more complex problem.

4. Summary

The application of electrokinetic sonic amplitude measurements in moderately concentrated ceramic powder suspensions was discussed in relation to the advantages and limitations of this technique. Examples were presented from a variety of ceramic particulate systems in both aqueous and non-aqueous media. ESA has proven to be a rapid and reliable method for characterizing the electrokinetic properties of suspensions, and has potential for on-line monitoring applications.

References

1. F.F. Lange, "Powder Processing Science and Technology for Increased Reliability," *J. Am. Ceram. Soc.*, **72** [1] 3-15 (1989).

2. L. Bergstròm, C.H. Schilling, and I.A. Aksay, "Consolidation Behavior of Flocculated Alumina Suspensions," *J. Am. Ceram. Soc.*, **75** [12] 3305-3314 (1992).

3. H.W. Lee and M.D. Sacks, "Pressureless Sintering of SiC-Whisker-Reinforced Al_2O_3 Composites. II. Effects of Sintering Additives and Green Body Infiltration," *J. Am. Ceram. Soc.*, **73** [7] 1894-1900 (1990).

4. T.M. Shaw and R.A. Pethica, "Preparation and Sintering of Homogeneous Silicon Nitride Green Compacts," *J. Am. Ceram. Soc.*, **69** [2] 88-93 (1986).

5. D.W. Fuerstenau, R. Herrera-Urbina, and J.S. Hanson, "Adsorption of Processing Additives and the Dispersion of Ceramic Powders"; pp. 323-351, *in* Ceramic Transactions, Vol. 1, Ceramic Powder Science. Edited by G. L. Messing, E. R. Fuller Jr., and H. Hausner, American Ceramic Society, Westerville, OH, 1988.

6. Y.K. Leong, P.J. Scales, T.W. Healy, and D.V. Boger, "Interparticle Forces Arising from Adsorbed Polyelectrolytes in Colloidal Suspensions," *Colloids Surf. A*, **95**, 43-52 (1995).

7. R.J. Hunter, Zeta Potential in Colloid Science, Academic Press, NY, 1981.

8. *Electroacoustics for Characterization of Particulates and Suspensions*, NIST Special Publication 856. Edited by S.G. Malghan, U.S. Department of Commerce, Technology Administration, National Institute of Standards and Technology, Gaithersburg, MD, 1993.

9. V.A. Hackley and S.G. Malghan, "Investigation of Parameters and Secondary Components Affecting the Electroacoustic Analysis of Silicon Nitride Powders," in reference 8, pp. 161-179.

10. R.W. O'Brien, "The Electroacoustic Equations for a Colloidal Suspension," *J. Fluid Mech.*, **212**, 81-93 (1990).

11. F.N. Desai, H.R. Hammad and K.F. Hayes, "Background Electrolyte Correction for Electrokinetic Sonic Amplitude Measurements," *Langmuir*, **9**, 2888-2894 (1993).

12. V.A. Hackley, U. Paik, B.H. Kim, and S.G. Malghan, "Aqueous Processing of Sintered Reaction-Bonded Silicon Nitride. I. Dispersion Properties of Silicon Powder," *J. Am. Ceram. Soc.*, **80** [7] 1781-1788 (1997)

13. V.A. Hackley, "Colloidal Processing of Silicon Nitride with Poly(acrylic acid). I. Adsorption and Electrostatic Interactions," *J. Am. Ceram. Soc.*, **80** [9], 2315-2325 (1997).

14. T. Sata and R. Ruch, Stabilization of Colloidal Dispersions by Polymer Adsorption, Marcel Dekker, NY, 1984.

15. R.E. Johnson, Jr. and W.H. Morrison, Jr., "Ceramic Powder Dispersion in Non-aqueous Systems," pp. 323-348, *in* Advances in Ceramics, Vol. 21, Ceramic Powder Science. Edited by G.L. Messing, K.S. Mazdiyasni, J.W. McCauley and R.A. Haber, American Ceramic Society, Columbus, OH, 1987.

16. K. Mikeska and W.R. Cannon, "Non-Aqueous Dispersion Properties of Pure Barium Titanate for Tape Casting," *Colloid Surf.*, **29**, 305-321 (1988).

17. B.I. Lee and U. Paik, "Dispersion of Alumina and Silica Powders in Non-Aqueous Media: Mixed-Solvent Effects," *Ceram. Int.*, **19**, 241-250 (1993).

18. U. Paik, V.A. Hackley, S. Choi, and Y. Jung, "The Effect of Electrostatic Repulsive Forces on the Stability of $BaTiO_3$ Particles Suspended in Non-Aqueous Media" *Colloids Surf. A*, in press (1998).

19. R.W. O'Brien, "The Dielectric Response of a Dilute Suspension of Semiconducting Particles," *J. Colloid Interface Sci.*, **177**, 280-281 (1996).

Acoustic Attenuation Spectroscopy for Particle Sizing of High Concentration Dispersions

Tõnis Oja
Malvern Instruments Inc., Southborough, MA 01772 USA

Felix Alba
Felix Alba Consultants Inc., Murray, Utah 84107 USA

Abstract

The experimentally determined acoustic attenuation spectrum of a concentrated dispersion, together with the theory of Epstein, Carhart, Allegra, and Hawley, is utilized to obtain mean particle size, a particle size distribution, and, in addition, the dispersion concentration. An acoustic attenuation spectrum is in general dependent on a variety of properties of the dispersed phase as well as on the properties of the continuous phase. As long as there exists a contrast between at least one thermomechanical property of the particle phase and the continuous phase, an acoustic attenuation spectrum relatable to the particle phase will result. Acoustic attenuation spectroscopy is a bulk measurement that can be carried out on any dispersed system of any particle density and surface charge (even zero) and over a wide range of particle sizes ranging from 10 nanometers to 1000 micrometers. Concentrations of the dispersed phase can range typically from 0.5 to 50 volume %. In this paper we will present particle size analysis results using acoustic attenuation spectroscopy on a variety of dispersed systems of interest to industry.

1. Introduction

As an acoustic wave travels through a pure liquid medium a certain amount of energy will be absorbed by the medium resulting in loss of energy and hence in an attenuation of the acoustic wave. If the liquid medium also contains suspended particles of a different material there will occur an additional attenuation due to scattering of the acoustic wave by the particles. The fundamental problem of scattering of acoustic energy by particles suspended in a fluid medium was first considered by Lord Rayleigh at the end of the 19th century and finally completed by Eptsein and Carhart[1] and by Allegra and Hawley.[2] The details of the Epstein Carhart Allegra and Hawley (ECAH) theory have been discussed in a previous paper in this publication[3] and in many other references. It is sufficient to note that if the particle phase has at least one thermomechanical property (density, specific heat, thermal conductivity, thermal expansion coefficient, sound attenuation, sound speed, and shear modulus if solid, or viscosity if liquid) different from the liquid phase, sound waves will be scattered by the particle phase. The amount of scattering will depend on the amount of contrast in the thermomechanical properties between the liquid phase and the particle phase, on the diameter of the particles, on the frequency of the sound wave and on the concentration of particles.

The complete solution of the scattering problem for a system of one type of particles suspended in a liquid medium, be they oil droplets in water or titania pigment dispersed in hexane, is summarized in the ECAH theory by a three dimensional surface of the logarithm of the attenuation plotted against particle size and the frequency of the applied acoustic wave as shown in Figure 1. Figure 1 is actually the sound attenuation surface for a system of silica (SiO_2) particles dispersed in water.

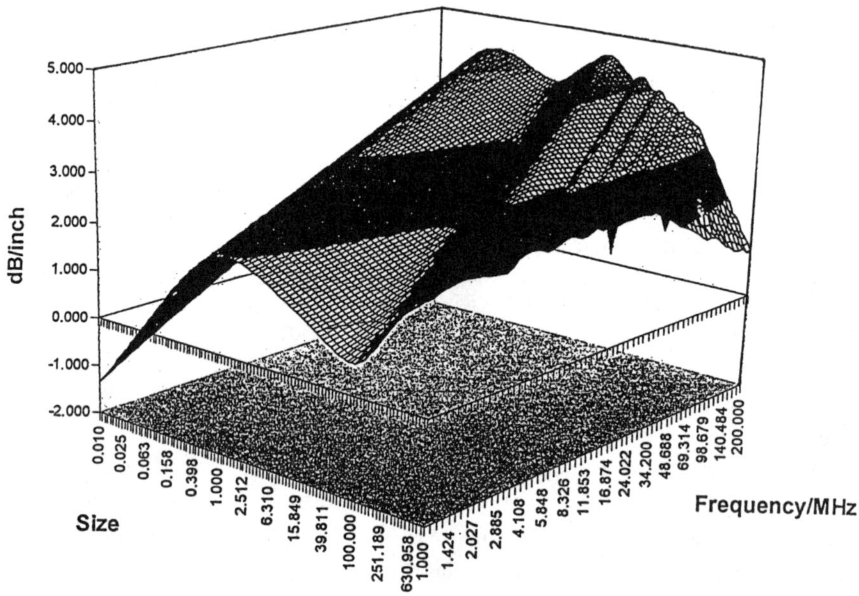

Figure 1. The sound attenuation surface for silica particles.

The problem of determining particle size by using acoustic attenuation spectroscopy can then be summarized as follows. First, it is a given that the type of particle and the nature of the liquid phase is known and that we have available the thermomechanical properties of both the particle phase and the liquid phase.† Secondly, we make a high accuracy measurement of the attenuation as a function of frequency over the largest possible frequency range. Such an attenuation measurement is shown in Figure 2 for two different concentrations of silica (5.0 and 10.0 volume %). Finally, we ask, to which particle size does the measured attenuation curve belong? Alternatively, equivalently, the measured attenuation vs. frequency curve cuts the size axis at which particle size values? This "inversion" calculation requires the latest desktop computing capabilities. The particle size obtained from "inverting" either of the two curves of Figure 2 is shown in Figure 3. It is possible to measure two different concentrations of the same particle system since a higher concentration simply shifts the surface uniformly upwards along the attenuation axis if multiple scattering effects are not present. It is also possible to obtain a particle size distribution width and to distinguish bimodal distributions.

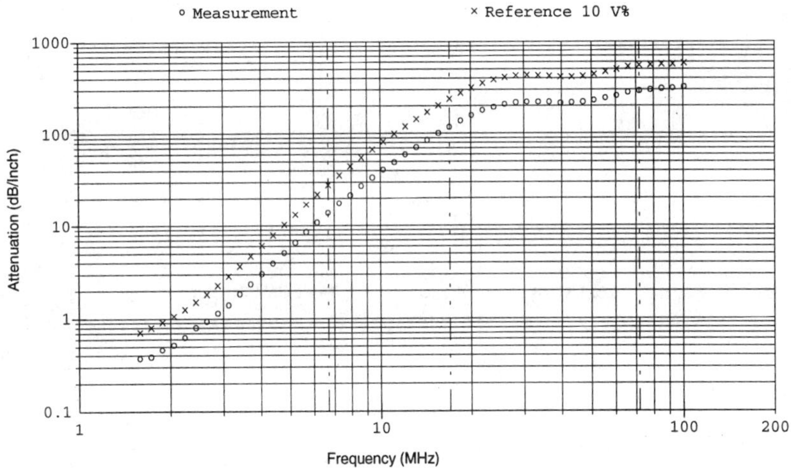

Figure 2. A cross section of the attenuation plot at fixed particle size and at two different concentrations for a silica dispersion.

Acoustic attenuation spectroscopy offers many appealing advantages for particle sizing. Most current sizing techniques employ laser light scattering methods, which can only be used on very dilute systems. Samples whose particle size is to be measured require first a severe dilution. Dilution may often change completely the nature of the particle size distribution; this is especially true for

† There are many instances where most of the thermomechanical properties are not needed and knowledge of the density of both phases is sufficient.

emulsion systems. Acoustic attenuation spectroscopy permits particle size measurements in systems at process concentrations, and hence permits one to monitor the onset of aggregation, flocculation, nucleation and particle growth, or conversely, to monitor deaggregation or dissolution. To make measurements at full process concentration also implies no sample handling, which in turn implies measurements can be made by less highly trained personnel. Critical results are obtained more rapidly permitting a quicker response to out-of-specification conditions in the process, which in turn results in tighter process control and a saving of materials and money.

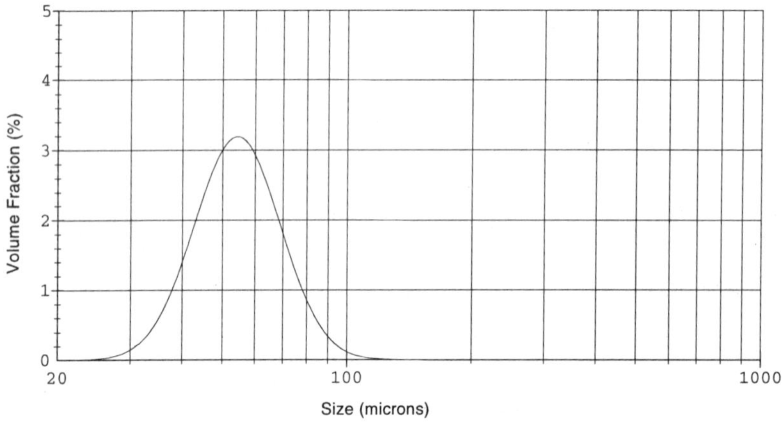

Figure 3. Particle size distribution for the silica dispersion in Figure 2.

2. Experimental

2.1 Apparatus

The apparatus used in all acoustic attenuation measurements was the Malvern Ultrasizer™ shown schematically in Figure 4. Since attenuation measurements must be made over the widest possible frequency range and since a range of materials from 0.5 volume % emulsions to 50 volume % of fine high density particles such as titania must be accommodated, a range of acoustic frequencies from 1 MHz to 150 MHz, using a set of high frequency and a set of low frequency transducers and variable transducer separations are used. For highly attenuating materials, the measurement separation is small whereas for low attenuating materials, such as low concentration emulsions, acoustic propagation paths of up to 10 cm may be required. The largest frequency ranges, and the corresponding transducer separations, are automatically determined by the Optimal Strategy™ function of the software. The measuring tank also contains a stirring propeller, a level sensor, and a drain and overflow valve, and heating elements with a temperature sensor.

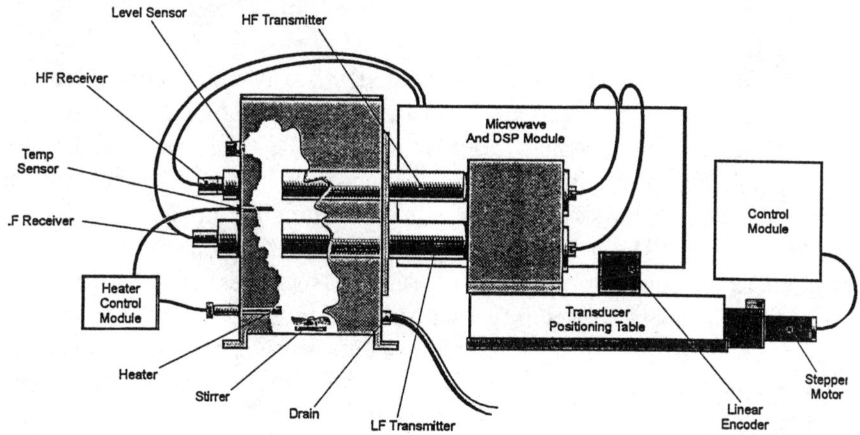

Figure 4. Instrument diagram for Ultrasizer acoustic attenuation spectrometer.

2.2 Procedure

Often materials to be measured are obtained already in the dispersed state. The material is poured, without further handling, into the Ultrasizer measuring tank and equilibrated at the desired temperature. Stirring is maintained at all times to avoid settling and to present a representative sample for measurement. Ultrasonic attenuation measurements are made in large volumes of material and not at a point or at a surface on the side of the container. The inversion matrices are selected depending on the range of particle sizes expected and a measurement "optimal strategy" is obtained for the particular sample. The attenuation spectrum measurement is then initiated which is followed by the particle size distribution calculation, the whole process taking at the most a few minutes, depending to some extent on the complexity of the particle size distribution. If the material is obtained in the dry, undispersed state, it must be dispersed according to good colloid science practices. In aqueous systems, pH and conductivity values must be monitored and set to desired values. pH values too close to the isoelectric point of the material dispersed in water will result in an unstable, hence aggregating, system which may lead in turn to distorted particle sizing results. The Ultrasizer can also be used to measure the onset of aggregation as measured by an increase in particle size of the system as the system is brought close to the isoelectric point (the point of zero zeta potential).

3. Results and Discussion

3.1 Particle Sizing in CMP Materials

Chemical - mechanical polishing (CMP) is a widely accepted technology for the planarization of semiconductor wafer surfaces following the deposition of layers of insulators and metals. CMP is expected to become a mainstream method in the manufacturing of advanced semiconductor devices. CMP polishing slurries are typically nanometer sized, fumed silica or alumina particles dispersed in an aqueous medium. The particle size and size distribution must be carefully controlled to avoid the production of defects (scratches) on the planarization surfaces. The dispersion stability and shelf life of CMP slurries is also an important concern. Nanometer sized particles at typical concentrations used (5-30 weight %), may aggregate and the particle size distribution may thus shift upwards with time resulting in an increase in defects produced. The technical challenge is to be able to rapidly measure the mean particle size and size distribution of CMP slurries at the full concentrations used in the planarization process.

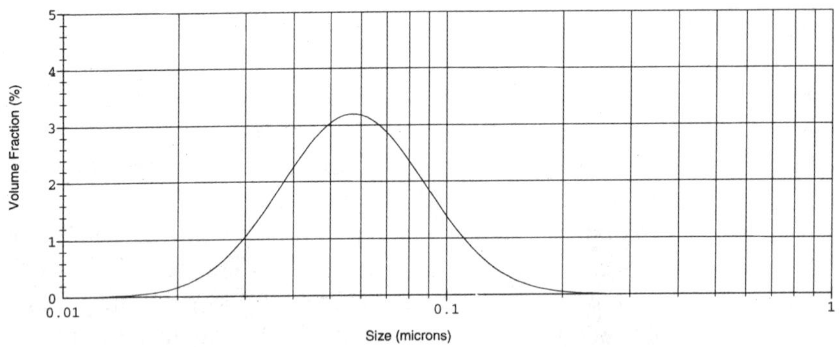

Figure 5. Particle size distribution of a CMP silica with no aggregation evident.

In Figure 5 is shown the mean particle size and distribution of a fine CMP silica (SiO_2) at a measured concentration of 4.0 volume %. The largest particle size does not extend beyond 200 nanometers. These data were taken in a matter of minutes close to the process line and indicates that the slurry system can be used for planarization without further concern. Figure 6 shows the data from an alumina (Al_2O_3) CMP slurry. Although the mean particle size of the main distribution is centered at about 35 nanometers, there is a significant quantity of material, about 10 %, centered at 150 nanometers, with an extended distribution tail reaching to above 300 nanometers. Independent verification of the particle size distribution in this material is shown in Figure 7, which shows results from a measurement using photon correlation spectroscopy (PCS).

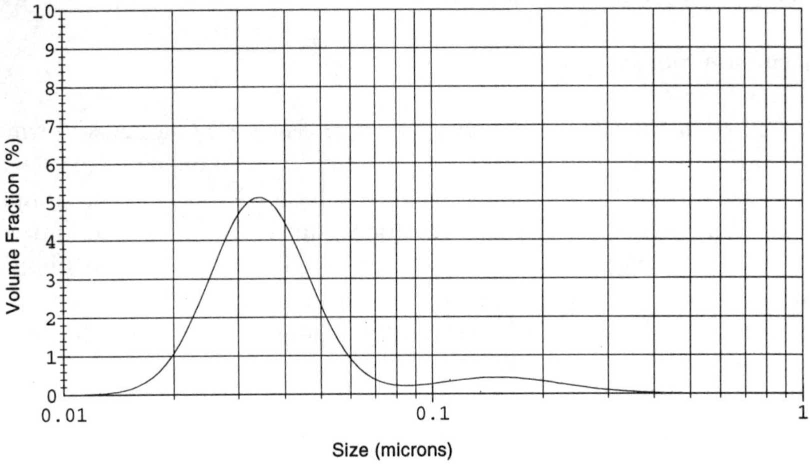

Figure 6. Particle size distribution of a CMP alumina with about 10 % by volume of aggregates at around 150 nm.

Peak Analysis by Intensity

Peak	Area	Mean(nm)	Width
1	11.4	35.2	5
2	88.6	192.5	31

Peak Analysis by Volume

Peak	Area	Mean(nm)	Width
1	92.1	35.4	10
2	7.9	192.3	60.2

Peak Analysis by Particle Number

Peak	Area	Mean(nm)	Width
1	99.9	35.3	10

Figure 7. Photon Correlation Spectroscopy results on the CMP alumina from Figure 6.

We can judge that this material is in the early stages of aggregation, and as time goes by more of the system will shift to larger particle sizes raising potential problems with the planarization process. Vigorous stirring or shearing in an attempt to redisperse the aggregated portion of the particle size distribution is time consuming and often not very successful.

3.2 Zirconium Silicate

Zirconium silicate is a widely used white wares glazing material which is ground from its initially coarse size to a much finer size. In Figure 8 are shown the results obtained from a 4 volume % material using the Ultrasizer, and comparing the results to that obtained with the Malvern Mastersizer™ using laser diffraction on very dilute samples. The bimodal nature of the particle size distribution is evident with both methods, and derives from the fact that the initially coarse material is mixed with some of the recycled ground material which still contains a coarse component.

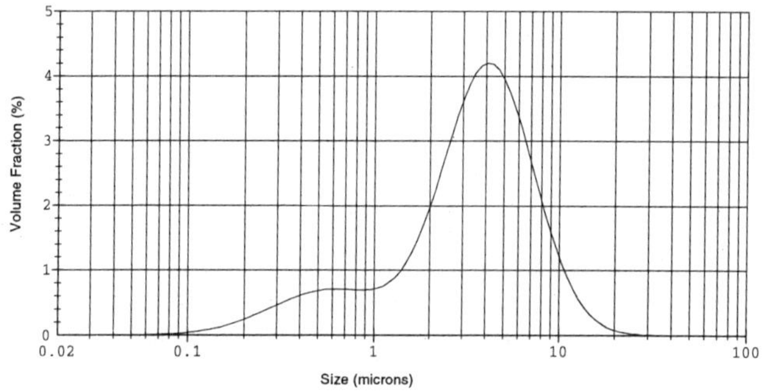

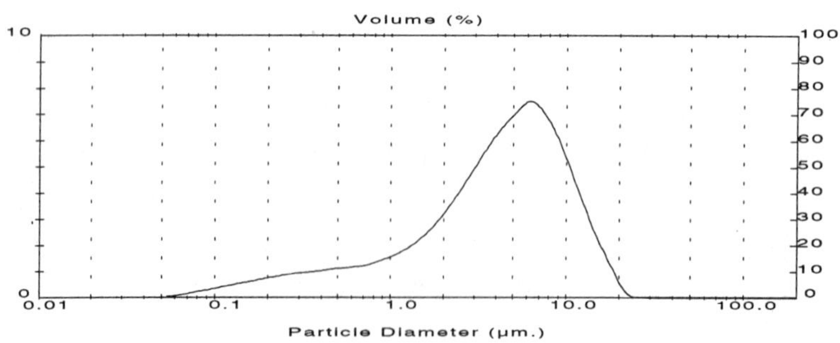

Figure 8. The particle size distribution in zirconium silicate glazing material comparing (top) concentrated (Ultrasizer) and (bottom) dilute (Mastersizer) measurements.

Also, since the separation process is centrifugal, and not perfect, a small quantity of the fine, end component is also included. This result shows the potential of the Ultrasizer measurement to faithfully reproduce the more established dilute laser diffraction results.

3.3 Bitumen Emulsions

An emulsion of bitumen (high viscocity crude oil) in water at 70 volume % was measured in both the Ultrasizer and the Mastersizer. The viscosity of the emulsion, and its consequent flowability, strongly depends upon the droplet size distribution. The viscosity of the pure bitumen (around 90 poise at 25°C) is reduced more than 10 times by the proprietary emulsification process. The final viscosity of the emulsion decreases as the droplet size distribution width increases. The optimal amount of surfactant also depends upon the droplet size distribution (surface area to be protected). Quality control through droplet size measurement and subsequent modification of droplet size are consequently paramount to the manufacturer. If the highly concentrated emulsion is carefully diluted, preserving the surfactant concentration, and measured immediately in the Mastersizer, a remarkable agreement between light scattering and ultrasound spectroscopy is achievable as shown in Figure 9.

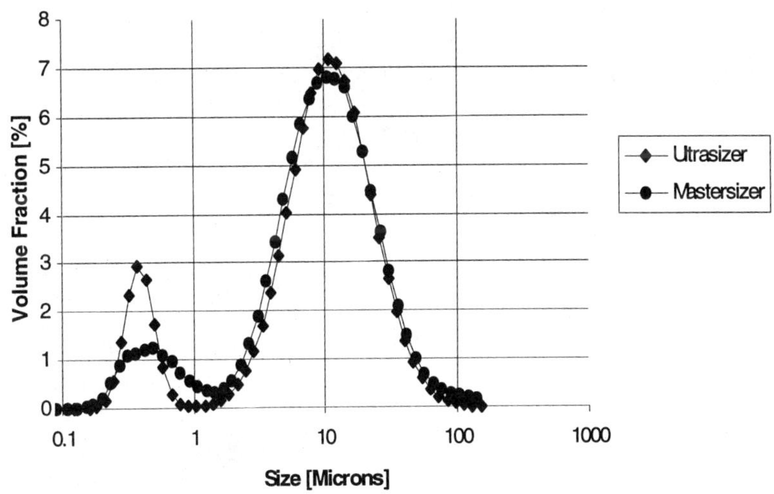

Figure 9. Comparing Ultrasizer and Mastersizer results from a bitumen emulsion.

3.4 pH Titrations

The material investigated is actually an alumina coated titania pigment widely used in paint formulations and for paper coatings dispersed at 1.7 volume %. The the particles will aggregate to larger sizes. Far from the i.e.p., that is, at pH 5.0 or at pH 11.0, the isoelectric point (i.e.p.), or the point where zeta potential is zero, is at around pH 9.0. In the neighborhood of the i.e.p, the system becomes unstable and particles are expected to exhibit sufficiently high electrostatic repulsion to prevent aggregation. It is of general interest to measure the particle size distribution as a function of pH in a concentrated system close to the i.e.p. In Figure 10 are summarized the concentrated particle sizing results at three pH values. At pH 5.0 the system is far removed from the i.e.p. and remains stable and well dispersed for an indefinite time. The measured particle size is about 0.30 micrometers, very close to the widely accepted size value. At pH 9.0 the system is in the neighborhood of the i.e.p. and is expected to aggregate to form larger particles. This is borne out by the spectrum in the middle where we observe the particle size distribution centered at around 3 micrometers, as might be expected, *plus* the remaining distribution centered about 0.2 micrometers. The ultrasonic results seem to imply a substantial distribution of fine unaggregated particles. This is an unexpected result. At pH 11.0 the system shows a broader and slightly shifted (to higher sizes) particle size distribution. All titrations are performed *in situ* on the original system.

The observation is that a system which has gone through the i.e.p. never returns to the original mean size or size distribution (i.e., there is some irreversible aggregation). On the other hand, if a system is prepared initially at pH 11.0, the original (pH 5.0) size and size distribution are measured. All the above observations were confirmed using laser light scattering measurements (Malvern Mastersizer) on dilute systems taking great care to maintain the original solution conditions at each measurement point.

4. Summary and Conclusions

The practical use of high accuracy ultrasonic attenuation spectra to determine the mean particle size, the particle size distribution, and the concentration of undiluted dispersions has been demonstrated. A variety of dispersions of industrial importance can now be characterized without resorting to extensive handling. Particle sizes can be measured from the nanometer range to hundreds of micrometers without any alterations or modification to the measuring instrument over the whole range of particle sizes. Although many practical details

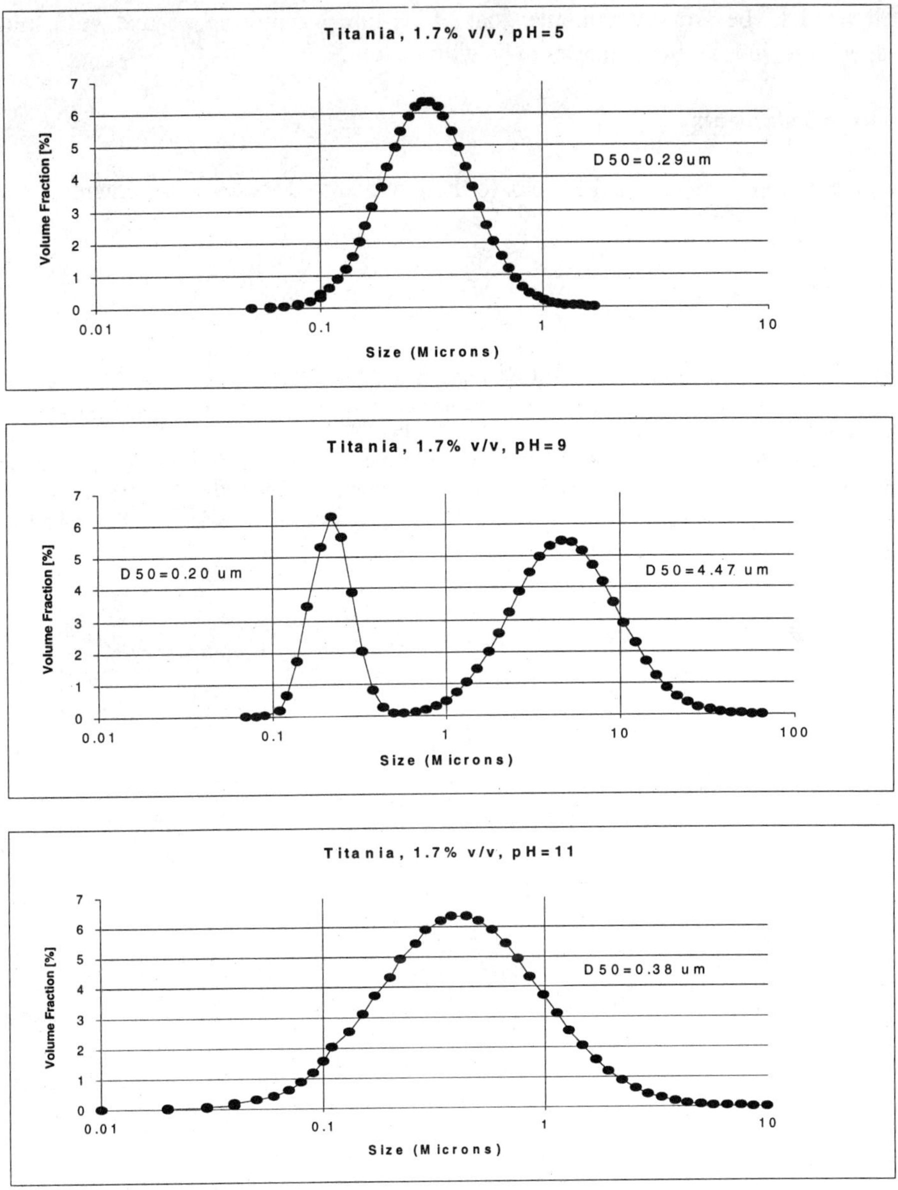

Figure 10. The particle size distribution of a titania based pigment as a function of pH: (top) pH 5, (middle) pH 9, (bottom) pH 11.

still need to be worked out, thc goal of on-line monitoring of real industrial dispersions, in real time, appears to be within sight.

Acknowledgments

We would like to thank Paul Kippax for help in many of these measurements.

References

1. P.S. Epstein and R.R. Carhart, "The Absorption of Sound in Suspensions and Emulsions I Water Fog in Air", *J. Acoust. Soc. Am.*, **25** [3] 553 (1953).
2. J.R. Allegra and S.A. Hawley, "Attenuation of Sound in Suspensions and Emulsions: Theory and Experiment", *J. Acoust. Soc. Am.*, **51**, , 1545 (1972).
3. F. Alba, D. Higgs, R. Jack, and P. Kippax, "Ultrasound Spectroscopy: A Sound Approach to Sizing of Concentrated Particulates", in *Handbook on Ultrasonic and Dielectric Characterization Techniques for Suspended Particulates.* Edited by V.A. Hackley and J. Texter. American Ceramic Society, Westerville, OH , 1998.

Applications in Inorganic Particulate Suspensions

Particle Size Characterization of Inorganic Colloids by Ultrasonic Attenuation Spectrometry

Michael Stintz, Frank Hinze and Siegfried Ripperger
Technische Universität Dresden, Institut für Verfahrenstechnik und Umwelttechnik, D-01062 Dresden, Germany

Abstract °

The particle or agglomerate size measurement at solids concentrations more than 5 vol % was investigated with the Acoustophor 8000 from PenKem Inc. Furthermore, limits of application to coarse disperse slurries and the sensitivity to agglomerates or flocks were quantified. Using alumina, titania, silica, and other inorganic test materials the particle size measurement by ultrasonic attenuation spectrometry was verified by comparison with SEM and optical methods. To test the accuracy three nearly monodisperse distributions of silica spheres and a bimodal mixture were measured. The size measurement of particles below 10 μm, down to about 10 nm was proved. In particular the upper limit must be regarded because of the growing influence of scattering effects. Simultaneous measurement of particle size and zeta potential opens up a new field for characterization of agglomeration and dispersion, e.g. by a dissolver or ultrasonic treatment. The stability of agglomerates can be studied under variable mechanical or chemical conditions. With growing time and power of dispersion the mean particle size becomes smaller and the standard deviation reduces. The spectra show a significant increase in the attenuation in the higher frequency range due to the growing amount of small primary particles over the time of dispersion. Electrostatic destabilizing or adding of a polymer agent are sensitively monitored by the attenuation spectra. But the particle size of a full flocculated system could not be calculated with the existing theory.

1. Introduction

Very fine particles in suspensions, especially in the submicron range, tend to agglomerate because of the dominating surface effects and the small interparticle distances. A lot of processes like flocculation, crystallization, dispersion, and filtration, and many materials like ceramic slurries, pigments, coatings, disperse reagents, and others, need the characterization of the state of agglomeration, i.e. the degree of dispersion. The complex relations between particle size, surface behaviour and rheological properties of the liquid cause a certain state of the dynamic equilibrium between agglomeration and deagglomeration. This state will change if the concentration of solids, the types of ions, or the polyelectrolytes are altered. Therefore, simultaneous determination of several characteristics, like agglomerate size, zeta potential, pH value, conductivity, and concentration has become important.

Different mechanisms of agglomeration, like solid bridges, adhesion, or shape effects, cause agglomerates with a different strength of their bonds and different porosity. Agglomerates cover the wide range between aggregates of partially molten primary particles and flocks, which can be destroyed by low shear forces. In addition to particle interaction, chemical reactions often change the properties of the particle surface and the type of particle interaction dramatically; so a flock may be converted to an aggregate. For particles with nonspherical shape, mechanical interactions or mechanical fastening are possible. Often, even very small amounts of additives or impurities change the properties of the suspension by adsorption on the surface of the particles. Thus the history of a particulate system in the production process and sample preparation may strongly influence the state of agglomeration.

Ultrasonic spectrometry should be tested for its ability to determine the particle size distribution or the state of agglomeration in non-diluted particulate systems.

2. Ultrasonic Spectrometry

In ultrasonic spectrometry, the attenuation of sound and dispersion of phase velocity over a wide frequency range is measured and analyzed. The physical background of these particle size dependent effects is understood to be scattering of the sound waves on the particles and certain dissipation processes that are inherent to sound propagation in disperse systems.[1]

The dissipation processes belong inevitably to the sound propagation, as sound waves by their nature are nothing less than propagating equilibrium disturbances. In disperse systems one distinguishes five different loss mechanisms:

- the particle size independent sound absorption in the fluid and disperse phase(s), the so-called "intrinsic losses",

- the energy loss of particle deformation and a non-stationary heat transfer across the phase interface due to different thermo-physical properties - "thermal losses",

- the energy loss of a viscous deformation of the flow field around the particle due to different inertial properties and the viscosity of real fluids - "visco-inertial" or "viscous losses",

- the energy loss of alternating electric currents due to the dipole behaviour of charged particles in acoustic fields - "electrokinetic losses", and

- the energy loss in inelastic interparticle links - "structural losses".

In most cases the last two dissipation mechanisms can be neglected.

The influence of both scattering and dissipation on the acoustic behaviour of disperse systems depends essentially on the operating conditions: if the dimensions of the particles are much smaller than the wavelength of the sound signal - this is called the long wavelength regime (LWR) and can be expressed mathematically in terms of the wave number by $ka \ll 1$ (k = wave number, a = particle radius) - the usual scattering effects are negligible, only pulsation and oscillation occur and the attenuation is mainly a result of dissipation. The system behaves in a totally different way in the intermediate wavelength regime (IWR), i.e., when the particle size is of the same magnitude as the wavelength. Dissipation plays here almost no role and scattering, in particular by diffraction, increases enormously. The remaining regime, the so-called short wave length regime, belongs already to the geometrical acoustics.

The ultrasonic spectrometer AcoustoPhor 8000 from Pen Kem is designed for the long wavelength regime with a frequency range from 1 to 100 MHz in the particle size range from 10 nm to 10 μm. The measured attenuation spectra are transformed into a log-normal particle size distribution as well as into monodisperse particle size and a bimodal log-normal distribution. A statistical measure for the coincidence with the experimental values allows the selection of the best hypothesis. The sample volume of 100 cm^3 with up to 50 vol % solids is pumped continuously through the measurement cell. For the calculation of a particle size from the attenuation spectra, a number of physical constants of the phases must be known: density, sound speed, specific attenuation, viscosity, heat capacity, heat conductivity, and thermal expansion coefficient. In the case of hard solid particles the thermal losses can be neglected and therefore the thermal properties are not required.

3. Verification of the Particle Size Measurement

Titania is often used in industrial applications, for example in the paint industry. As a result of the AcoustoPhor measurement (with 5 vol % titania) we obtained a median particle size of 233 nm with a standard deviation of 0.325 (see Figure 1).[2] For comparison with a similar non-diluting measurement technique the particle size was determined with the AcoustoSizer (Matec) in the Fraunhofer Institute IKTS-Dresden at 5 vol % giving 245 nm as mean size and 0.35 as standard deviation.[3]

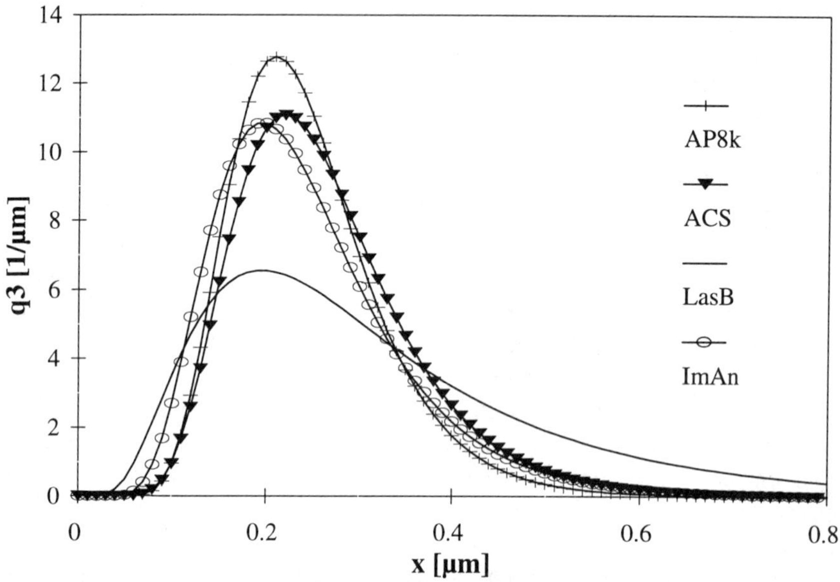

Figure 1. Particle size distribution of a 5 vol % titania suspension measured with AP8k: AcoustoPhor, ACS: AcoustoSizer, LasB: laser diffraction and ImAn: SEM-image analysis.

This relatively narrow distributed material was measured by laser diffraction after strong dilution and dispersion. The result was a broader distribution of a slightly larger particle size. The analysis with the scanning electron microscope (SEM) shows for Selectipur™ (Merck) a non-spherical convex shape and a broad size distribution with a median size of 142 nm (number-based) and a standard deviation of 0.4. The transferred value into a median size by weight was 229 nm.

The measured attenuation spectrum of these particles is shown in Figure 2. The different components of attenuation are additionally calculated.[4] The visco-inertial losses dominate and only a small amount of internal loss occurs at higher frequencies. For this well prepared sample with small particle sizes there are no scattering losses. Also, thermal losses do not participate in the total sound attenuation for this suspension.

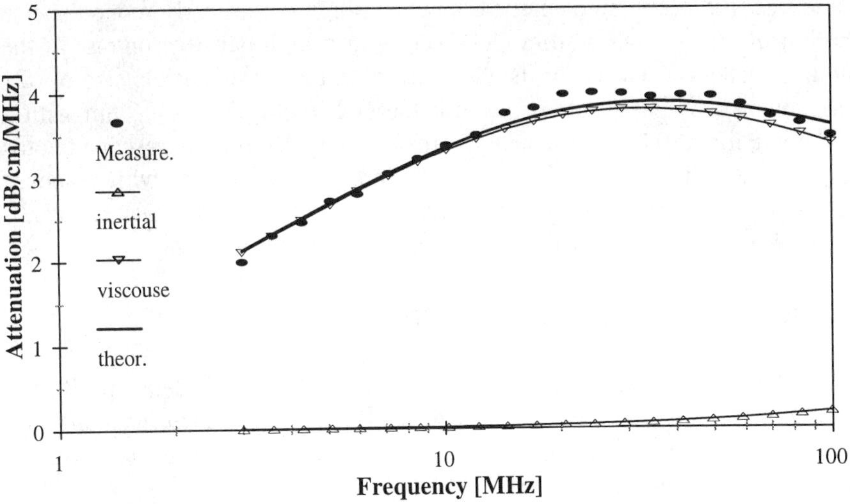

Figure 2. Attenuation spectra of a 5 vol % titania suspension.

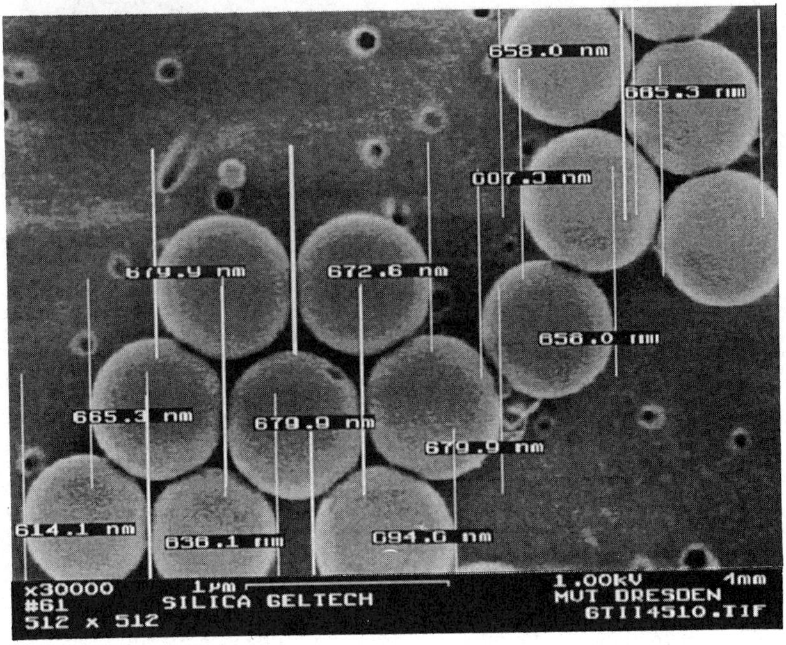

Figure 3. SEM micrograph of Gel Tech 500 silica monospheres on a membrane filter.

To test the accuracy of particle size measurement a nearly monodisperse distribution of silica spheres from Gel Tech with a high density contrast to the surrounding water was used. For the Gel Tech 500, the mean particle size of 621 nm was obtained by AcoustoPhor and for the Gel Tech 1500, 1486 nm with a standard deviation of 0.04. Image analysis of scanning electron microscope photos (see Figure 3) delivers mean sizes of 685 nm and 1510 nm with standard deviations of about 0.02. Continuously the solids concentration was increased from 1 vol % up to 20 vol % without big changes in the particle size.[5]

4. Characterization of the Dispersion State

To test a real agglomerated material system, the titania Selectipur™ was used in Figure 4. First the powder was suspended in a 5 vol % suspension without any sample preparation. Then a dissolver and after that an ultrasonic disintegrator were applied for dispersing.

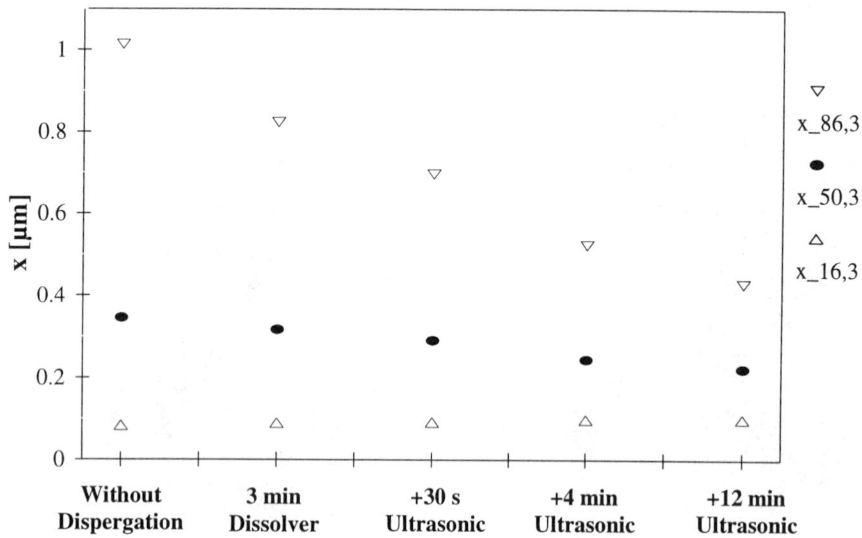

Figure 4. Quantiles of the particle size distribution of Selectipur™ titania over the dispersion time.

With growing time and power of dispersion the mean particle size becomes smaller and the standard deviation reduces. The spectra (see Figure 5) show a significant increase in the attenuation in the higher frequency range due to the growing amount of small primary particles over the time of dispersion.

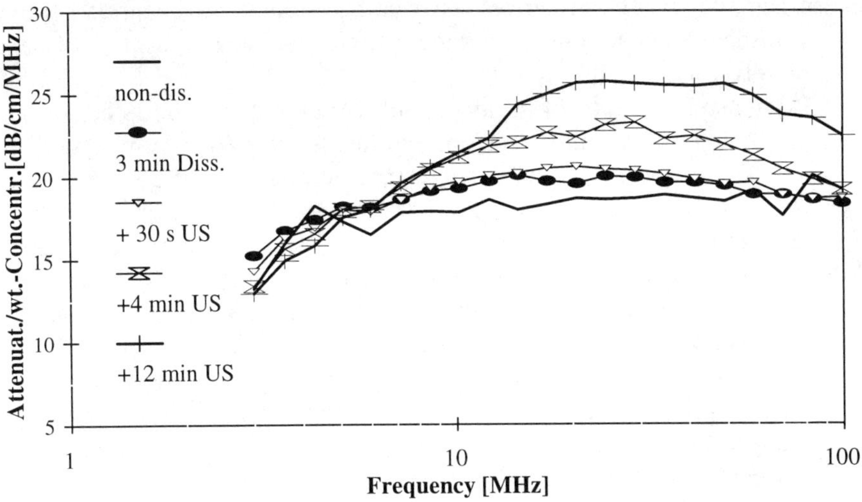

Figure 5. Influence of dispersion on the attenuation spectra of the 5 vol. % Selectipur™ titania suspension.

With a laser diffractometer, after dilution at constant pH and conductivity, the dispersion effect of only the dissolver could be remarked. The dispersion effect in the submicron range by the ultrasonic disintegrator was dominated by re-agglomeration in the first 20 seconds of dilution.

It seems that drastic dilution for scattering methods causes a loss of dispersion stability because of desorption of stabilizing ions from the particle surface, even if pH changes are avoided. So it was not possible with a laser diffractometer to document the dispersion in the slurry in the smaller particle size range.

5. Limits of the Measuring Range

The influence of both scattering and dissipation on the acoustic behaviour of disperse systems depends essentially on the operating conditions - the hardware of a measuring technique. As mentioned in section 2, with growing particle size the acoustic behaviour will be dominated by scattering effects. Consequently a maximum limit particle size of about 10 μm must be accounted for the AcoustoPhor 8000, which is designed mainly for the measurement of acoustic dissipation.

However, in some cases very large and stable agglomerates or aggregates can scatter a high amount of energy in the high frequency range. This attenuation can be wrongly interpreted by the software as viscous losses of very fine particles smaller than 100 nm. High fit errors between theory and experiment (larger than 5 %) or a drastic reduction of the high frequency attenuation as a result of dispersion with a dissolver indicate this effect.

Systematic variation of the amount of particles above 10 μm deliver limits for the beginning of false interpretation. The fine fraction (0.3-5 μm) of a mixture of glass ballotini spheres is measured without errors even if 25 vol % coarse spheres (10-15 μm) are added. This limit depends on the particle size distribution. So large silica agglomerates of silica powder had to be dispersed to an amount fewer than 10 vol % larger than 10 μm for correct interpretation of the attenuation spectra. Such coarse particle size limits are well known for established measuring techniques, like photon correlation spectroscopy.

Near the lower limit particle size of theAcoustoPhor 8000 (10 nm), fumed alumina with a primary particle size of 13 nm (TEM analysis by the producer) was measured. The obtained mean size of 32 nm agrees well with our SEM image analysis, which showed aggregated particles up to 100 nm.

6. Flocculated Particle Systems

Systems flocculated by polymers, or electrostatically non-stable material systems, are often used or handled in industry. For simulation a polymer agent was added to a well dispersed suspension. It came to a strong agglomeration and visible change in the suspension structure. The attenuation spectrum changed significantly as shown in Figure 6. For redispersion a dissolver at 4000 min^{-1} was applied for 30 s and also the attenuation of low and middle frequencies became greater.

Without polymer agents, flocculation also can be obtained by electrostatic destabilization. This state exists near the isoelectrical point, i.e. for titania Selectipur™ at pH 3.6. This destabilization tendency was observed by continuous measurement of the zeta potential. The change of the particle size could be confirmed as a change in the corresponding attenuation spectrum.

The particle size of the fully flocculated systems could not be calculated with the existing theory. The large particle size of the flocks does not explain the deviation problem sufficiently as simulation experiments have shown. It is assumed that additional mechanisms affect ultrasonic attenuation, which are not included in the existing theory.

7. Concentration Effects

For this effect, a concentration series of the rutile R900 suspension was measured. The calculated PSDs are shown in Figure 7 as x_{84}, x_{50}, and x_{16} values

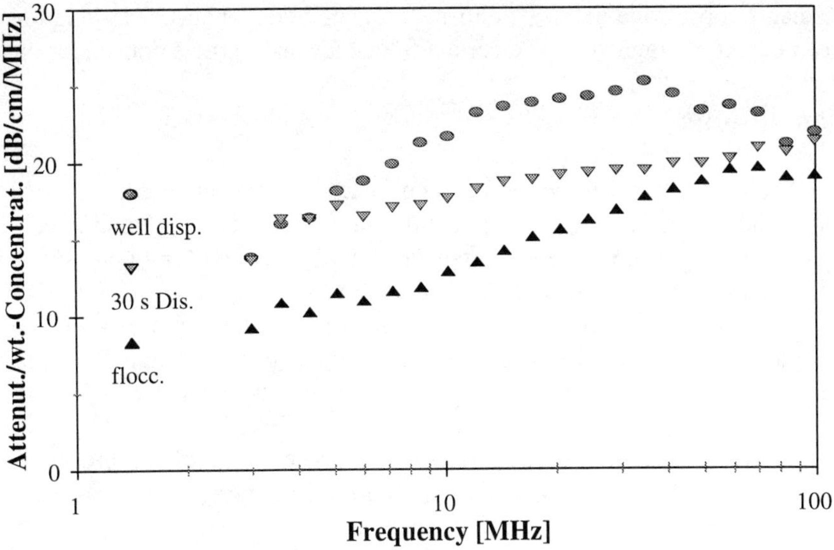

Figure 6. Flocculation of a 5 vol % titania suspension by a polymer and redispersion with a dissolver.

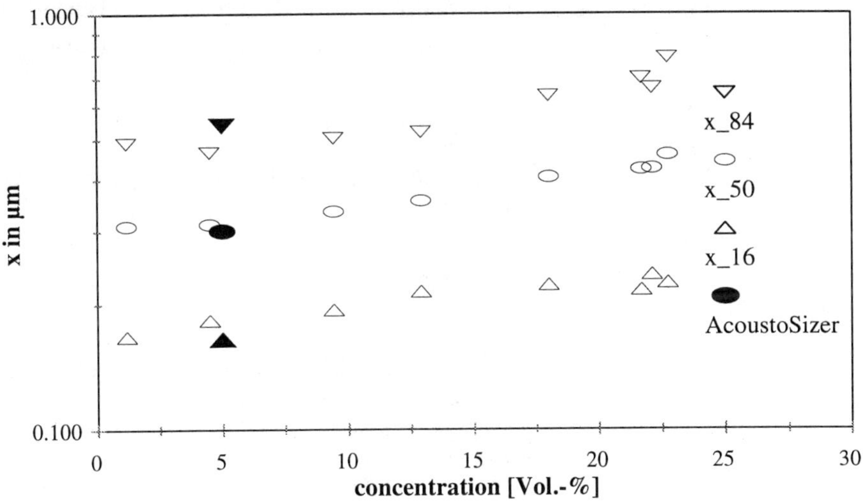

Figure 7. Quantiles of the particle size distribution for Rutile R900 titania.

over the concentration. Additionally, the PSD obtained by the AcoustoSizer is marked in the diagram. The agreement is quite well at lower concentrations. However, from about 10 vol % onward the calculated PSD is getting coarser. The experiments do, nonetheless, confirm the predicted non-linear relationship between extent of attenuation at a certain frequency and particle concentration.[1]

8. Conclusions

The experiments in dense submicrometer suspensions confirmed that the characterisation of the dynamic equilibrium between agglomeration and deagglomeration requires the simultaneous measurement of particle size, zeta potential, solids concentration, pH, and conductivity.

Ultrasonic spectrometry can be used to measure the size of particles below 10 μm, down to about 10 nm. In particular, the upper limit must be regarded because of the growing influence of scattering effects not detectable with the AcoustoPhor 8000.

Online measurement for the study and optimization of the processing of dense slurries is possible without sample dilution in contrast to optical methods. Additionally, the solids concentration must be determined if the particles are generated or reduced.

Flocculation changes the attenuation spectrum significantly. The interpretation in terms of particle size distributions is not possible with the existing theory. Further investigations concerning modeling and experiments are necessary. Concentration effects of very dense slurries should be studied also.

Acknowledgements

The authors wish to thank the German Science Foundation (DFG) for financial support.

References

1. A. Dukhin and P. Goetz, "Acoustic Spectrometry for Concentrated Polydisperse Colloids with High Density Contrast", *Langmuir*, **12**, 4987-4997 (1996).

2. F. Hinze, M. Stintz, and S. Ripperger, "Ultraschallspektrometrie zur Partikelgrößenanalyse", *CLB*, **5**, 193-196 (1997).

3. C. Knösche, Diploma Thesis, TU Dresden, 1995.

4. F. Babick, Diploma Thesis, TU Dresden, 1997.

5. F. Hinze, M. Stintz, and S. Ripperger, "Particle Characterisation in Highly Concentrated Submicron Suspensions by Ultrasonic Spectrometry", First European Congress on Chemical Engineering ECCE 1, Florence, 1997.

6. F. Babick, F. Hinze, M. Stintz, and S. Ripperger, "Ultrasonic Spectrometry for the Particle Size Analysis in Dense, Submicron Suspensions", submitted to *Part. Part. Sys. Charact.*, July 1997.

Electroacoustic Studies of Aqueous Ce-ZrO_2 and Si_3N_4 Suspensions

Lennart Bergström, Eric Laarz and Richard Greenwood*
Institute for Surface Chemistry, P.O. Box 5607, S-114 86 Stockholm, Sweden

Abstract

We have investigated the suspension properties of concentrated, aqueous suspensions of zirconia and silicon nitride powders using electroacoustics in combination with other techniques, e.g. rheology. Initial problems of achieving reproducible electroacoustic data when the suspension passed through a flocculated stage was solved by using ultrasonication; this treatment also gave a good correlation between electroacoustic and rheological measurements at moderate volume fraction of solids resulting in an isoelectric point at pH 8 for the Ce-ZrO_2 powder. Microelectrophoresis measurements of the slightly soluble zirconia powder resulted in a significantly lower isoelectric point. The studies have focused on the effect of powder pre-treatment and addition of dispersants where the rheological and electrokinetic information have been combined with adsorption data to elucidate more information of the relation between particle interactions and rheology.

1. Introduction

Control of the suspension properties is of great importance for industrial processes involving ceramic slurries, e.g. mixing, moulding, and forming. For example, a concentrated suspension with an attractive particle network may show a viscoelastic, gel-like rheological behavior while a colloidally stable suspension of equal powder volume fraction is very fluid with a low viscosity; these differences greatly affect e.g. the casting rate or spray drying performance. Since the macroscopic behaviour of colloidal suspensions, e.g. rheology, is directly linked to the total interparticle forces, either attractive or repulsive,[1,2] there is of interest to control and characterize the parameters determining these forces. The particle surface charge is one of the key parameters that controls the forces

* present address: Birchall Centre, Department of Chemistry, Keele University, Staffordshire ST5 5BG, UK

acting between particles in aqueous ceramic suspensions.

Electroacoustic methods are capable of measuring particle mobility in opaque concentrated suspensions. The method is based on the application of a high-frequency electric field to a colloidal suspension. The alternating field causes a relative movement of particles and the continuous liquid phase, which leads to a measurable acoustic signal, the electrokinetic sonic amplitude (ESA).[3] The measured signal as a function of frequency defines the dynamic mobility spectrum of the investigated material.

Fitting the dynamic mobility data to the theory developed by O'Brien,[4,5] assuming a unimodal log-normal particle size distribution of non-agglomerated hard spheres, the zeta potential and the particle size distribution can be obtained. Recent theoretical and experimental work have illustrated the influence of particle shape,[6] semi-conducting particle properties,[7] and ionic strength[8,9] on zeta potential measurements. The use of electroacoustics for the characterization of inorganic materials has been demonstrated in a number of ceramic systems, e.g. silica, yttria, silicon nitride, aluminium hydroxide, titania, and clay minerals.[8-13]

The work presented in this paper emphasizes the use of electroacoustics as a sensitive tool to study effects of polymer adsorption and powder surface treatment in concentrated ceramic suspensions. It will be shown how electroacoustics can be combined with complementary methods such as rheology and adsorption isotherms to obtain a more complete picture of the details and possible mechanisms for stabilization or flocculation.

Two aqueous systems: a Ce-ZrO_2 system containing a commercial anionic polyelectrolyte (Dispex A40), and Si_3N_4 suspensions with oxidized and leached powder surfaces were studied. Initial studies with the zirconia system were focused on developing a procedure for reproducible electrokinetic measurements at high solids loading. We will illustrate how the slight solubility and other possible effects makes an interpretation from dilute microelectrophoresis experiments difficult to perform. We will also briefly discuss the effect of polyelectrolyte adsorption on the suspension properties.[14]

Concentrated aqueous silicon nitride suspensions were studied with respect to the influence of changes in surface oxygen content on the zeta potential. For the Si_3N_4 system it is known that microstructure and mechanical properties after sintering depend on the oxygen content of the powder.[15] Equally important, the surface oxygen content affects the dispersion behavior in colloidal processing as shown in other reports.[16-19]

2. Materials and Methods

The ceria-stabilized ZrO_2 powder (CEZ 10, Zirconia Sales Inc., GA, USA) used in this study contained 13 mol % ceria and belongs to the class of partially stabilized, polycrystalline tetragonal zirconia powders known as TZP's. The powder has a specific surface area of 21.4 m^2g^{-1} and a mean particle size of 0.4 μm with a significant portion (18 wt %) of the particles smaller than 0.1 • m. The powder was slightly soluble at acidic conditions,[14] i.e. cerium ions were dissolved at pH<6. The additive, Dispex A40 (Allied Colloids, UK), is a polyacrylic acid neutralized with ammonia; fully ionized at pH>8. The dispersant has a mean molecular weight of 10,000 and a polydispersity of M_W/M_N=1.56.[14] Double distilled water was used and 1 M HCl and 0.5 M NaOH was used to adjust the pH.

The Si_3N_4 powder (SN-E10, UBE Industries, Japan) has a mean particle size of 0.5 μm and a specific surface area of 10.4 m^2g^{-1}. The leaching pre-treatment consisted of boiling 25 g as-received powder for 24 hours in 500 ml deionized water at pH 11 in a polypropylene container. The hot suspension was filtered after boiling and the filter cake was washed twice with hot deionized water, dried for 10 hours at 120 °C, stored in an desiccator and re-dispersed in deionized water by ultrasonication for subsequent experiments. Oxidation pre-treatment was done by heating the as-received powder for 8 hours at 800 °C in air.

Suspensions with 10 or 25 vol % solids content in 0.01 M NaCl electrolyte were prepared for rheological experiments by mixing powder, electrolyte and dispersant, followed by adjustment of pH and ultrasonication. Steady shear experiments were carried out with a controlled shear rate rotational rheometer (Bohlin VOR, Bohlin Rheology, Sweden). A bob and cup measuring geometry with 1.25 mm gap width and 1 min preshearing at the highest obtainable shear rate was used.

Electroacoustic measurements (AcoustoSizer™, Colloidal Dynamics, Australia) were also carried out on 10 vol % samples in 0.01 M NaCl electrolyte, prepared the same way as described above for the rheology experiments and then kept stirring over night. After adjusting the pH, the sample was either subjected to simple stirring or ultrasonic treatment for 2x3 min using an ultrasonic horn of diameter 9.5 mm (Soniprep 150, MSE Scientific Instruments, UK) before the next measurement was carried out. The type of suspension pre-treatment will be described in more detail in relation to each experiment.

Microelectrophoresis experiments, (Zetamaster 4, Malvern Instruments) were also used to determine the electrophoretic mobility, from which the zeta potential was calculated using the Smoluchowski equation. The dispersion had to be fairly dilute since these measurements are based on light scattering; volume fractions of 0.005 were used. Here, all samples were measured as quickly as possible, within 20 - 25 minutes after preparation. The powder was added to the electrolyte solution (0.01 and 0.005 M NaCl) at a pre-adjusted pH, and then stirred and settled for 20 minutes.

XPS studies (Kratos, AXIS-HS) were carried out using a Mg Kα radiation source (1253.6 eV). Intensity peaks at 281.8 eV (C 1s), 529.0 eV (O 1s), 394.1 eV

(N 1s) and 98.6 eV (Si 2p) were evaluated with the corresponding sensitivity factors 0.25, 0.66, 0.42 and 0.27. More detailed descriptions of the above experimental procedures can be found elsewhere.[14]

3. Results and Discussion

3.1 The Isoelectric Point of Ce-ZrO_2

Steady shear measurements on 25 vol % samples revealed a strong shear thinning behaviour and comparably high viscosities in the range pH 4-11. This indicates that the samples were flocculated and that the surface charge density acquired through dissociation of surface groups is not high enough to provide electrostatic stabilization. The pH-dependence of the low-shear viscosity– the shear-range where the particle interactions have a strong influence on the steady shear properties– , is shown in Figure 1. There is a definite viscosity maximum at about pH 8, indicating strong flocculation. If we assume that only electrostatic forces can reduce the magnitude of the van der Waals attraction between the particles, the point of maximum viscosity at pH≈8 should correspond to the point of zero charge (pzc).

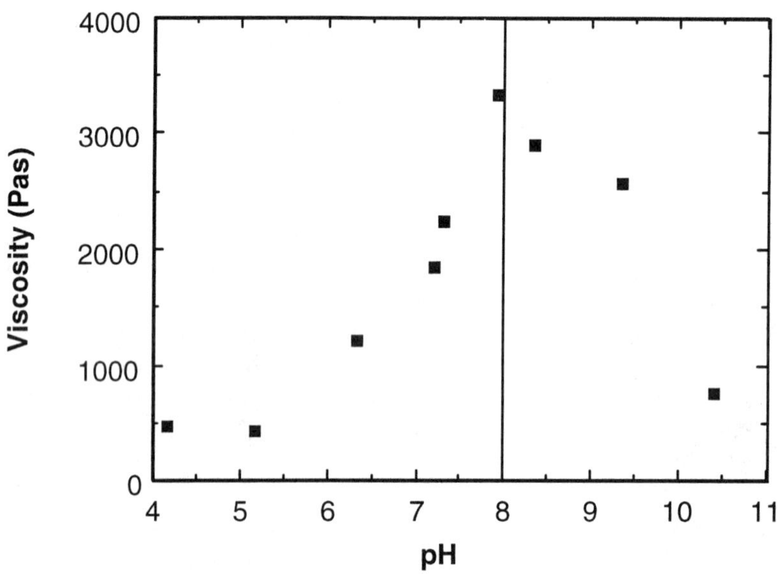

Figure 1. Steady shear viscosity versus pH for a Ce-ZrO_2 aqueous suspension at ϕ=0.25 at a shear rate of 0.184 1/s.

An alternative way of determining the pzc is by electrokinetic measurements in an inert electrolyte where the isoelectric point (iep) and pzc should coincide. We have used both microelectrophoresis in dilute suspensions ($\phi \approx 0.005$) and electroacoustics for suspensions of intermediate concentration ($\phi = 0.1$) and compared the results.

Figure 2 shows the results from the microelectrophoresis measurements which displayed an isoelectric point for the Ce-ZrO_2 powder at pH 5.5. The curves for both electrolyte concentrations (10 mM and 5mM NaCl) display an identical isoelectric point, showing that the electrolyte is inert. However, this $pH_{iep}=5.5$ is significantly lower than the pzc indicated by the rheology experiments at high solids loading.

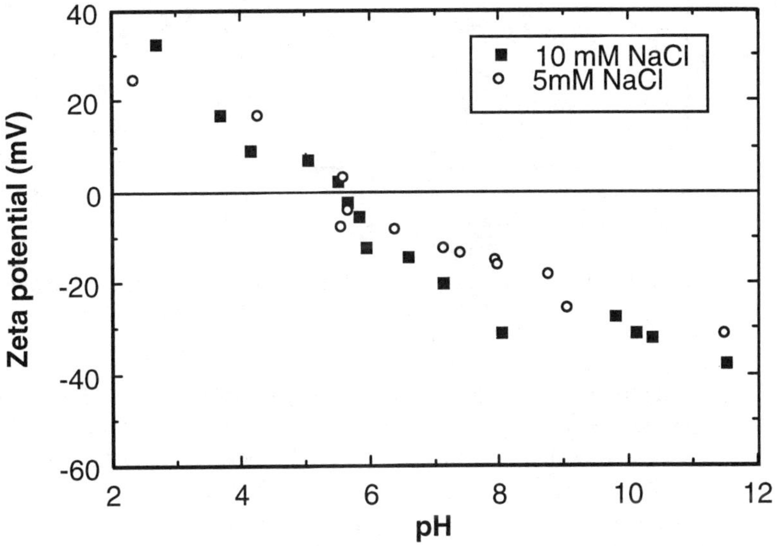

Figure 2. Microelectrophoretic measurements of a dilute Ce-ZrO_2 suspension at two different background electrolyte strengths.

Electrokinetic studies at much higher powder concentrations using electroacoustics are shown in Figures 3 and 4. Figure 3 shows the result from pH titrations of two separate concentrated suspensions; one prepared at a low pH (pH≈4) and one prepared at high pH (pH≈12). After preparation, the sample were left stirring overnight before the pH titration started were acid were added to the high pH suspension and base was added to the low pH suspension directly in the measuring container with the rotating paddle, rotating at 300 rpm, as the only means of homogenizing the suspension. Figure 3 clearly shows that the two separate runs display a significant hysteresis with apparent iep's at pH_{iep}=9.1 and 7.5, for the acid and base titration, respectively.

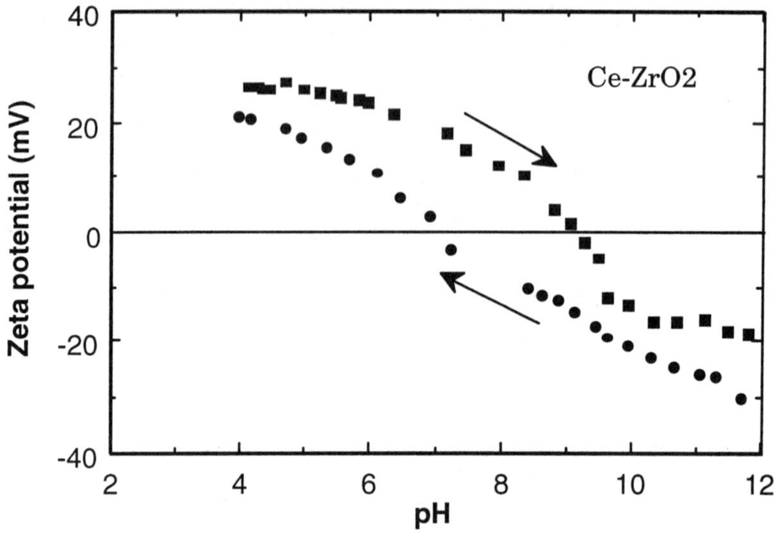

Figure 3. Electroacoustic measurements over a pH range performed by *in situ* titration, starting from a basic (circles) and acidic (squares) pH, respectively, and using only the stirrer for mixing and deagglomeration of the two suspensions.

The hysteresis may be attributed to several different effects. The relatively weak stirrer may be unable to break up the flocs that are formed when the surface charge density is low. However, at high or low pH, where the zeta potential attains the highest absolute values, the particle flocs should be weak enough to be broken down by the hydrodynamic forces introduced by stirring. Since we find a significant difference in zeta potential also at these high and low pH values, far from the apparent iep, this explanation is less likely.

Another, more likely cause of the observed hysteresis is related to the solubility of the powder. Since cerium will dissolve at pH values below pH 6, it is possible that the suspension prepared at acidic pH will contain a significant amount of dissolved cerium ions which might adsorb as cationic species or even reprecipitate on the powder surfaces as cerium hydroxide when the pH is increased. These effects would lead to an increase in the zeta potential and an upward shift of the iep, just as we can see in Figure 3. For the suspension prepared at pH 12, it is possible that zirconia will dissolve and form zirconate ions ($Zr(OH)_5^-$).[20] In parallel to the acidic case, this anionic specie might adsorb or reprecipitate, thus resulting in a decrease in the zeta potential and a downward shift of the iep.

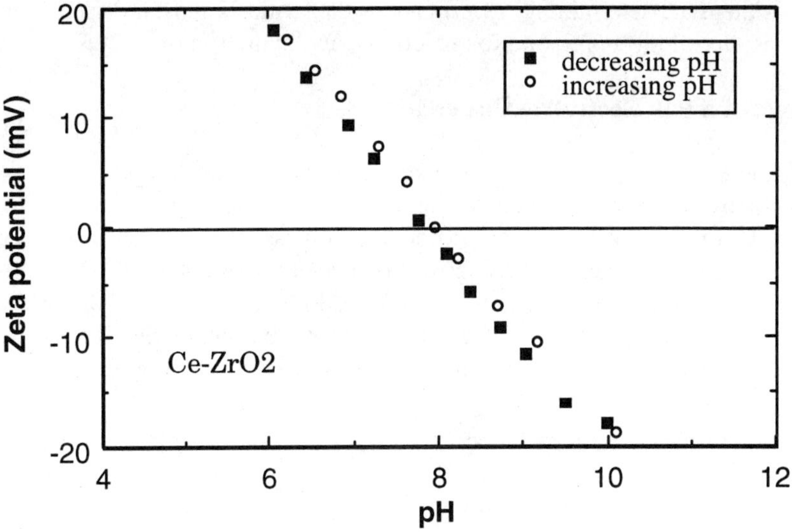

Figure 4. Electroacoustic measurements at different pH values where the suspension was subjected to ultrasonic treatment after each pH adjustment. Reprinted from Ref. 14 with kind permission of ©Elsevier Science Ltd.

To remedy these equilibration and dissolution problems, we have established an alternative experimental procedure to obtain a reproducible state of the suspension. Ultrasonication for 3 times 2 minutes was used to break any possible flocs being formed after each pH adjustment and we performed electroacoustic measurements over a more narrow pH range (pH=6-10). This treatment resulted in a good correlation between electroacoustic and rheological measurements at moderate volume fraction of solids. In Figure 4, the electroacoustic results using the ultrasonic treatment after each pH change is shown. The measurements show no hysteresis when increasing or decreasing the pH and the iep at pH≈7.9, coincides very well with the maximum in viscosity (Figure 1). Further evidence for an iep near pH 8 was noted in settling experiments where the sample at pH=8 displayed the highest settling rate. Hence, we define a $pH_{iep}≈8$ for the Ce-ZrO_2 powder in equilibrium with an electrolyte solution at relatively high solids loading.

It appears that microelectrophoresis experiments with dilute suspensions of a slightly soluble powder like Ce-ZrO_2 powder may result in different values for the iep as observed in this study (pH_{iep}=5.5) and as reported by Bleier *et al* (pH_{iep}=6.9).[21] Both of these iep-values are substantially lower than the iep obtained from the electroacoustic measurements in this study ($pH_{iep}≈8$). A possible explanation to this discrepancy is the much lower surface to volume ratio in the dilute suspensions which may result in longer equilibration times for surface potential determining reactions at the solid-liquid interface, such as hydrolysis, dissolution-precipitation reactions, and hydration of solutes.

3.2 Effect of a Polyelectrolyte Dispersant

The effect of polyelectrolyte dispersant addition on the properties of Ce-ZrO_2 suspensions were investigated at basic pH values; the acidic range was avoided due to dissolution of the powder. Previous studies have shown that a low molecular weight polyacrylic acid (PAA) dispersant (Dispex A40) adsorbs at the zirconia-water interface and may impart colloidal stability.[14] The effect of polyelectrolyte adsorption can be easily shown by combining electroacoustic and rheological measurements at different additions of dispersant (Figure 5).

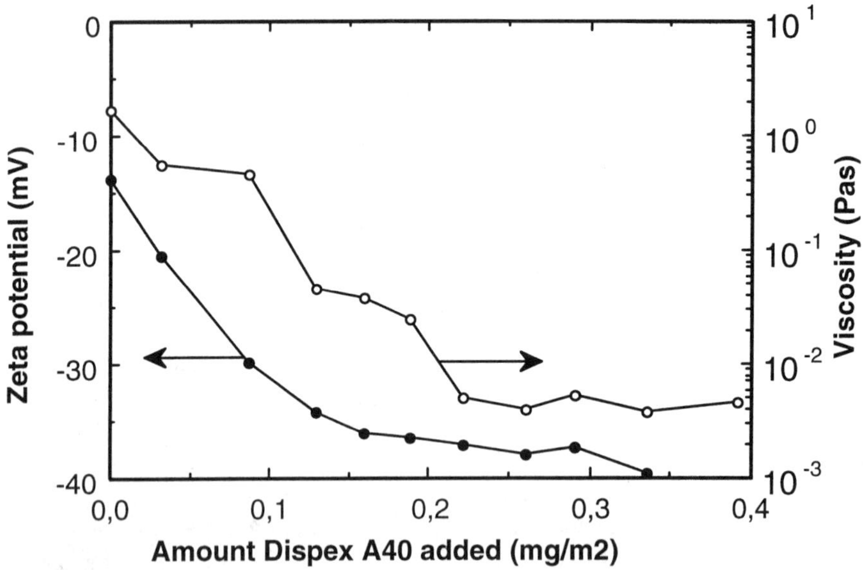

Figure 5. The effect of dispersant addition on the zeta potential and steady shear viscosities at a shear rate of 1.84 1/s; pH=9.3 and ϕ=0.10. Reprinted from Ref. 14 with kind permission from ©Elsevier Science Ltd.

The results show that the suspension viscosity decreased and the zeta potential increased in magnitude, with increasing added amounts of Dispex A40; both quantities approaching plateau values at higher dispersant concentrations. Extended rheological characterization (not shown here) displayed an abrupt change from a shear thinning behaviour at low additions to a Newtonian behavior with a very low viscosity (η=4 mPa s) at higher additions of the dispersant.[14] The decrease in viscosity and degree of shear thinning with the addition of more

Dispex A40 implies that the degree of flocculation decreases. At a critical addition of dispersant the viscosity is at a minimum and the suspension becomes Newtonian, implying that the suspension is now colloidally stable.

The increase in the magnitude of the zeta potential (i.e, it became more negative) with addition of polyelectrolyte can be explained by adsorption; the surface charge density at the solid/liquid interface increases as the highly charged anionic polyelectrolyte adsorbs. The zeta potential gradually increases in magnitude with addition until a plateau region is reached where further addition of the dispersant did not affect the zeta potential, i.e. the surface is fully covered with polyelectrolyte.

We can compare the values of the amount of dispersant corresponding to monolayer coverage from adsorption isotherms (0.21 mg/m^2),[14] with the additions required to reach minimum viscosity (0.25 mg/m^2) or maximum zeta potential (0.25 mg/m^2) using steady shear and electroacoustic measurements, respectively. The good agreement between the amount needed for monolayer coverage and the minimum amount of dispersant needed to reach minimum viscosity and maximum zeta potential for the different suspensions indicate that full coverage is needed to reach colloidal stability. It also points to the usefulness of electroacoustic measurements as a quick and simple method to determine the optimum dosing of charged dispersants. However, this is complicated by the complex nature of the interactions between surfaces carrying adsorbed polyelectrolyte layers which is presently being investigated in more detail using an Atomic Force Microscope to directly probe the interparticle forces in this polyelectrolyte system.[22]

3.3 The Aqueous Si_3N_4 System

Leaching in boiling water at high pH was used to reduce the surface oxygen content of the as-received silicon nitride powder. Dissolution of silica is catalyzed by hydroxide ions and the reaction (Equation 1) is accelerated in boiling water.

$$SiO_2 + 2H_2O \longleftrightarrow Si(OH)_4(aq) \quad (1)$$

The XPS results of decreasing O/Si and increasing N/Si atomic ratios of the leached powder compared to the as-received confirms that the leaching was successful (Table I).

Atomic ratios of the oxidized powder were calculated by using the XPS and Thermogravimetric analysis data reported by Butt *et al.*[23] for UBE SN-E10 powder and by adopting the proposed model of a diffusion-rate-limited oxidation reaction:

$$Si_3N_4 + 3O_2(g) \longleftrightarrow 3SiO_2 + 2N_2(g) \quad (2)$$

The O/Si ratio decreases by a factor of 0.5 due to leaching, whereas the thermal oxidation at 800 °C results in an increase by a factor of 2.4. Previously, it was suggested that the ratio of number of silanol (Si-OH) groups to number of secondary amine (Si_2-NH) groups at the silicon nitride surface correlates with the O/Si and N/Si ratios determined by ESCA.[17] Thus, thermal oxidation results in an increased surface density of silanol groups whereas leaching decreases the number of silanol groups compared to the secondary amine groups. Since the potential-determining reactions of the silanol and secondary amine groups controls the surface charge and thus the isoelectric point, this suggests that electrokinetic measurements can be used as a tool to detect changes in the composition of the functional surface groups.

Table I. Atomic ratios in the surface layer of UBE SN-E10 powders.

	O/Si (own measurements)	N/Si (own measurements)	O/Si (Butt et al. measurements)*	N/Si (Butt et al. measurements)*
as received	0.59	1.03	0.50	0.95
Leached	0.30	1.20	-	-
Oxidized	-	-	1.20**	0.19**

* Data taken from Butt et al.[23]
** Calculated values, based on data from Butt *et al.*[23]

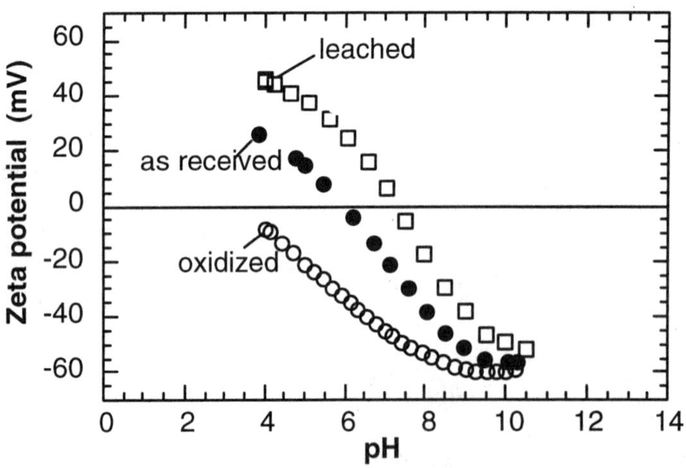

Figure 6. Electroacoustic measurements of as-received and pre-treated (leached and oxidized) Si_3N_4 suspensions.

The electroacoustic results in Figure 6 illustrate the sensitivity of electrokinetic measurements. The isoelectric points of the three different powders: as-received, leached and oxidized, displayed significant differences. Leaching resulted in an increase in the iep to $pH_{iep}=7.3$ compared to the as-received powder ($pH_{iep}=6.0$) which can be directly related to the low relative surface concentration of silanol groups (or high relative concentration of secondary amine groups). The oxidized powder displayed a low iep, $pH_{iep}<3.5$, since the surface is dominated by the amphoteric silanol groups.

4. Summary and Conclusions

Rheology and electroacoustics experiments on concentrated Ce-ZrO_2 suspensions resulted in a $pH_{iep}\approx8$. Microelectrophoretic measurements showed a significantly lower iep. This difference in iep with solids concentration– and thus surface-to-volume ratio of the system– may be related to the difference in the kinetics and equilibrium of complex surface reactions, e.g. dissolution, dissociation, adsorption and possibly hydration.

We have established an experimental procedure, using ultrasonication to obtain a reproducible state of the suspension, which gave good correlation between electroacoustic and rheological measurements at moderate volume fraction of solids. These results illustrate the importance of electroacoustics as an excellent technique for studying complex suspensions of slightly soluble powders at high volume fractions. It is often problematic to use microelectrophoresis data alone and extrapolating the results to higher volume fractions when studying ceramic suspensions.

We found that electroacoustics can be used as a convenient and simple method to obtain information regarding the change in zeta potential with the addition of dispersant. The results in this study also illustrate how the powerful combination of rheology and electroacoustics can be utilized in the investigation and optimization of the effect of different dispersants on the properties of concentrated ceramic powder suspensions. The work on silicon nitride, pre-treated by leaching or oxidation, illustrates that electrokinetic measurements can be used as a sensitive tool to detect small changes in the composition of the functional surface groups.

Acknowledgements

This work was partly financed by the European Commission, CEC Human Capital and Mobility Programme, contract CHRX CT 94-0574, and partly by "The Brinell Centre- Inorganic Interfacial Engineering", a jointly funded research initiative by NUTEK and an industrial consortium.

References

1. J.N. Israelachvili, *Intermolecular & Surface Forces*, 2nd ed., Academic Press Limited, London, 1991

2. W.B. Russel, D.A. Saville, and W.R. Schowalter, *Colloidal Dispersions*, Cambridge University Press, Cambridge, 1989

3. T. Oja, G.L. Petersen, and D.C. Cannon, United States Patent No. 4497208, 1985

4. R.W. O'Brien, "Electro-Acoustic Effects in a Dilute Suspension of Spherical Particles", *J. Fluid Mech.*, **190**, 71-86 (1988).

5. R.W. O'Brien, D.W. Cannon, and W.N. Rowlands, "Electroacoustic Determination of Particle Size and Zeta Potential", *J. Colloid Interface Sci.*, **173**, 406-418 (1995).

6. M. Loewenberg and R.W. O'Brien, "The Dynamic Mobility of Non-Spherical Particles", *J. Colloid Interface Sci.*, **150**, 158-168 (1992).

7. R.W. O'Brien, "The Dielectric Response of a Dilute Suspension of Semiconducting Particles" (Note), *J. Colloid Interface Sci.*, **177**, 280-281 (1996).

8. W.N. Rowlands, R.W. O'Brien, R.J. Hunter, and V. Patrick, "Surface Properties of Aluminium Hydroxide at High Salt Concentration", *J. Colloid Interface Sci.*, **188**, 325-335 (1997).

9. V.A. Hackley and S.G. Malghan, "Investigation of Parameters and Secondary Components Affecting the Electroacoustic Analysis of Silicon Nitride Powders", pp 161-179 in *Electroacoustics for Characterization of Particulates and Suspensions,* NIST Special Publication 856, Edited by S.G. Malghan. USDC Technology Administration, Gaithersburg, MD, 1993 .

10. J.P. Pollinger, "Oxide and Non-Oxide Powder Processing Applications using Electroacoustic Characterization", pp 143-160 in *Electroacoustics for Characterization of Particulates and Suspensions,* NIST Special Publication 856, Edited by S.G. Malghan. USDC Technology Administration, Gaithersburg, MD, 1993 .

11. V.A. Hackley and S.G. Malghan, "The Surface Chemistry of Silicon Nitride Powder in the Presence of Dissolved Ions", *J. Mater. Sci.*, **29**, 4420-4430 (1994).

12. W.N. Rowlands and R.W. O'Brien, "The Dynamic Mobility and Dielectric Response of Kaolinite Particles", *J. Colloid Interface Sci.*, **175**, 190-200 (1995).

13. M. Rasmusson, W.R. Rowlands, R.W. O'Brien, and R.J. Hunter, "The Dynamic Mobility and Dielectric Response of Sodium Bentonite", *J. Colloid Interface Sci.*, **189**, 92-100 (1997).

14. R. Greenwood and .L. Bergström, "Electroacoustic and Rheological Properties of Aqueous Ce-ZrO_2 (Ce-TZP) Suspensions", *J. Europ. Ceram. Soc.*, **17**, 537-548 (1997).

15. S. Natansohn, A.E. Pasto, and W.J. Rourke, "Effect of Powder Surface Modifications on the Properties of Silicon Nitride Ceramics", *J. Amer. Ceram. Soc.*, **76**, 2273-2284 (1993).

16. H. Stadelmann, G. Petzow, and P. Greil, "Effects of Surface Purification on the Properties of Aqueous Silicon Nitride Suspensions", *J. Europ. Ceram. Soc.*, **5**, 155-163 (1989).

17. L. Bergström and E. Bostedt, "Surface Chemistry of Silicon Nitride Powders: Electrokinetic Behaviour and ESCA Studies", *Colloids Surfaces*, **49**, 183-197 (1990).

18. P. Greil, R. Nitzsche, and H. Friedrich, "Evaluation of Oxygen Content on Silicon Nitride Powder Surface from the Measurement of the Isoelectric Point", *J. Europ. Ceram. Soc.*, **7**, 353-359 (1991).

19. V.A. Hackley, P.S. Wang, and S.G. Malghan, "Effects of Soxhlet Extraction on the Surface Oxide Layer of Silicon Nitride Powders", *Mater. Chem. Phys.*, **36**, 112-118 (1993).

20. L. Bergström, "Surface Chemical Characterization of Ceramic Powders"; pp. 71-126 *in Surface and Colloid Chemistry in Advanced Ceramics Processing*, Edited by R.J. Pugh and L. Bergström, Marcel Dekker, New York, 1994.

21. A. Bleier, P.F. Becher, K.B. Alexander, and C.G. Westmoreland, "Effect of Aqueous Processing Conditions on the Microstructure and Transformation Behaviour in Al_2O_3-ZrO_2(CeO_2) Composites", *J. Amer. Ceram. Soc.*, **75**, 2649-2658 (1992).

22. H. Guldberg-Pedersen and L. Bergström, "Forces Measured between Zirconia (TZ3Y) Surfaces in Poly(acrylic acid) Solutions", *J. Amer. Ceram. Soc.*, submitted.

23. D.P. Butt, D. Albert, and T.N. Taylor, "Kinetics of Thermal Oxidation of Silicon Nitride Powders", *J. Amer. Ceram. Soc.*, **79**, 2809-2814 (1996).

Characterization of Ceramic Slurries by Ultrasonic Attenuation Spectroscopy

Shin-ichi Takeda

Department of Applied Chemistry, Faculty of Engineering, Okayama University, 3-1-1 Tsushima-naka Okayama, 700 Japan

Abstract

Aqueous slurries of ceramic powders are used in a variety of powder forming techniques for electronic and structural ceramics. This paper presents experimental proof that ultrasonic attenuation spectroscopy is capable of characterizing particle or agglomerate size distributions in highly concentrated ceramic slurries. Well characterized alumina powder was used to test the theory by using the Pen Kem-8000. This commercial instrument measures ultrasonic attenuation spectra over a wide frequency range from 1 to 100 MHz, and converts them into particle size distributions. Topics to be presented include: 1) evaluation of the particle size distribution in binary systems, such as an alumina/alumina systems with different sizes and volume fractions. 2) evaluation of the particle and agglomerate size distribution of nano-sized particles. 3) monitoring the change of particle size distribution in slurries by changing the dosage of dispersant.

1. Introduction

It is important to control inhomogeneities that originate in the particle-packing structures during shape forming of ceramic powders, since they can produce fracture origins in sintered materials or they can lead to shape distortion and cracking during drying, pyrolysis and sintering.[1-4] Such inhomogeneities are closely related to the particle size distribution (PSD) in the prepared slurries. In general, slurries used in actual ceramic processing contain several components such as sintering aids, dispersants, and binders which interact in complex ways and as a result, proper control of the particle-packing structure is a difficult problem.[5] These components can alter the interfacial chemistry during processing.[6] In addition, the mixing process itself, where these components react with the powder surface in aqueous solutions, may change the state at the powder/solution interface. Wide variations in PSD and surface chemical properties of powders thus occur between manufacturers and even between batches from the same source.

By measuring properties of a green body, such as density and porosity, it is possible to determine whether an employed condition of processing is proper or not, and then this indirect information can be applied to compensate for property fluctuations during or after sintering. Once one can estimate a PSD during processing or prior to sintering, the analysis and control of slurry behavior will be facilitated and problems will be more easily avoided before casting and firing.

Recently, ultrasonic attenuation spectroscopy has developed rapidly as an alternative to the light scattering method for the determination of particle size distributions.[7-9] Many new works related to ultrasonic attenuation spectroscopy have appeared since 1991, when McClement's review of ultrasonic attenuation spectroscopy was published.[10,11] The development of automatic measuring instrumentation using ultrasonic attenuation or electroacoustic spectroscopy has also been performed and the instruments are now out on the market. The biggest advantage of ultrasonic attenuation spectroscopy compared with other techniques, such as light scattering, is the capability to characterize intact concentrated dispersed systems. That is, we do not have to dilute the slurries when we estimate the dispersed/flocculated particle size distribution in slurries.

The purpose of the present paper is to confirm the capability to distinguish the dispersed and flocculated size in highly concentrated slurries using ultrasonic attenuation spectroscopy.

2. Materials and Experimental Procedure

2.2 Powders and Prepared Slurries

Commercial-grade high purity powders of Al_2O_3 (Sumitomo Chemical Co., Ltd., Tokyo, Japan) were used for all measurements other than the nano-sized particle system. For the measurements of nano-sized particles, alpha alumina (Degussa and Nano Tek), anatase (Nano Tek), and hematite (Nano Tek) were used. All slurries were prepared by dispersing dry powders in distilled water or prepared aqueous solutions using a high intensity ultrasonic probe. The dispersant used in the present study was sodium hexametaphosphate (SHP). The percentage of the concentration of SHP used in this study represents the mass fraction of SHP against the mass of distilled water.

2.3 Measurement Technique

A precise and detailed background of the theory for ultrasonic attenuation spectroscopy has been given elsewhere.[9] The attenuation spectra of the sample slurries were measured using the Acoustophor PK-8000 (Pen Kem Inc., Mount

Kisko, NY, USA). The attenuation was measured over a frequency range from 1 to 100 MHz. The variable gap technique makes it possible to exclude calibration procedures. There is no restriction on the volume fraction but the sample must be fluid in order to pump the slurry through the measuring chamber using a peristaltic pump.

Analysis software converts attenuation spectra into the particle size distribution by minimizing the difference between theoretical and measured spectra. The absolute error of the theoretical fit is the measure of this difference. In the case of the log-normal distribution the absolute error is minimized by adjusting median size and standard deviation. In the case of the bimodal distribution, the absolute error is also minimized by adjusting median sizes and standard deviations of each mode.

3. Results and Discussion

3.1 Single Powder Systems

Figure 1 shows the measured attenuation spectrum in decibels per centimeter per MHz (dB/cm/MHz). Experimental points at the low frequencies (<3 MHz) are excluded because of large errors; those frequencies lie too close to the transducer piezo crystal harmonics). It can be seen that the log-normal distribution provides a good fit to the measured attenuation spectrum for 10 vol% alumina slurry (AKP-30). The corresponding particle size distribution is shown in Figure 2.

The median size obtained shows good agreement with that reported by the manufacturer as shown in Table I. In order to verify the capability of the ultrasonic attenuation spectroscopy for determining the median size of the particles in single powder slurry systems, alumina powders with different sizes were used. Table I represents a comparison of median size between measured and reported values. The latter values were determined for diluted slurries by using the light scattering technique. All of the obtained data show good agreement, which demonstrates the capability to determine the size of single powder slurry systems in the well-dispersed state.

Table I. Comparison of median size of particle distribution obtained by the PK-8000 and that reported by the manufacturer.

Sample	AKP-15	AKP-20	AKP-30	AKP-50	AA-07	AA-2
	Median size (μm)					
PK-8000	0.59	0.50	0.28	0.20	0.75	2.1
Reported value	0.62	0.58	0.32	0.23	0.66	2.0

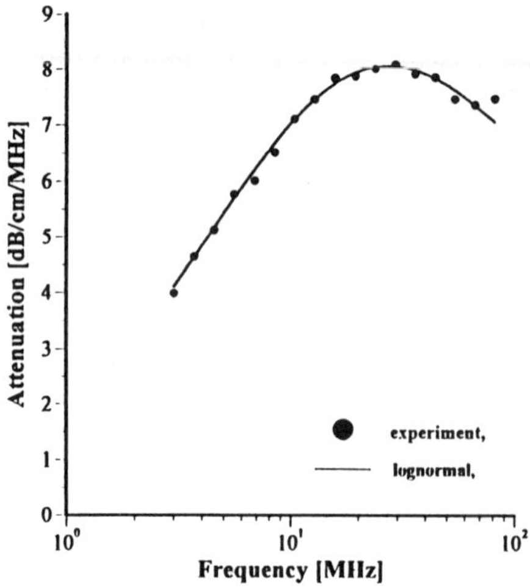

Figure1. Experimental and theoretical attenuation spectra for alumina slurry (AKP-30), 10 vol %.

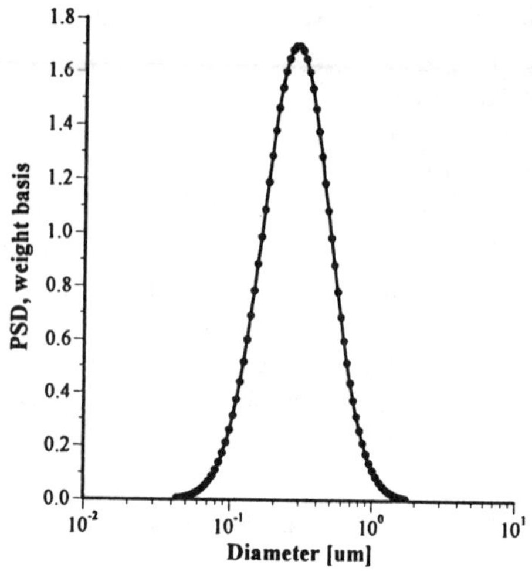

Figure 2. Particle size distribution for alumina slurry (AKP-30), 10 vol %.

3.2 Binary Powder Systems

In aqueous-based ceramic processing such as slip-casting and tape-casting, control of dispersion properties is required. In particular, avoiding formation of agglomerates is important. Since current theory provides the capability to characterize the median particle sizes in binary systems, we tested this capability in binary powder systems which are taken as a model system for slurries containing some agglomerates.

To illustrate the capability, we present data for binary systems which involved two kinds of alumina powders with different volume fractions and different sizes. The median sizes of each powder used here are summarized in Table I. Figures 3(a) and (b) show the calculated particle size distributions assuming a bimodal distribution, which consists of 10 vol% slurries of AA-2 and AKP-30 with different mixing ratio of two powders at a well-dispersed state. Figures 3(c) and (d) show the calculated and measured attenuation spectra at the corresponding conditions shown in Figure 3 (a) and (b). A log-normal distribution curve was fit to the experimental data points for single powder systems, while a bimodal distribution, shown in Figure 3 (a) and (b), were fit very well to the measured points for binary systems. Results clearly demonstrate the ability of this technique to determine the original component distributions from mixtures and to detect the presence of another component with a fraction of about 1vol% (these slurries contain 10 vol% total powder and hence this technique can detect a change of about 1 vol% for the system with the ratio of 90/10). On the other hand the measured attenuation spectra were also analyzed to obtain the best fit by recalculating attenuation spectra based on the theory. For recalculating the attenuation spectra of bimodal distribution, the median size and the standard deviation of the larger particle and its volume fraction were changed as variables (the median size of the smaller particle (AKP-30) was fixed to 0.28 μm). In this case the obtained median size for the larger-size particle was 2.1 μm, which corresponds to the reported value shown in Table I. Hence the analyzing procedure also gives the correct value of the particle size, which suggests this technique has a capability to determine the particle size correctly if the primary size is known.

Figure 4 (a) and (b) show the measured attenuation spectra for other binary systems of well-dispersed 10vol% slurries containing AA-2, AA-07, AKP-15, AKP-20, AKP-30 and AKP-50, and for 50/50 mixtures of two powders. The spectra are also dominated by viscous losses, and clearly show the variations of the shape with the sizes of each mixed powder. Figure 4(c) and (d) show the corresponding particle size distributions calculated from model fits of the experimentally obtained spectra. The analyzing procedure adopted here was the same as that mentioned above. These binary systems were also fit assuming a bimodal distribution and results clearly demonstrate the ability of this technique to resolve the two original component distributions from mixtures for the case where the difference between the median sizes was greater than 1.3 μm. On the other hand, when the difference was below 0.3 μm, two peaks did not appear in the distribution curve, but still the bimodal distribution gives a good fit to the measured attenuation spectra.

3.3 Nano-Sized Particle Characterization

It can be seen from Figure 5 that the bimodal distribution provides a good fit to the measured attenuation spectrum of Nano Tek alumina powder, which suggests that there are some agglomerates existing in the measured slurry. The corresponding PSD is shown in Figure 6, where two peaks appear in the distribution curve. Since the median sizes of this particle reported by the manufacturer was 37 nm, estimated by surface area, the median size for the lower size peak (34.8 nm) shows a good agreement with the reported value. Hence, the median size for the higher size peak corresponds to the size of the agglomerates.

In order to verify the capability of ultrasonic attenuation spectroscopy for determination of PSD in the nano-size range, PSDs were measured by different techniques such as dynamic light scattering and surface area analysis. The

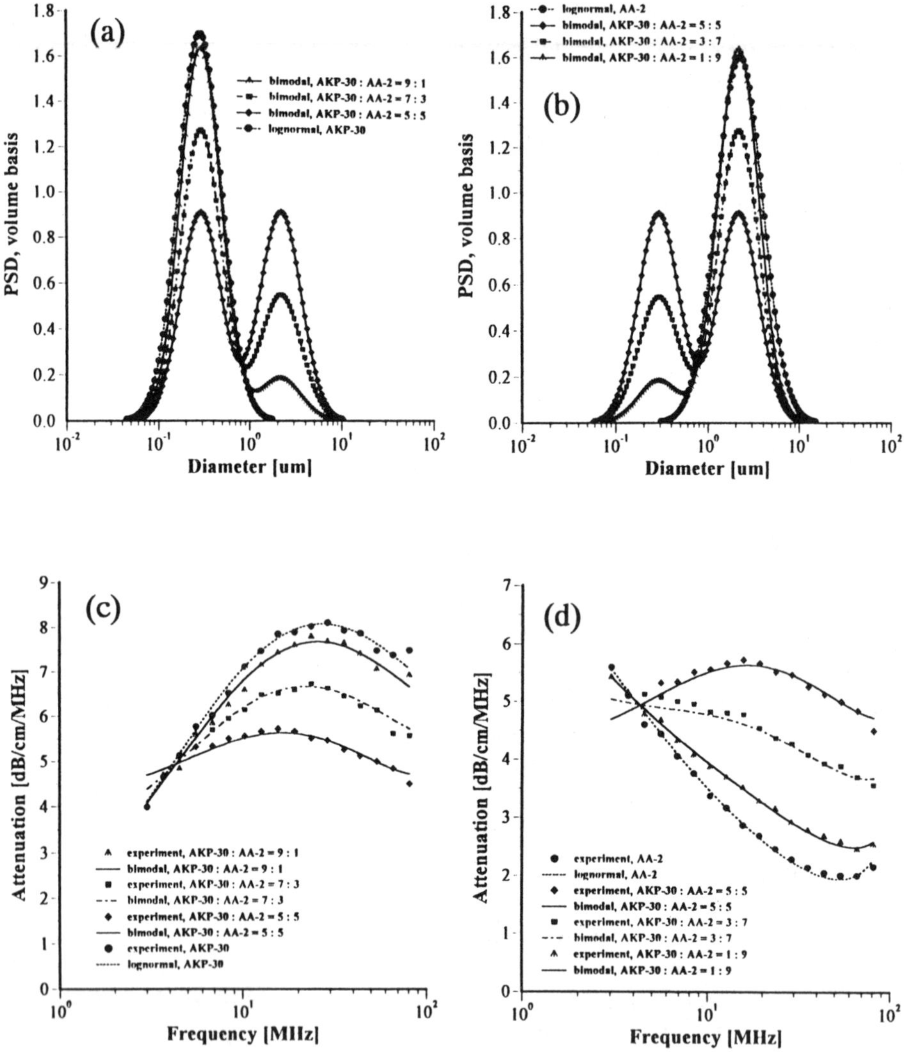

Figure 3. (a)-(b) Particle size distribution for well-dispersed 10 vol % alumina slurry of AA-2 and AKP-30, and for different mixtures of these two powders. (c)-(d) Experimental and theoretical attenuation spectra for well-dispersed 10 vol % alumina slurry of AA-2 and AKP-30, and for different mixtures of these two powders.

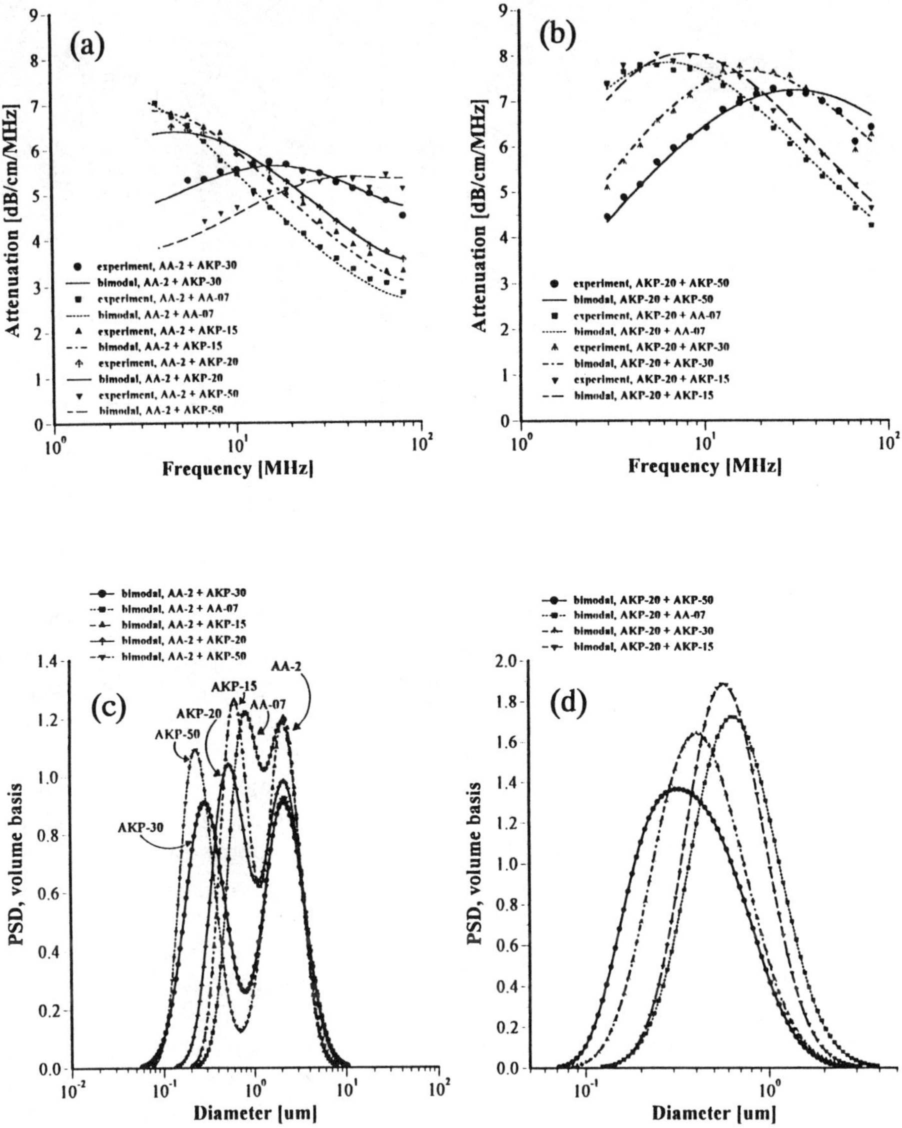

Figure 4. (a)-(b) Experimental and theoretical attenuation spectra for well-dispersed 10 vol % alumina slurry of AA-2, AA-07, AKP-15, AKP-20, AKP-30, and AKP-50, and for 50/50 mixtures of two powders. (c)-(d) Particle size distribution for well-dispersed 10 vol % alumina slurry of AA-2, AA-07, AKP-15, AKP-20, AKP-30, and AKP-50, and for 50/50 mixtures of these two powders.

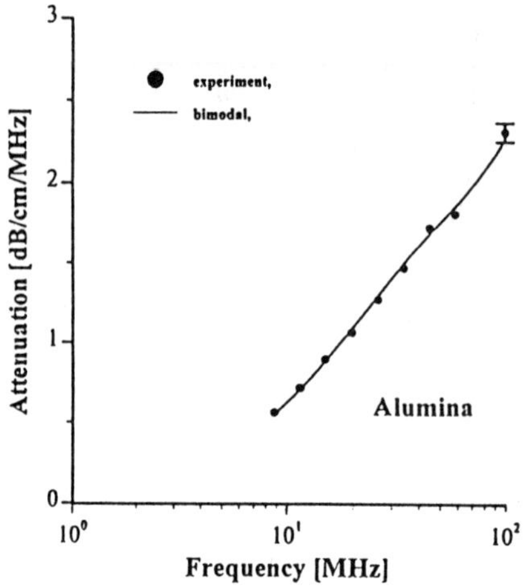

Figure 5. Experimental and theoretical attenuation spectra for alumina (Nano Tek), 5 vol %.

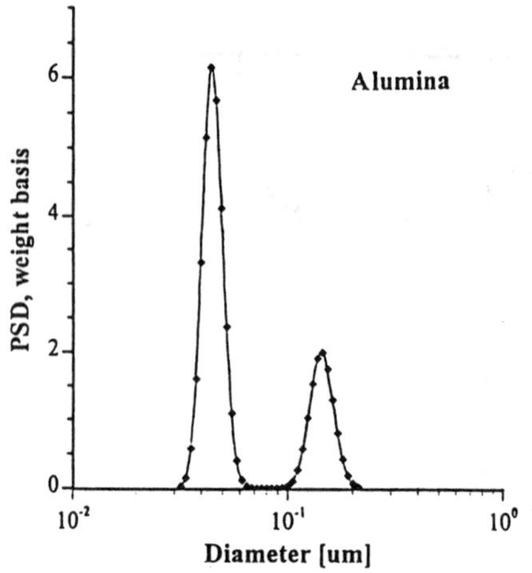

Figure 6. Particle size distribution for alumina (Nano Tek), 5 vol %.

obtained PSD by the light scattering technique also gives a bimodal distribution, but in this case, the volume fraction of the larger-sized particles (agglomerates) involved in the measured slurry was much higher than that measured by ultrasonic attenuation spectroscopy. In general, the intensity of scattered light coming from the larger-sized particle is much stronger than that coming from smaller ones even if the volume fractions of each size particles are the same. Hence, the measured intensity does not reflect the exact volume fraction of the larger-sized particles existing in the slurry. As a result, we cannot estimate how many larger-sized particles exist. On the other hand, the extent of attenuation also depends on the density, the concentration, and the size of the particles, but in this case the theory can account for each contribution to the measured attenuation. For this reason, we can compare only the sizes of the two peaks of the bimodal distribution, but not the volume fraction of them.

The median sizes in distribution curves obtained by light scattering were 34 nm for the lower size peak and 138 nm for higher one. On the other hand, the median sizes for the lower and higher size peak measured by the PK-8000 were 34.8 nm and 133 nm, respectively. Therefore, it is clear that these results show good agreement and the ultrasonic spectroscopy gives an adequate estimation of the particle size.

To further confirm the capability of ultrasonic spectroscopy to determine PSDs with sizes less than 100 nm, another size and density of nano-sized particles was used. Figures 7(a)-(c) show the measured attenuation spectra for Degussa alumina, Nano Tek anatase, and Nano Tek hematite. It can be seen that there are distinctive differences among the spectra. These differences reflect the differences in sample properties, such as particle size and density. Particle size distributions calculated for each sample are shown in Figures 8(a)-(c). A log-normal distribution provides a good fit to the measured attenuation spectra of Degussa alumina, and bimodal distributions provide a good fit to the others. The PSDs obtained from Nano Tek anatase and hematite are broader than that from Degussa alumina. The reason for this is thought to be as follows: the isoelectric point (IEP) of the alumina should be near pH 9. On the other hand the IEP of anatase and hematite is around pH 5. When the pH of the slurry is adjusted to pH values far from the IEP, the slurry should be well-dispersed. In this case, all slurries were adjusted to pH 4, and as a result the difference from the IEP was not sufficiently large to give the slurries a well-dispersed state. The median particle sizes are summarized in Table II. The data for Degussa alumina shows reasonable agreement of the median size, because the estimated value of average size reported by the manufacturer is based on TEM observation of primary particle size. The median sizes of anatase and hematite for lower size peaks in the PSD curve which were obtained by the PK-8000, also show good agreement with the primary particle size estimated by the manufacturer. But, those values demonstrate a large difference from that measured by the light scattering technique. This is also because the higher size peak in the PSD curve obtained by light scattering suggests that the slurry contains agglomerates to some extent even in the very diluted (25 ppm) system, and as a result, the intensity of the scattering was dominated by that coming from larger particles.

3.4 Effect of Dosage of Sodium Hexametaphosphate on PSD

To see the effect of the dosage of SHP on the PSD in slurries, ultrasonic attenuation spectra were measured at various concentrations of SHP. Test solutions containing from 0 to 2.0 mass% of SHP were prepared and then mixed with 10 vol% alumina powder.

It is evident from Figure 9 that there are distinct changes in attenuation spectra with SHP concentration, which suggests that the PSD is changed depending on the concentration of SHP. The log-normal distribution provides a good fit to the slurries with concentrations of 0 and 0.2 mass%, and the samples with higher concentrations of SHP (1.0 and 2.0 mass%) were fitted well using bimodal distributions. These differences imply the presence of agglomerated or flocculated particles in the slurries with concentrations of 1.0 and 2.0 mass%. The corresponding PSDs are shown in Figure 10. The PSD for slurries with concentrations of 0 and 0.2 mass% SHP show log-normal distributions, where the median size of the slurry with 0 mass% was 0.40 μm with a standard deviation

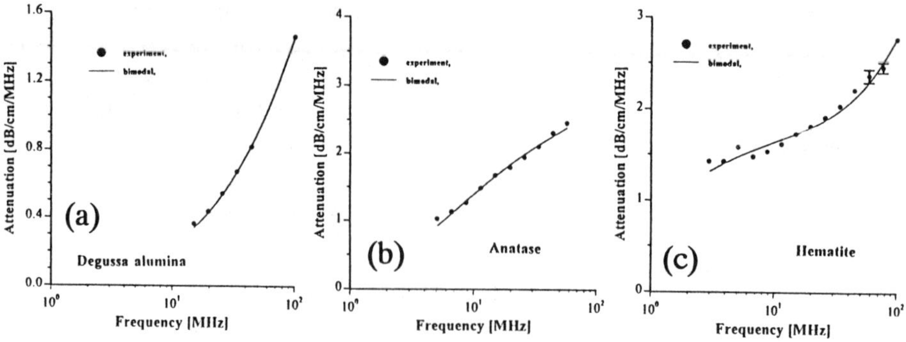

Figure 7. (a)-(c) Experimental and theoretical attenuation spectra for alumina (Degussa), anatase (Nano Tek), and hematite (Nano Tek), 5 vol%.

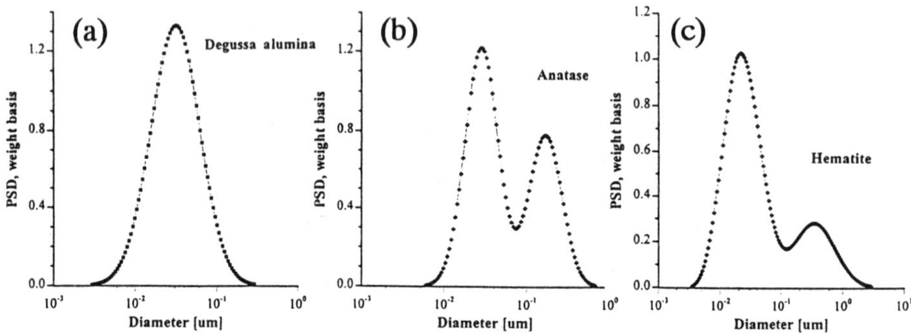

Figure 8. (a)-(c) Particle size distribution for alumina (Degussa), anatase (Nano Tek), and hematite (Nano Tek), 5 vol%.

Table II. Comparison of median size of particle distribution obtained by the PK-8000 and light scattering method for 5 vol % of alumina (Degussa), anatase (NanoTek), and hematite (NanoTek).

Determining Method	TEM (nm)	Surface Area (nm)	PK-8000 (nm)	Light Scattering (nm)
Alumina (Degussa)	13		30.5	
Anatase (NanoTek)		32	40.6 217.6	1172
Hematite (NanoTek)		23	20.4 347.1	444

(log[d_{16}/d_{50}]) of 0.545, and that with 0.2 mass% was 0.29 μm with a standard deviation of 0.235. The result obtained for the latter slurry suggests that the slurry was at the well-dispersed state. Since the median size measured by the light scattering method, as reported by the manufacturer, was 0.32 μm, this result shows a good coincidence. On the other hand, for the slurries with concentrations of 1.0 and 2.0 mass%, it is seen that the PSD curves have two peaks. The median size of the lower peak was 0.28 μm for both slurries and that of the higher one was 2.6 and 2.4 μm, respectively. These results demonstrate that these slurries contained some amount of agglomerates or flocculated particles. All these results demonstrate a capability to characterize the submicron-sized PSD in 10 vol% slurry systems without dilution.

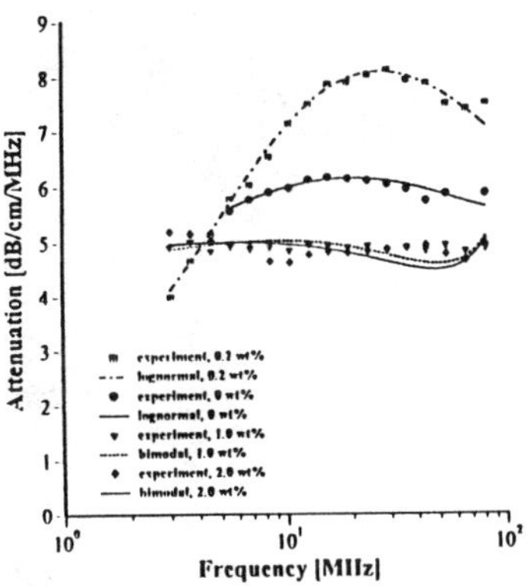

Figure 9. Experimental and theoretical attenuation spectra for 10 vol % alumina slurries with various concentrations of sodium hexametaphosphate (SHP). The test solutions contain from 0 to 2.0 mass % of SHP.

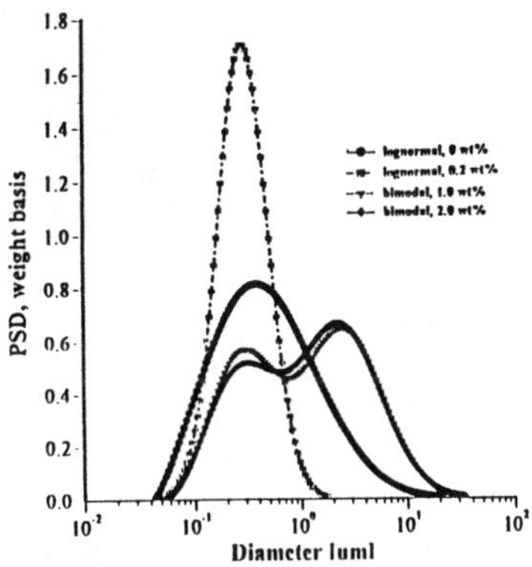

Figure 10. Particle size distribution for 10 vol % alumina slurries with various concentrations of sodium hexametaphosphate (SHP). The test solutions contain from 0 to 2.0 mass % of SHP.

It is well known that mixing procedures for powder and media affect slurry behavior. To confirm the sensitivity of this technique for monitoring the change in PSD caused by changes in the mixing procedure, a different procedure from that adopted for the above results was used. In the above measurements we added SHP to the test solution, but in this measurement we added SHP solution *after* mixing the powder with distilled water. In both cases, we applied ultrasonication subsequent to preparing the slurries. Figure 11 shows the calculated PSD. At a concentration of 2.0 mass%, two peaks appear and the height of the greater size one is much larger than that shown in Figure 10. This difference demonstrates the presence of a higher proportion of agglomerates in the slurry. It is clear that the mixing procedure affects the PSD. The slurry, and thus this technique, has the sensitivity to distinguish the difference in PSDs.

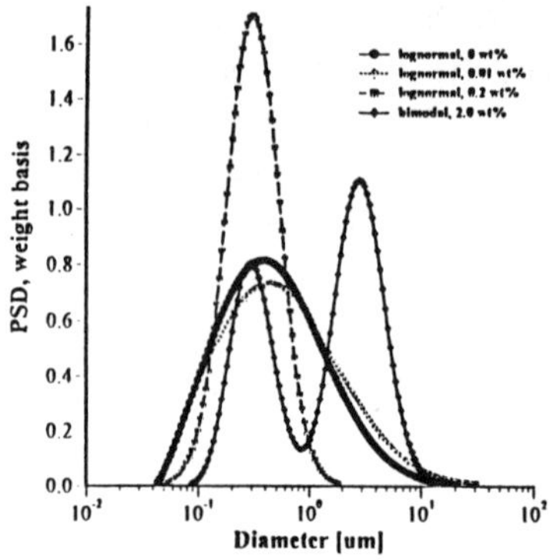

Figure 11. Particle size distribution for 10 vol% alumina slurries with various concentrations of sodium hexametaphosphate (SHP). The test solutions contain from 0 to 2.0 mass% of SHP. SHP solution was added to the slurry after mixing of powder and distilled water.

4. Conclusion

Particle size distributions in various kinds of slurries with 10 vol% solids concentration were determined by conversion of the measured ultrasonic attenuation spectra. All of the data were found to be almost identical to that measured independently by the light scattering method or that reported by the manufacturer. These agreements on PSD's demonstrate that ultrasonic attenuation spectroscopy makes it possible to characterize ceramic slurries without dilution.

Acknowledgment

The author gratefully acknowledges Sumitomo Chemical Co., Ltd. for supplying the Al_2O_3 powder.

References

1. F.F. Lange, "Processing-Related Fracture Origins: I, Observations in Sintered and Isostatically Hot-Pressed Al_2O_3/ZrO_2 Composites", *J. Am. Ceram. Soc.*, **66**, 396-98 (1983).

2. F.F. Lange and M. Metcalf, "Processing-Related Fracture Origins: II, Agglomerate Motion and Crack-like Internal Surfaces Caused by Differential Sintering", *J. Am. Ceram. Soc.*, **66**, 398-406 (1983).

3. F.F. Lange, B.I. Davis and I.A. Aksay, "Processing- Related Fracture Origins: III, Differential Sintering of ZrO_2 Agglomerates in Al_2O_3/ZrO_2 Composite", *J. Am. Ceram. Soc.*, **66**, 407-08 (1983).

4. I.A. Aksay, F.F. Lange and B.I. Davis, "Uniformity of Al_2O_3/ZrO_2 Composites by Colloidal Filtration", *J. Am. Ceram. Soc.*, **66**, C-190 92 (1983).

5.V.A. Hackley and S.G. Malghan, "Investigation of Parameters and Secondary Components Affecting the Electroacoustic Analysis of Silicon Nitride Powders"; pp.161-79 in *Electroacoustics for Characterization of Particulates and Suspensions*, Edited by S.G.Malghan, National Institute of Standards and Technology Special Publication 856, US Department of Commerce (1993).

6. J. Cesarano III, I.A. Aksay and A. Bleier, "Stability of Aqueous α-Al_2O_3 Suspensions with Poly(methylacrylic acid) Polyelectrolyte", *J. Am. Ceram. Soc.*, **71**, 250-55 (1988).

7. T.A. Strout, "Attenuation of Sound in High-Concentration Suspensions: Development and Application of an Oscillatory Cell Model", PhD Thesis, The University of Maine, 1991.

8. A.K. Holmes, R.E. Challis and D.J. Wedlock, "A Wide-Bandwidth Ultrasonic Study of Suspensions: The Variation of Velocity and Attenuation with Particle Size", *J. Colloid Interface Sci.*, **168**, 339-48 (1994).

9. A.S. Dukhin and P.J. Goetz, "Acoustic Spectroscopy for Concentrated Polydisperse Colloids with High Density Contrast", *Langmuir*, 12, 4987-4997 (1996).

10. J.D. McClements, "Ultrasonic Determination of Depletion Flocculation in Oil-in-Water Emulsions Containing a Non-Ionic Surfactant", *Colloids Surf.*, **90**, 25-35 (1994).

Effect of Polyacrylate Adsorption on ESA of Aqueous Dispersions of Phosphors

Arunava Dutta and Leonard V. Dullea
OSRAM SYLVANIA Development Inc., Research & Development, 75 Sylvan Street, Danvers, MA 01923 USA

Abstract

This investigation concerns the adsorption of a low molecular weight polyelectrolyte on the surface of different phosphor types. The profiles of ESA (electrokinetic sonic amplitude), pH and conductivity of the phosphor dispersions are monitored as a function of the concentration of the ammonium polycarboxylate. The viscosity profile is also measured. In addition the intrinsic IEP (isoelectric point) of each phosphor, obtained by potentiometric titration, is reported. The information is interpreted to point out characteristics of any particular phosphor type and identify differences, if any, between phosphors belonging to the same type.

1. Introduction

Anionic dispersants are often added as a component in the formulation of aqueous coating suspensions of inorganic solids in order to achieve dispersion stability. This practice is prevalent in a wide variety of industries like ceramics, paper, and pigment. These dispersants are often sodium, potassium, or ammonium salts of polyacrylic acid or its derivatives.

Polyacrylates have a hydrocarbon backbone with alternate C atoms having branched sp^2 planar active COO^- groups.[1] In general, the spacing between the carboxylate groups that can interact with the solid surface is about twice that of polyphosphate based dispersants. The alternate COO^- groups adsorb on the positively charged sites of the solid surface while the intermediate carboxylate groups extend into the aqueous phase thereby providing molecular charge and a double layer.[1] This is how the polyacrylates function as dispersants. By preventing or minimizing particle agglomeration or flocculation, relatively stable phosphor coating suspensions can be made using these polyacrylate based dispersants.

The phosphors used in this study are photoluminescent inorganic materials that emit in the visible spectrum when subjected to a flux of UV photons, primarily of 254 nm wavelength. Depending on the matrix and the activator ions, a phosphor can be made to emit different wavelengths of visible light. The primary application of the phosphors in this study is in low pressure mercury vapor lamps.

Electrokinetic studies on phosphors were conducted using a Matec ESA-8000 system. In this apparatus[2] a high frequency electric field is applied to a phosphor dispersion between two parallel electrodes causing the phosphor particles to oscillate producing a coherent sound wave. ESA, or electrokinetic sonic amplitude, is the pressure amplitude of the sound wave per unit of applied electric field. The units of ESA are mPa.m/V. It has been shown[3] that ESA is proportional to the zeta potential of the particle which in turn is a measure of the stability of the material to intraparticle and/or interparticle flocculation. This study focuses on ESA rather than the zeta potential because of the uncertainty surrounding the applicability to phosphors of the equation used to compute zeta potential from the ESA.[4]

2. Experiment

2.1 Materials

Six phosphor families from three different manufacturers were investigated: $LaPO_4$:Ce,Tb (emits in the green), Y_2O_3: Eu (emits in the red), (Ce, Tb)$MgAl_{11}O_{19}$:Ce,Tb (emits in the green), $Ca_5(F,Cl)(PO_4)_3$:Sb:Mn (broad band white emitter), Zn_2SiO_4: Mn (green emitter) and $Sr_2P_2O_7$:Eu (emits in the blue). A total of eleven phosphors belonging to the six families comprised this study. The phosphors are assigned random arbitrary numbers by the authors, starting from unity, and reference is not made to them using the manufacturers' part numbers. While this makes it impossible to know which phosphor came from which manufacturer, it does not detract from the research objectives of this paper. The particle size distribution of the phosphors, measured using a Coulter LS130 laser diffraction unit, is shown in Table I. The phosphor samples were externally sonified for 5 minutes prior to the run using a Fisher Scientific Ultrasonic FS-9 unit and during sample loading and for 1 minute before each run. No surfactants were used and the obscuration target was 10%. Each of the eleven phosphors was subjected to the same sample preparation technique and the size was measured three times to ensure reproducibility.

The dispersant used is commercially available as Dispex A40 from Allied Colloids. The manufacturer describes it as a low molecular weight ammonium salt of a polycarboxylic acid. It has 40% active material and a density of 1160 kg/m^3. A 1% Dispex solution (1cm^3 of Dispex made up to 100 cm^3 of solution) is used

Table I. Particle size distribution of phosphors.

Coulter LS130 Laser Diffraction Particle Sizing System

Phosphor Type	Phosphor #	10%>	25%>	50%	75%>	90%>	S.A.
Green							
LaPO4:Ce,Tb	1	23.470	13.130	8.220	4.442	0.498	0.6879
LaPO4:Ce,Tb	2	6.670	5.040	3.462	2.080	0.316	1.0786
Red							
Y2O3:Eu	3	9.557	7.201	4.670	2.501	0.326	1.0273
Y2O3:Eu	4	12.290	9.832	6.974	4.512	2.799	0.4493
Y2O3:Eu	5	9.331	7.559	5.351	3.291	0.428	0.8229
Green							
(Ce,Tb)MgAl11O19:Ce, Tb	6	10.220	7.940	5.326	2.907	0.332	1.2430
(Ce,Tb)MgAl11O19:Ce, Tb	7	15.780	11.090	6.694	3.403	0.998	0.7153
Cool white							
Ca5(F,Cl)(PO4)3:Sb:Mn	8	24.270	18.170	12.330	6.951	0.849	0.8397
Ca5(F,Cl)(PO4)3:Sb:Mn	9	18.510	14.720	10.540	6.241	0.649	0.9661
Green							
Zn2SiO4:Mn	11	18.380	14.170	9.625	5.132	0.555	0.8067
Blue							
Sr2P2O7:Eu	12	21.300	15.080	9.872	5.181	0.587	0.8619

Note: Particle size micrometers

in all the experiments. This works out to 4.64 mg of active Dispex/cm^3 of titrant.

2.2 Methods

The following method was used to obtain the intrinsic IEP of each phosphor. A 6 vol % dispersion of the phosphor in 0.2 L DI water was made in a Matec ESA 8000 sample cell and the suspension allowed to reach a stable temperature of 298K. Next, the contents of the cell were allowed to age for a period of 48 to 72 hours depending on the sample, till stable values of both ESA and pH were attained. A phase calibration was then performed. Finally, a potentiometric titration was conducted using either 0.1N HNO_3 or 0.1N NH_4OH as appropriate. The ESA vs pH profile and hence the IEP of the phosphor (where ESA reached zero) were noted.

The concentration series using Dispex was conducted in the following manner. A 6 vol % dispersion of the phosphor in 0.2 L DI water was made in a sample cell and the suspension allowed to reach a stable temperature of 298K. Next, the contents of the cell were allowed to age for a period of 48 to 72 hours, depending on the sample, till both the ESA and pH had stabilized. A phase calibration was then performed. Next, the suspension was titrated with a 1% Dispex solution and the ESA, pH, and conductivity were recorded.

The rheology of the phosphor dispersion as a function of the polyacrylate concentration was studied using a Brookfield DVII digital viscosimeter using a temperature controlled (298K) small sample adapter and using a shear rate varying from 0.4 s^{-1} to 39.6 s^{-1}. Here too, as in the Matec experiments, the phosphor suspension was stabilized before any Dispex was added.

Since the phosphors used have appreciable material densities, some as high as 5000 kg/m^3, the phosphor particles tend to settle out in the viscosimeter sample cell. This problem was eliminated by adding 2% Polyox, a commercially available poly(ethylene oxide) from Union Carbide, which increased the viscosity of the medium and decreased the settling of the dense powders. The use of Polyox is supported by the fact that no detectable rheological interactions between Dispex and Polyox were found. There is negligible change in viscosity of a 2% Polyox solution upon addition of Dispex over the entire range of shear rates investigated and up to the maximum Dispex concentration that was used in the ESA studies.

3. Results

3.1 $LaPO_4$:Ce,Tb Phosphor Family

Two phosphors were used in this study. It is seen from Figure 1 that both the phosphors have an identical IEP of 7. This points to an almost neutral surface for this phosphor type as far as interactions of hydroxyl and hydrogen ions with the surface are concerned. The reason for using ESA rather than computing the zeta potential lies mainly in the uncertainty of the equations relating ESA to the zeta potential when broad particle size distributions are present as is the case for most phosphors.[4]

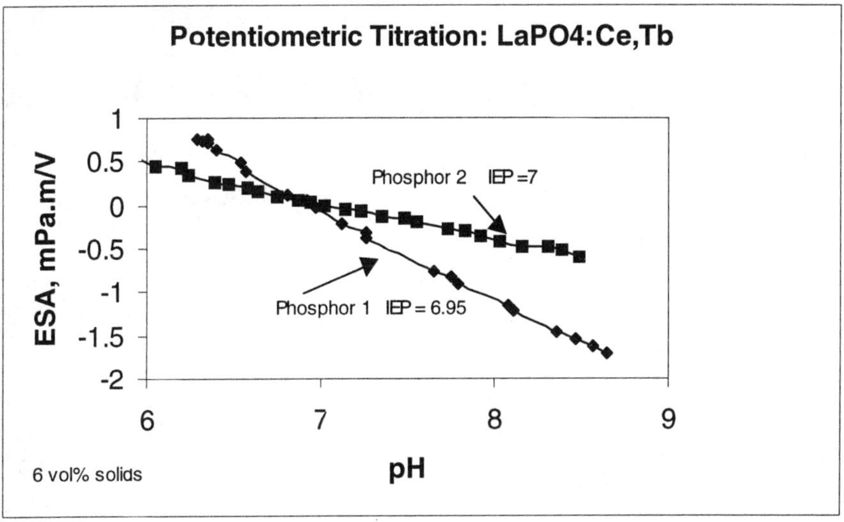

Figure 1. Potentiometric titration of lanthanum phosphate phosphors.

Although the IEP for the two phosphors is identical, the slope of the ESA vs pH profile for phosphor 1 is about twice that of phosphor 2. A review of the particle size distribution of these two phosphors from Table I indicates that phosphor 1 is much larger in size than phosphor 2. Hence, the inertial correction factor for 1 should be larger than that for 2 which should reduce the magnitude of the ESA for the former. The fact that phosphor 1 still has a higher ESA magnitude than phosphor 2 is attributed, therefore, to an intrinsically higher zeta potential for phosphor 1. It is also noted that phosphor 1 would have an overall smaller surface area than 2 which would mean fewer active sites for interaction with the hydroxyl and hydrogen ions and the need for less titrant.

The Dispex concentration series is shown in Figure 2. For both phosphors, which start out positive initially, adsorption of Dispex rapidly reduces the positive surface charge and reduces it to zero at a pH which is lower than the IEP of this phosphor type. Further adsorption of Dispex drives the phosphor negative until an equilibrium value of negative ESA is reached. By this time, the pH has also substantially reached a plateau. This point in the ESA profile reflects a maximum in Dispex adsorption and corresponds to some sort of a fractional coverage of the phosphor surface by the polyacrylate anions which are the anchoring groups for the adsorption process.

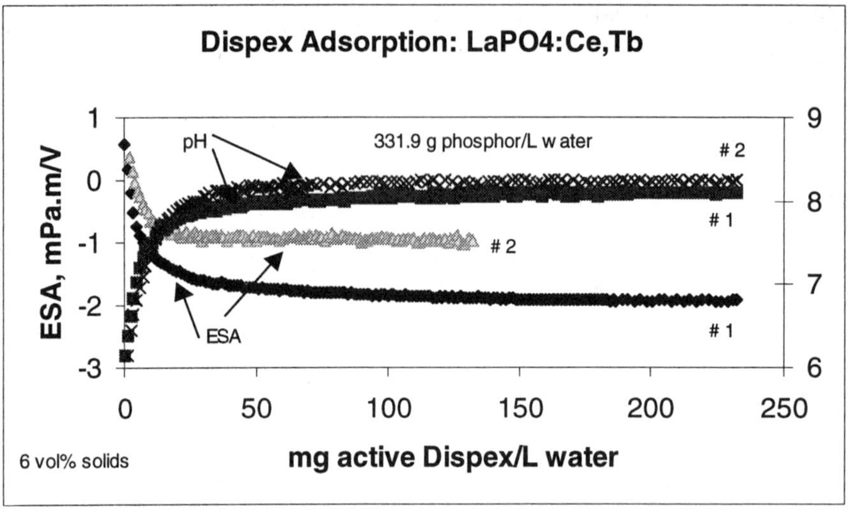

Figure 2. Dispex adsorption on lanthanum phosphate phosphors.

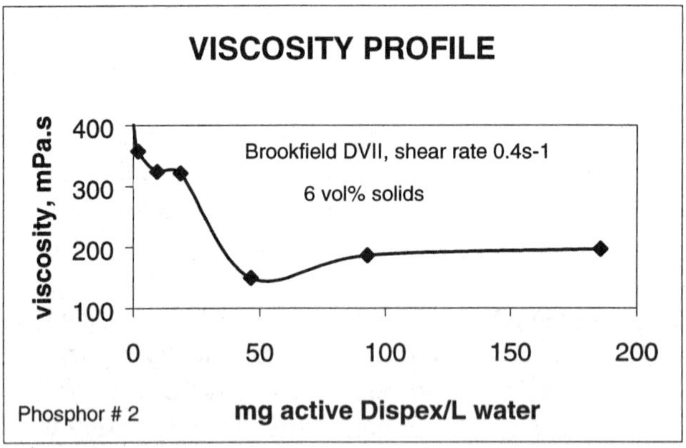

Figure 3. Viscosity profile-Dispex adsorption on lanthanum phosphate phosphor.

Further additions of the polyacrylate make little change in ESA and pH of this system, but can be detected in a progressive rise in conductivity of the contents of the cell. While the conductivity before Dispex addition is about 15 μS/cm, it reaches about 105 μS/cm at the maximum addition level of Dispex. The equilibrium ESA for phosphor 1 is about twice that for phosphor 2. This is again attributed to an intrinsically higher zeta potential for phosphor 1.

The viscosity profile for one of the phosphors is shown in Figure 3. The viscosity drops rapidly as Dispex is added. The decrease in viscosity is explained by the increase in charge on the particle surface as more Dispex is adsorbed which leads to a progressively more stabilized dispersion since the volume fraction of floccs is reduced. A minimum in viscosity, corresponding to a stable dispersion, is reached at a polyacrylate concentration higher, by about a factor of two, than that at which ESA is stabilized for this phosphor. Beyond this level of Dispex, the viscosity increases quickly but remains substantially flat thereafter.

3.2 Y2O3: Eu Phosphor Family

The intrinsic ESA vs pH profiles for phosphors from three different manufacturers are shown in Figure 4. It is evident from the wide variation in IEP values that a substantial variation in surface chemistry prevails among the three phosphors. The IEP ranges from a low of 7 (neutral surface) to a basic 8.4. Due to the differences in surface chemistry, the titrations had to be done with acid for phosphor 3 and with base for phosphors 4 and 5. The slope of the profiles is substantially similar for phosphors 3 and 5 which differ in IEP by about 1.4 pH units.

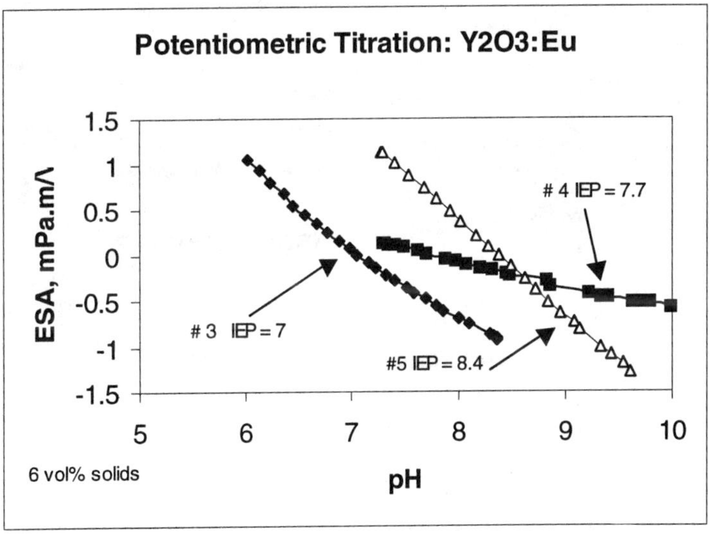

Figure 4. Potentiometric titration of yttrium oxide phosphors.

It is seen from Table I that these phosphors also have very comparable particle size distributions. The slope for phosphor 4 is very shallow and about a quarter of that for the other two: phosphor 4 is also significantly larger in size than the other two, which leads to a larger inertial correction factor than that for the other phosphors. Also, the smaller surface area of phosphor 4 would imply fewer active sites for interaction with H^+ and OH^- ions and the need for less titrant.

The Dispex concentration series is shown in Figure 5. While phosphor 5 is markedly positive before the onset of polyacrylate adsorption, phosphor 4 is much less positive while phosphor 3 is negative to start with. This is consistent with the IEP and starting pH levels for these phosphors. It follows that neutralization of surface charge is not applicable to phosphor 3 while adsorption of Dispex rapidly neutralizes phosphors 4 and 5 and then drives them negative till an equilibrium value of negative ESA is reached. As stated before, this point corresponds to some sort of a fractional coverage of the phosphor surface by the polyacrylate anions. The equilibrium ESA at this maximum Dispex adsorption point is comparable for phosphors 3 and 5 and about twice that for phosphor 4 which is detectably larger in size compared to the other two phosphors.

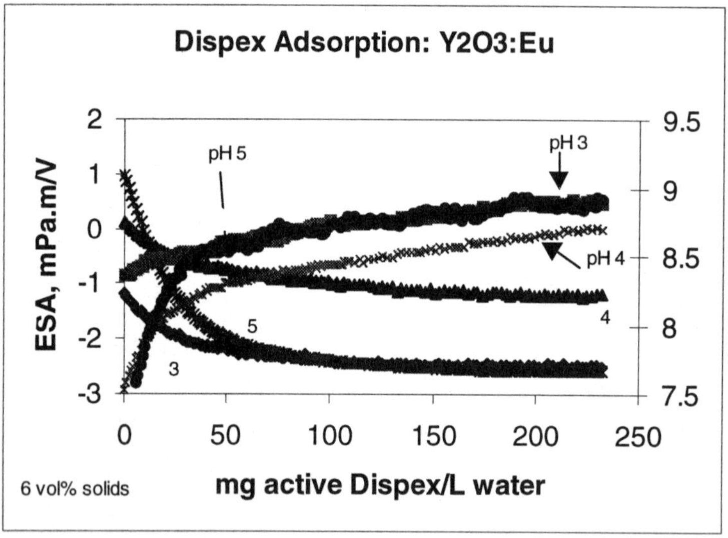

Figure 5. Dispex adsorption on yttrium oxide phosphors.

Further additions of Dispex leads to little change in ESA. There is a small but detectable increase in pH, and a progressive increase in the conductivity of the suspension. While phosphors 4 and 5 have comparable initial and final conductivities, about 25 μS/cm and 110 μS/cm respectively, phosphor 3 has a detectably higher initial and final conductivity of about 75 μS/cm and 165 μS/cm respectively.

The viscosity profile for phosphor 4 is shown in Figure 6. The viscosity drops rapidly as Dispex is added and a minimum is reached at a polyacrylate concentration comparable to that at which the ESA is subtantially stabilized for this phosphor. Beyond this level of Dispex, the viscosity remains unchanged and there is no detectable increase.

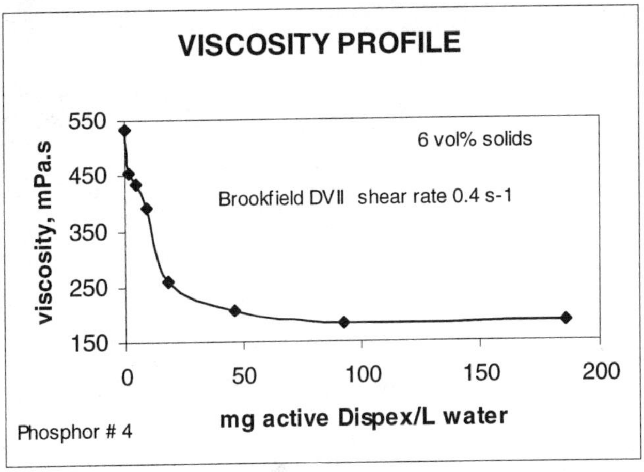

Figure 6. Viscosity profile-Dispex adsorption on yttrium oxide phosphor.

3.3 (Ce, Tb)$MgAl_{11}O_{19}$:Ce,Tb Phosphor Family

The intrinsic ESA vs pH profiles for phosphors from two different manufacturers are shown in Figure 7. The IEP of the two phosphors is very comparable, 8.35 vs 8.7, and reflects a basic surface chemistry. While phosphor 6 is smaller in size than phosphor 7, the slope of the two profiles is comparable for pH < IEP.

Regarding the Dispex concentration series, it is seen from Figure 8 that both the phosphors are distinctly positive to start with, and addition of Dispex rapidly neutralizes the phosphors and drives them negative until an equilibrium value of negative ESA is attained (beyond this point there is little change in pH too) at a Dispex concentration of about 100 mg active Dispex/L water. The equilibrium ESA corresponding to maximum polyacrylate adsorption on the phosphor surface is larger for phosphor 6 which is a smaller size than phosphor 7. Both the phosphors have comparable initial and final conductivities of about 25 μS/cm and 110 μS/cm respectively.

It is observed from Figure 9 that the viscosity of a suspension of phosphor 6 rapidly drops in viscosity upon addition of Dispex and that a minimum is reached at a Dispex concentration that is about half of that at which the ESA stabilizes for this phosphor. Beyond the minimum in viscosity, there is a very small gradual increase in viscosity as the Dispex concentration is increased.

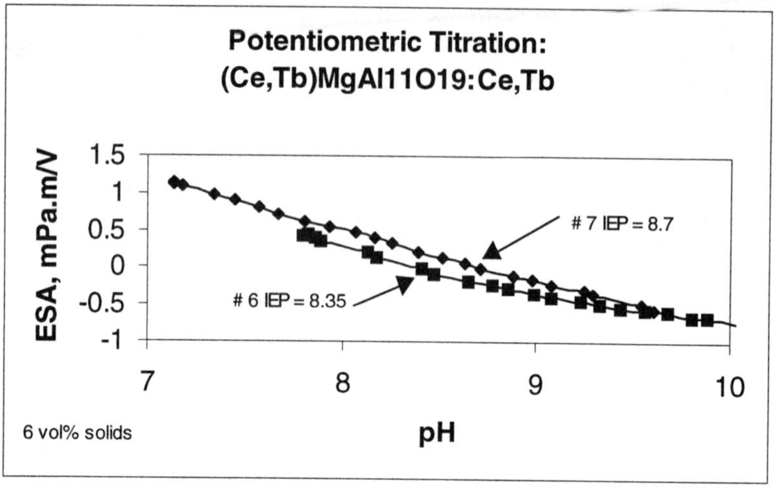

Figure 7. Potentiometric titration of Ce,Tb magnesium aluminate phosphor.

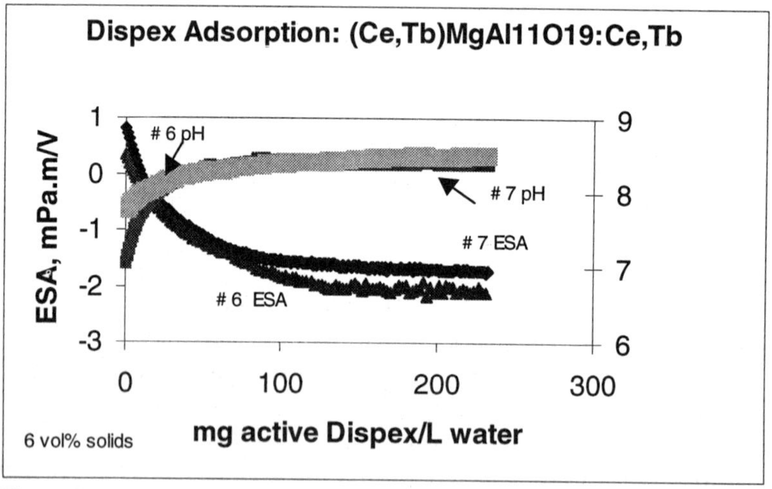

Figure 8. Dispex adsorption on Ce,Tb magnesium aluminate phosphor.

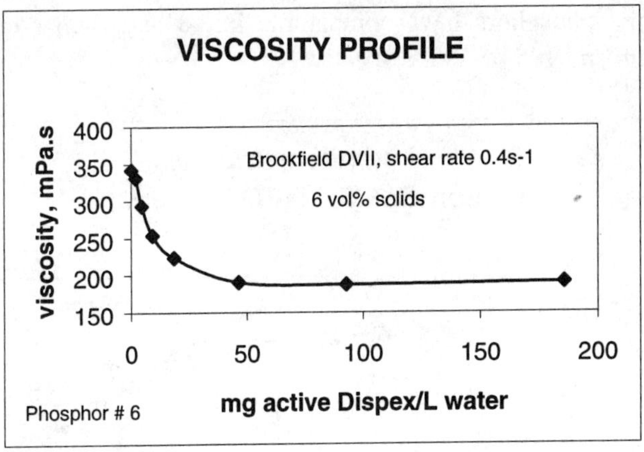

Figure 9. Viscosity profile: Dispex adsorption on Ce,Tb Mg aluminate phosphor.

3.4 $Ca_5(F,Cl)(PO_4)_3$:Sb:Mn Phosphor Family

The intrinsic ESA vs pH profiles for phosphors from two different manufacturers are shown in Figure 10. This phosphor type is acidic: the IEP of the two phosphors is identical: 5. While the potentiometric titration profiles are very similar for pH > 5, they differ greatly at lower pH values. In particular, the slope of ESA vs pH for phosphor 9 becomes much shallower compared to phosphor 8. This is attributed to higher dissolution in the case of phosphor 9 which is a smaller particle size material than phosphor 8 thereby providing a larger surface area to the liquid phase for reactions that enhance solubility. The formation of a larger concentration of various hydroxy phosphates of Ca in the liquid phase in the case of phosphor 9 which can potentially reprecipitate on the phosphor surface could adversely affect the adsorption of H^+ ions on the surface of the phosphor, stemming thereby an increase in magnitude of positive ESA as pH is lowered.

Both the phosphors are negatively charged (excess of negative sites over positive sites) at the beginning of the Dispex titration. This is consistent with the IEP and the starting pH levels for these phosphors. While no neutralization of surface charge due to Dispex adsorption is possible here, it is seen from Figure 11 that Dispex does increase the negative ESA for this phosphor till an equilibrium ESA is reached. It is believed that the Dispex is adsorbing on the positive sites of the phosphor and not forming a complex with neutral metal sites. The equilibrium ESA is almost the same for the two phosphors and is reached at almost identical levels of Dispex: about 50 mg active Dispex/L water. The pH does not, however, show signs of stabilization at this point and continues to increase with Dispex levels until a pH equilibrium is reached at Dispex levels of about 150 mg/ L water. Both the phosphors have comparable initial and final conductivities of about 15 μS/cm and 105 μS/cm, respectively.

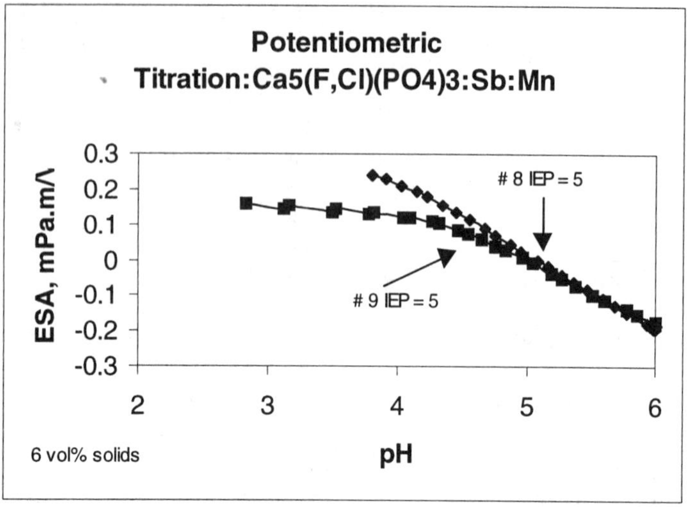

Figure 10. Potentiometric titration of Ca halophosphate phosphor.

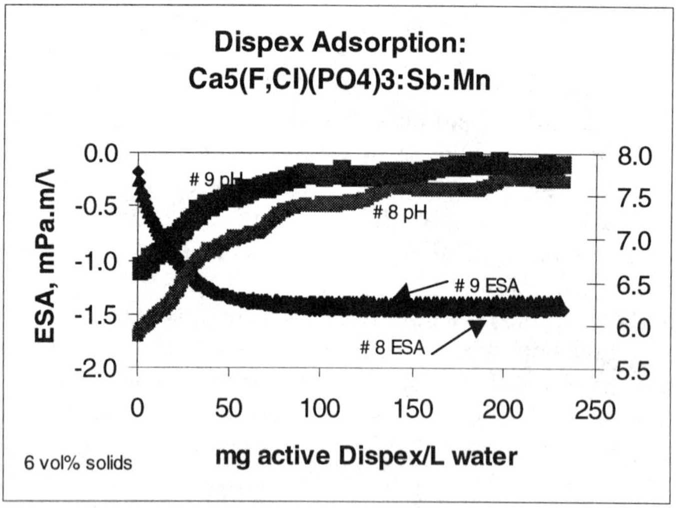

Figure 11. Dispex adsorption profile: Ca halophosphate phosphor.

Figure 12 shows the viscosity profile for phosphor 9. While a minimum in viscosity is observed at a Dispex concentration very similar to that at which the ESA stabilized for this phosphor, further addition of Dispex leads to a detectable increase in viscosity followed by a decline as the polyacrylate level is increased further. The viscosity at the highest concentration of Dispex is comparable to, if not lower than, the minimum observed at around 50 mg Dispex/ L water.

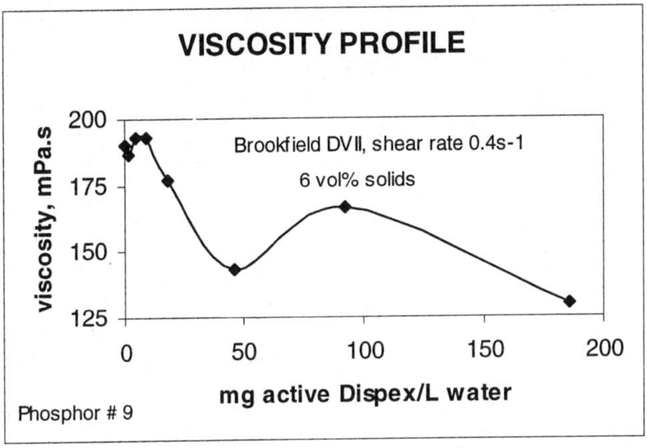

Figure 12. Viscosity profile-Dispex adsorption on Ca halophosphate phosphor.

3.5 Zn_2SiO_4: Mn Phosphor Family

The ESA vs pH profile is seen in Figure 13. The phosphor is initially negative to start with at pH of 7.5 and an acid titration is needed to decrease the magnitude of the negative ESA. While the ESA decreases progressively until a pH of about 6, there is very little change as further acid is added and the pH lowered. No IEP could, therefore, be determined for this phosphor within the conditions of this study. It can, however, be said that this is a highly acidic phosphor. The plateau like profile of the ESA at low pH is attributed to the dissolution of this phosphor that increases as the pH is lowered. In particular, various aqua metal cations of Zn are formed which can potentially interfere with interaction of H^+ with the phosphor surface thereby stemming the progress of the ESA towards a zero value.

Adsorption of Dispex on the phosphor surface is seen in Figure 14. Since the surface is negative to start with (excess of negative sites over positive sites) , a fact consistent with the IEP and starting pH for this phosphor, further adsorption of Dispex drives the surface more negative until an equilibrium ESA is reached at a Dispex concentration of about 150 mg/L water. It is believed that the Dispex is adsorbing on the positive sites of the phosphor and not forming a complex with neutral metal sites. The pH also starts to level out at about this concentration of the polyacrylate. The initial and final conductivity are about 30 and 120 μS/cm, respectively.

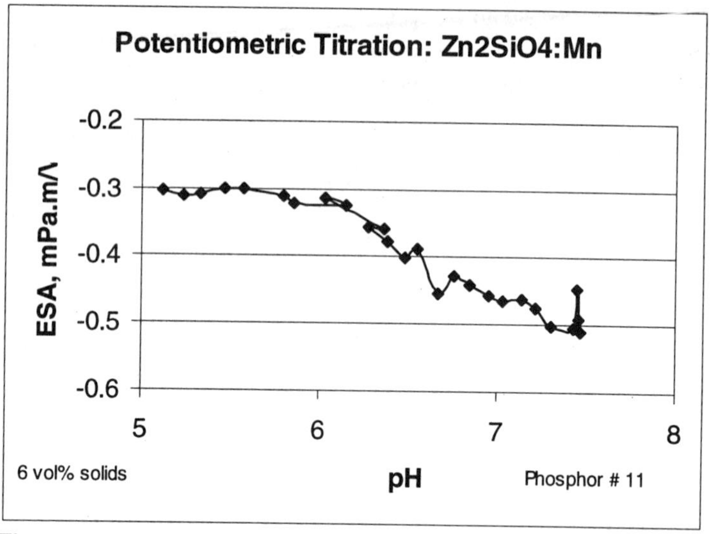

Figure 13. Potentiometric titration of zinc silicate phosphor.

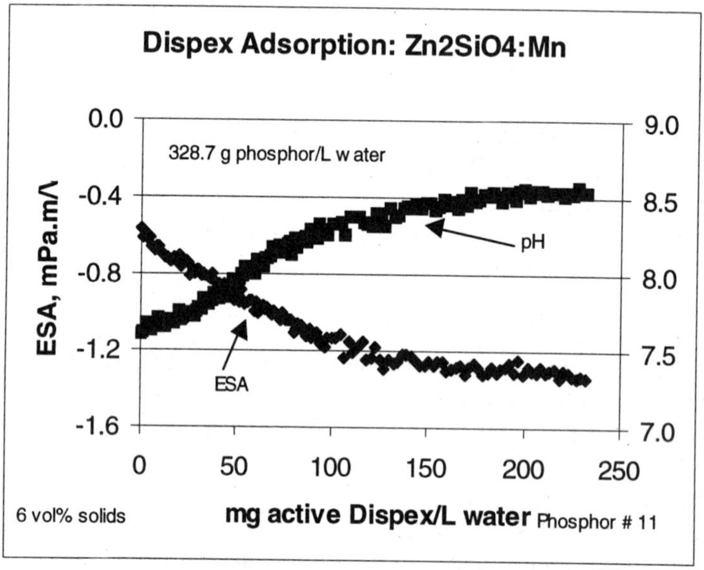

Figure 14. Dispex adsorption profile- zinc silicate phosphor.

The viscosity profile of this phosphor suspension is seen in Figure 15. A minimum in viscosity is attained at a Dispex concentration of about a third of that at which the ESA stabilizes. There is also a detectable increase in viscosity as the Dispex concentration is increased beyond the point of minimum viscosity.

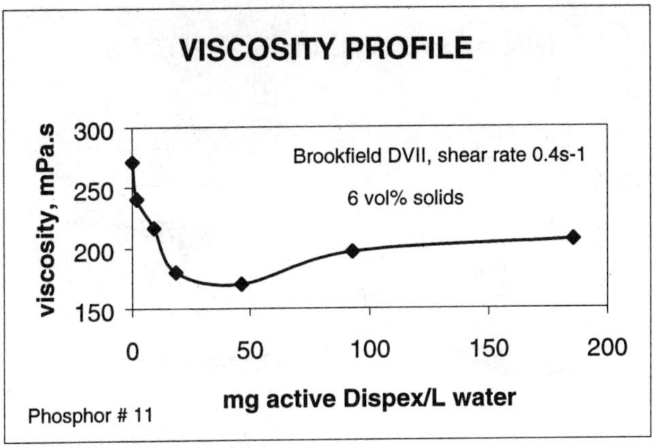

Figure 15. Viscosity profile-Dispex adsorption on zinc silicate phosphor.

3.6 $Sr_2P_2O_7$:Eu Phosphor Family

The intrinsic ESA vs pH profile for this phosphor is seen in Figure 16. This is a very acidic phosphor with an IEP of about 2.5. The potentiometric titration is done with acid in this case and for pH < IEP, the polarity of the ESA flips back to negative instead of crossing over to the positive territory. While the exact mechanism for this is not known presently, it is hypothesized that the formation of various phosphate anions in the liquid phase at low pH counteracts the H^+ ion interaction with the phosphor surface, leading to a flip back to negative polarity.

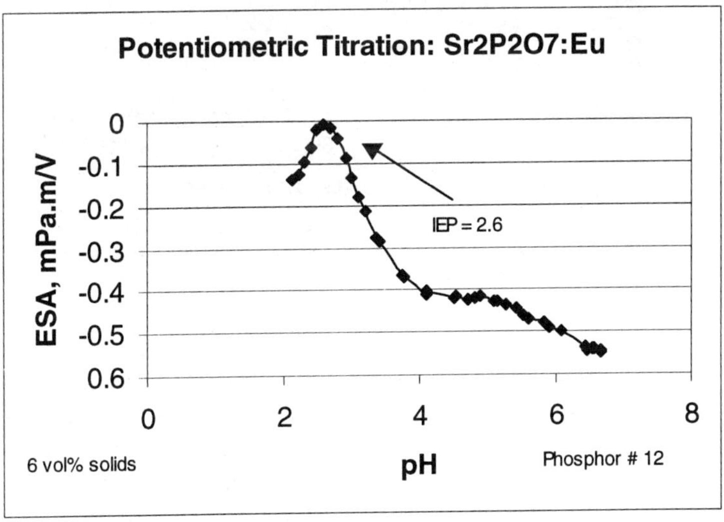

Figure 16. Potentiometric titration of Sr pyrophosphate phosphor.

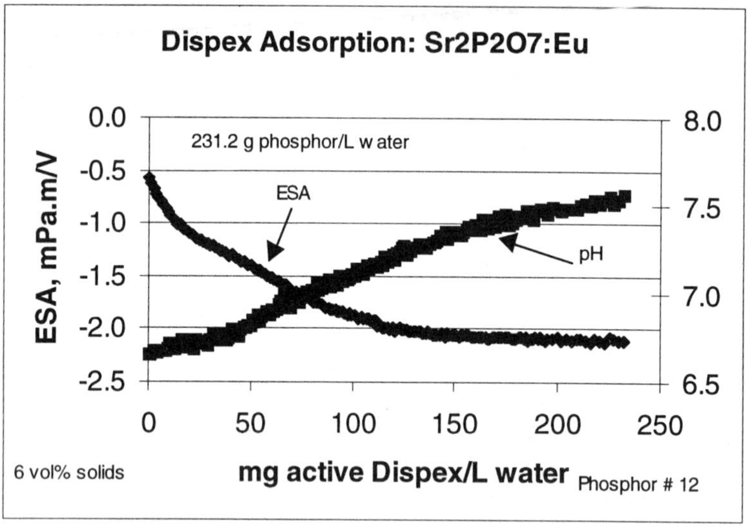

Figure 17. Dispex adsorption profile on Sr pyrophosphate phosphor.

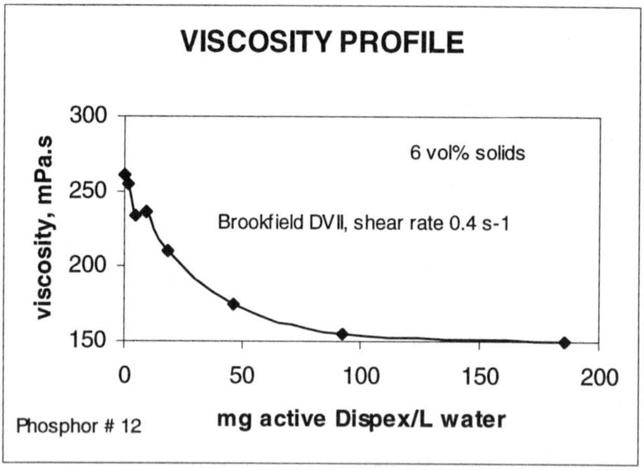

Figure 18. Viscosity profile-Dispex adsorption on Sr pyrophosphate phosphor .

This phosphor is negatively charged before the onset of Dispex addition. Addition of Dispex, Figure 17, drives the ESA more negative until an equilibrium ESA is reached at a Dispex level of about 150 mg/L water. The pH continues to increase and reaches a stable value at a Dispex concentration of about 200 mg/L water. The initial and final conductivity for the Dispex concentration series is about 30 and 120 μS/cm, respectively.

The viscosity of this phosphor suspension is shown in Figure 18. A minimum in viscosity is attained at a Dispex concentration very similar to that at which the ESA stabilizes for this phosphor. No increase in viscosity beyond the minimum is observed within the experimental conditions of this investigation.

4. Conclusions

This investigation shows the use of ESA measurements to characterize the surface chemistry of various phosphors in aqueous dispersions. The phosphors vary widely in intrinsic surface chemistry and it is shown how well the ESA technique can differentiate between the different phosphor systems. The IEP values range from a low of about 2.5 for $Sr_2P_2O_7$:Eu to about 8.7 for (Ce, Tb)$MgAl_{11}O_{19}$:Ce,Tb. Depending on the IEP of each phosphor and the starting pH level, the phosphor can either be positively or negatively charged before the onset of polyacrylate adsorption. If the phosphor is positive to start with, adsorption of Dispex rapidly neutralizes the phosphor surface. In either case, an equilibrium ESA is reached at a certain Dispex level. This point in the ESA profile corresponds to maximum adsorption of Dispex on the phosphor surface and reflects some sort of fractional coverage of the phosphor surface by the polyacrylate anions which are the anchoring groups for the adsorption process. Differences are observed between phosphors made by different manufacturers regarding both the intrinsic IEP and the Dispex adsorption profiles. The viscosity of the phosphor suspension drops rapidly on addition of Dispex, and in all cases a minimum is reached. The decrease in viscosity is explained by the increase in charge on the particle surface as more Dispex is adsorbed which leads to a progressively more stabilized dispersion since the volume fraction of floccs is reduced. The minimum in viscosity corresponds to a stable dispersion. For some phosphors, there is a small but detectable increase in viscosity as more Dispex is added. Depending on the phosphor type, the Dispex concentration at which the minimum in viscosity is attained is lower than, comparable to, or higher than the polyacrylate concentration at which an equilibrium ESA is reached.

References

1. R.F. Conley, *Practical Dispersions: A Guide to Understanding and Formulating Slurries*, VCH Publishers, Inc., New York, 1996.

2. Matec Applied Sciences Inc., Hopkinton, MA, ESA-8000 System Manual, 1992.

3. R.W. O'Brien, "Electro-Acoustic Effects in a Dilute Suspension of Spherical Particles", *J. Fluid Mech.*, **190**, 71 (1988).

4. A. Dutta, "Electrokinetics of Phosphors", pp. 274-285 in *Electroacoustics for Characterization of Particulates and Suspensions*, NIST Special Publication 856. Edited by S.G. Malghan. NIST, Gaithersburg, MD, 1993.

Reaction Chemistries in Dynamic Colloidal Systems

Jorge L. Valdes, Sanjay S. Patel, and Y. L. Chen
Bell Labs, Lucent Technologies, 600 Mountain Ave 1A-355, Murray Hill, New Jersey 07974 USA

Abstract

The interfacial chemistry of sub-micron particles is of great practical importance in determining the stability, dynamics, and reactions of colloidal dispersions used for sol-gel processing. We discuss an alternative approach to sol-gel process chemistry that relies upon the use of quasi-"multi-valent" amines (QMVA) to serve as both the dispersing and flocculating agent. Processing advantages of this new approach include a reduction of additives and the potential for designing formulations in a systematic way to achieve gels with tailored physico-chemical properties. We have conducted a wide range of process analytical experiments, using single-point and variable frequency electroacoustics, to derive a conceptual framework about sol-gel process chemistry that permits specification of a broad class of chemical agents that afford new pathways for generating sol-gel reactions. These agents are characterized by having a multiplicity of ionizable nitrogen groups within the same molecular structure. When these agents are used in conjunction with colloidal silica materials, they provide two essential process functions by: 1) precluding the aggregation of particles thus effecting dispersion stability and maintaining sol fluidity, and 2) enabling a shift of the iso-electric point (i.e.p.) to within a prescribed range suitable for the formation of a gel body using an appropriate pH-reducing agent. An understanding of the process chemistry has led to a universal correlation of the i.e.p. parameter and the thermodynamic ionization value, pKa, of the QMVA. Such a relationship permits the precise prediction, replication, and control of conditions for sol-gel processing and supports a postulated theory for the mechanism of gelation in these complex systems based upon effective surface charge compensation by multiply-ionized chemical agents.

1. Introduction

Sol-gel technology for producing silica glass has been recognized for many years as having the potential for low cost and low temperature processing in the production of overcladding tubes used in optical fiber manufacture. Despite this great potential, little success has been achieved in applying sol-gel technology for the large-scale production of silica glass tubes. This fact alone is a testament to the highly complex nature of sol-gel processing. As such, the ability to predict, control, and model sol-gel process chemistry is of critical importance to development of an optimized manufacturing process. In this work, we describe an alternative chemical reaction approach that may have potential in the fabrication of sol-gel bodies for a wide range of technological applications.

As part of a comprehensive research and development program devised for understanding the process chemistry and engineering of sol-gel phenomena, we have two clear objectives: 1) to establish a strong knowledge base about the chemical formulation upon which the sol-gel process is based, and 2) to examine and exploit alternative colloidal reaction pathways that may afford processing advantages.

Using a very powerful and state-of-the art radio-frequency electro-acoustic technique,[1-4] we have been able to conduct a wide range of process analytical experiments designed to elucidate and further enhance our understanding about the interfacial chemistry and dynamics of sol-gel processing. Of particular importance is the materials characterization and chemical behavior in the first three stages of this process: 1) dispersion of silica particles, 2) stabilization of the resultant sol material, and 3) the conversion (by ester hydrolysis) of the fluid-like sol into a solid-like gelled body with appropriate structural integrity to withstand tube launching and overcome stresses incurred during drying.

Sol-gel methods for generating silica sols fall into two major categories: 1) polymerized alkoxides and 2) colloidal dispersions. In the colloidal dispersion approach, fumed silica is dispersed uniformly in distilled water resulting in a dispersion whose pH is typically in the range of 2 to 3 pH units. In order to ensure stability of the dispersion and prevent the agglomeration of particles, it is necessary to increase the pH of suspensions to between 9 and 10 pH units by the use of an appropriate base (typically ammoniacal). In the this pH regime, silica particles take on a net negative surface charge as a result of the ionization of silanol groups on the surface according to the following reaction:

$$\text{-Si-OH} + \text{OH}^- \text{ -> -Si-O}^- + H_2O \quad (1)$$

After a stable colloidal silica dispersion is produced, additional agents (such as binder and excess base) may be introduced into the suspension to achieve formulations capable of yielding gels that can be cast, dried, and fired into glass tubes. The task of deriving a specific formulation capable of producing large bodies is unquestionably an arduous one. However, there exists at least two fundamental guidelines for developing an alternative sol-gel formulation. The first one involves particle stabilization by electrostatic repulsion using an appropriate base. The second requires that when the pH of the dispersion is reduced by the addition of a suitable gelling agent, such as methyl formate, neutralization of the surface charge on silica promotes the coalescence of silica particles with concomitant formation of a gel body. A useful parameter in assessing the tendency of a given colloidal silica dispersion to coalesce and produce a gel is the iso-electric point (i.e.p.). The i.e.p. is defined generally as the pH condition at which the effective electrokinetic charge or zeta potential on the particles is zero Conventional silica dispersions typically exhibit an i.e.p. around pH 2. Consequently, gelation depends critically on shifting the i.e.p. to a sufficiently high pH such that this value is accessible upon reducing the pH by hydrolysis of an ester. In certain formulations, this condition is achieved by introducing the tetramethylammonium cation (TMA^+) which is shown, in this work, to exhibit a strong adsorptive capacity for silica. As a result, the adsorption of the TMA^+ cation by the particles provides for a shift of the i.e.p. to a higher pH than would otherwise be required for a non-adsorptive or weakly adsorptive ion such as the ammonium cation (NH_4^+). Hence, the use of quaternary ammonium hydroxides serves an important dual function in sol-gel processing, albeit this use defines a fairly narrow pH window over which a gel body may be produced. In this work, we exploit our understanding about the surface chemistry of silica particles to develop alternative reaction chemistries for the processing of large sol-gel bodies.

2. Theoretical Hypothesis

The present hypothesis relates to a generally applicable processing method for the production of silica-based sol-gel bodies which may have potential in a wide range of technological applications. The new process provides an effective method whereby gelling agents are generated *in situ*, and homogeneously throughout a colloidal suspension of silica, to yield a large (e.g. > 1 kg) sol-gel body that is capable of being further processed by known ceramic methods to produce monolithic glass structures.

The new process of producing large sol-gel bodies involves the addition of an appropriate quantity of an amine-based chemical compound to a suspension of sub-micrometer silica particles in water. These chemical compounds are defined here as quasi-"multi-valent" amine (QMVA) agents which are broadly characterized by their ability to possess a multiplicity (i.e., > 1) of ionizable nitrogen (Lewis base) groups within a given unit molecular formula. When these agents are introduced into an aqueous silica dispersion, they serve, first, as stabilizers by increasing the pH such that aggregation of particles is minimized by electrostatic repulsion forces. Addition of the agent thus ensures the establishment of a stable, fluid-like, and homogeneous silica sol for processing. The formation of a gel body is accomplished in a homogeneous and controlled manner by the *in situ* generation of multiply-charged moieties as a result of the ionization of the

QMVA agent whence the pH of thc system is reduced using, for example, the hydrolysis of an ester component such as methyl formate. Multiply-charged moieties produced *in situ* by the ionization of QMVA provide effective and powerful agents capable of producing large, wet sol-gel bodies which in turn can be further processed to produce glass overcladding tubes. One of the criteria for determining optimal utility of a given QMVA agent for practical sol-gel processing is its capacity for the stabilization of the colloidal silica through electrostatic repulsion. Typically, this requires the agent to be a sufficiently strong base to permit raising the pH of a colloidal silica and water dispersion to a level adequate for producing a stable (non-agglomerating), fluid-like (low viscosity) silica sol. Stable colloidal silica dispersions are usually formed by adjusting the pH of the system to values above pH 7. In general, a broad range of QMVA agents may be used to impart stabilization to a silica dispersion. These include but are not limited to primary, secondary, and tertiary amines and combinations thereof. Examples of QMVA agents that have been used to successfully produce silica gel bodies include: diethylenetriamine (DETA), ethylenediamine (EDA), hexamethylenetetramine (HMTA), N,N'-dimethylethylenediamine (DMEDA), hexamethylenediamine (HMDA), and tris(2-aminoethyl) amine (STAR). For those QMVA agents that may not exhibit the requisite strength in basicity, combination of the QMVA with a strong, fully dissociated organic base such as tetramethlyammonium hydroxide (TMAH) may be employed to achieve the desired level of sol stabilization. The second most important criteria for determining the ability of a QMVA to generate, *in situ*, an effective agent for gelation is related to the degree of charge a single moiety can attain and the conditions necessary to bring about this charge. The ionization of a given amine-based system is typically brought about by controlling the pH of the solution. The partition between a singly ionized state and its corresponding un-ionized state can be represented by the following equilibrium reaction:

$$-R_3N^+H = H^+ + -R_3N \qquad (2)$$

When the pH of a suspension is made substantially greater than the pK_a of the specific QMVA, the moiety is found principally in its un-charged state (i.e. $-R_3N$). Typically, this is accomplished by the addition of an appropriate quantity of the QMVA to the suspension. Only a small fraction of the QMVA agent is actually ionized in order to attain the desired high pH for stabilization of the suspension.

Multiple ionizations are brought about in the same manner and will depend upon the specific equilibrium pK_a values of the QMVA agent. The pK_a of a given amine base is defined as the equilibrium pH at which 50% of the compound has ionized or dissociated into its corresponding conjugate acid. At a given ionic strength and temperature, the pK_a is given by:

$$pK_a = pH + \log \frac{[\text{conjugate acid}, -R_3NH^+]}{[\text{a min e base}, -R_3N]} \tag{3}$$

The relative concentrations of the base and its conjugate acid will change most rapidly around the pH corresponding to the value of the pK_a. When the pH of a colloidal silica dispersion is brought to within the pK_a of the QMVA agent, effective neutralization of the surface charge by *in situ* ionization of the QMVA agent occurs to produce a gel network. The effectiveness of surface neutralization will depend upon the degree of charging on the QMVA moiety.

From the classical DLVO theory of colloidal stability, it is known that the adsorption or charge compensation of a given ionic species on a solid surface will depend upon both the solvation energy of the ion and its effective molecular diameter. So, for instance, Cs^+ ions are more likely to adsorb on the surface of silica than are K^+ or Li^+ cations. By the same token, DLVO theory also predicts that the critical coagulation concentration (c.c.c.) required to induce destabilization of a colloidal dispersion depends on the sixth power of the charge manifested on the cation or "gelling agent".

$$\text{c.c.c.} \sim Z^6 \tag{4}$$

This means that Ca^{+2} is 64 times more effective, and Al^{+3} 729 times more effective at causing gelation than simple monovalent cations. The use of alkali metals or even transition metal cations, of higher order valency, as gelation agents is obviously incompatible for producing high purity glass for optical transmission systems. Nonetheless, this framework of understanding can be applied to the development of an alternative class of chemical agents capable of producing the desired sol-gel product. Consequently, QMVA agents with a multiplicity of ionizations result in a demonstrably more efficient method of producing gel materials for sol-gel processing. In comparison to the use of fully ionizable quaternary ammonium monovalent cations (such as tetramethylammonium hydroxide or even ammonium hydroxide), considerably less QMVA agent is required to produce the same desirable effect. This approach thus has the distinct advantage of necessarily reducing the amount of agent used which in turn facilitates their removal during later stages in the process.

3. Experimental Methodology

Electroacoustic measurements were conducted with the MATEC MBS-8000 electronics system using an SP-80 probe. The electroacoustic measurement device consists of a piezoelectric transducer mounted on one end of a glass buffer rod. The opposing end of the glass rod has two gold electrodes placed at a fixed distance apart and configured as a disc in parallel to thin bar. This arrangement ensures precise and reproducible electrode gaps for tuning the applied frequency around its nominal value of 1 MHz to achieve the desired system resonance and maximum signal strength. The test colloid is placed between electrodes by partially immersing the electroacoustic device into the liquid dispersion. Experiments are performed by applying a series of radio-frequency voltage pulses (typically 5-10 cycles) between electrodes. Sound waves produced by colloidal particles in the suspension are then measured by a sensitive piezoelectric transducer. The glass rod acts a buffer or delay line eliminating electrical "cross talk" - an undesirable signal produced by the initial application of the high voltage pulse across the test electrodes.

Electroacoustic measurements were conducted in a Teflon vessel containing approximately 200 cm^3 of the test suspension. Colloidal dispersions are continuously stirred with an impeller-type mixer. The measurement vessel is equipped with several sensors for measuring the pH, temperature, and electrical conductivity, in addition to the electroacoustic response of the system. An automatic, computer-controlled reagent delivery system coupled to the cell provides the ability to conduct facile potentiometric titrations of colloidal dispersions. Reagents are delivered using a dual syringe, digital burette with 0.5 microliter resolution.

A wide-range of electroacoustic experiments were conducted on commercially available colloidal silica dispersions. Colloidal dispersions are known to contain approximately 46 % (w/w) fumed silica (with a nominal particle diameter of 50-100 nm) in water. To ensure adequate shelf-life, approximately 0.06M of TMAH is introduced into these systems at the time of sol preparation to produce an electrostatically stabilized silica dispersion. The resultant pH of these suspensions is thus typically around 10 pH units. In this range of pH, the solubility of silica is minuscule (~ 100 parts per million) and therefore is not expected to contribute in any significant way to the measured electrokinetic parameters. Experiments were conducted at room temperature (~21-23 ^{o}C) and the pH was controlled through the addition of reagent grade chemicals (HCl, formic acid). Solutions for titrations and dilutions of the concentrated colloidal silica were prepared with high purity, organic-free, de-ionized water having a nominal resistivity of 18 MΩ-cm.

4. Results and Discussion

Experimental data on the electroacoustic characterization of several silica-water dispersions are shown in Figure 1. The dependence of the zeta potential on pH is shown over a broad range of pH values ($2 \leq pH \leq 10$). A well-defined, common i.e.p. at a pH of 5.2 was measured for each dispersion. The i.e.p. delineates two distinct regions of interfacial chemistry on colloidal silica. For pH values higher than the i.e.p., a colloidal particle of silica has a net negative surface charge. For these conditions, the surface charge manifested in the silica-water system can be described on the basis of a single acid-site dissociation model according to the surface reaction described by Equation (1).

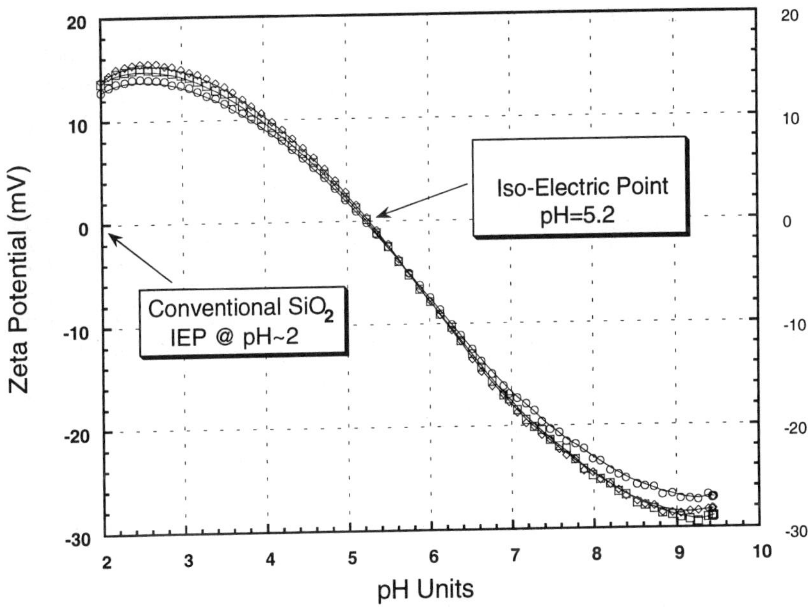

Figure 1. Electroacoustic characterizations of colloidal silica dispersions: zeta potential-pH titrations for 1 % (v/v) silica using 1M formic acid as the pH-reducing agent.

For pH values lower than the i.e.p., protonation of silanol groups on silica gives rise to a positively charged species by the following reaction:

$$Si\text{-}OH + H^+ = Si\text{-}OH_2^+ \qquad (5)$$

Although the results of Figure 1 are consistent with the ionization of existing amphoteric sites on the silica surface, the i.e.p. occurs at an appreciably higher pH value than is typically reported in the literature for the silica-water system.[5] The i.e.p. of the silica-water system is usually between a pH of 2 and 3. The positive electrokinetic charge on silica at pH 2 to 5 is related to the compensation of surface charge resulting from the affinity of specifically adsorbed cations on the particle surface.

An understanding of the surface chemistry required to establish a given i.e.p. is extremely useful in defining the conditions necessary for the processing of sol-gel materials. For example, gel formation is usually thought to occur when the pH of the system is brought close to the i.e.p. Commercially available silica containing approximately 0.06M TMAH (pH ~ 9.8), has sufficient negative surface charge to produce a highly stabilized dispersion. On the other hand, it has been reported that large organic monovalent cations (such as TMA^+) have the potential of exhibiting specific adsorption on these charged interfaces.[6] Specific adsorption of cations on a negatively charged silica surface would necessarily cause a displacement of the i.e.p. and thus lead to net positive surface charge on silica at pH values higher than those typically ascribed to colloidal silica.

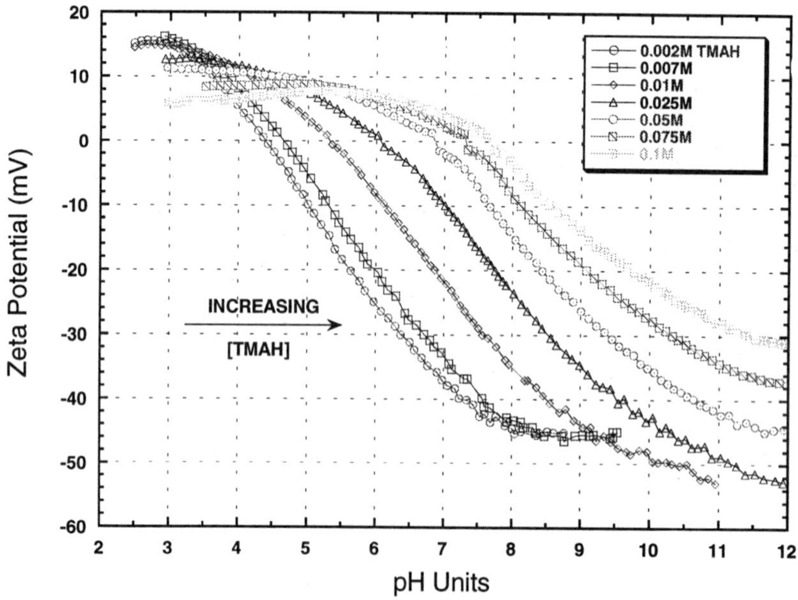

Figure 2. Concentration dependence of the zeta potential-pH curves for TMAH and silica. Iso-electric point (i.e.p.) shifts with increasing concentration of TMAH.

A series of experiments were conducted to determine the extent to which adsorption of the TMA^+ cation alters the i.e.p. of colloidal silica. The results of the zeta potential-pH curves for L-50 colloidal silica measured at various added concentrations of TMAH are shown in Figure 2. The concentration of the silica particles in the dispersion was maintained constant at 1 % (v/v). These results confirm that increasing the concentration of the TMA^+ cation in the suspension does indeed shift the i.e.p. to higher pH values and alludes to a possible adsorption mechanism for this process. Shifting of the i.e.p. to higher pH values is a critical component of the sol-gel process because it permits the "gelling agent" (such as methyl formate) to reach within a finite domain of the i.e.p. of the dispersion. Within this domain, particle repulsion forces are sufficiently diminished to allow for effective particle-particle interactions leading to a cohesive gel network. Gels made by using conventional, non-pH shifted, colloidal silica and a pH-reducing "gelling agent" will generally yield materials that are not sufficiently robust for subsequent processing steps.

The notion that appropriate control of the i.e.p. is an essential component of producing robust bodies provided the stimulus to explore alternative reaction chemistries for sol-gel processing. The development of a new chemical pathway for producing sol-gel materials necessarily requires a system that is: 1) capable of increasing the pH of the silica dispersion to ensure charge stabilization, and 2) has sufficient surface charge compensation (e.g. adsorption of cationic charge) for silica such that the effective i.e.p. is shifted to within a pH domain accessible to the gelling agent.

For the case of the "tri-functional" agent DETA, only two of the three ionization states (pK_1=9.94, pK_2=8.88, and pK_3=3.70) are attainable under normal processing conditions to produce a quasi-doubly charged moiety (~2). The third ionization state (pK_3=3.70) is inaccessible because the pH of the system does not usually drop below pH 7 whence a gel body is formed. For the case of STAR, all three ionization states ($pK_{1,2,3}$=10.29, 9.59, and 8.56, respectively) can be attained to produce a very potent quasi-triply charged gelling agent (~3). In the preferred processing mode, a gel body is produced upon introducing a pH-lowering additive (such as an ester) which has the effect of further catalyzing the ionization of the QMVA agent. In effect, this methodology serves to generate *in situ* charge compensation on silica that in turn provides for an effective thermodynamic free energy driving force for gelation to occur.

Experimental data on the zeta potential-pH curves for colloidal silica measured at different concentrations of added DETA are shown in Figure 3. The results suggest that DETA can indeed dramatically shift the i.e.p. of colloidal silica by almost 5 decades, toward higher pH values, when compared to an equivalent concentration of TMAH (see Fig. 2). An interesting feature in this family of curves is that for concentrations higher than about 0.02M, the zeta potential drops precipitously with decreasing pH around the i.e.p. The i.e.p. also happens to coincide approximately with the pH necessary to ionize the QMVA. This feature suggests a process mechanism whereby charge compensation of silica particles is driven by the particular ionization behavior of the QMVA agent.

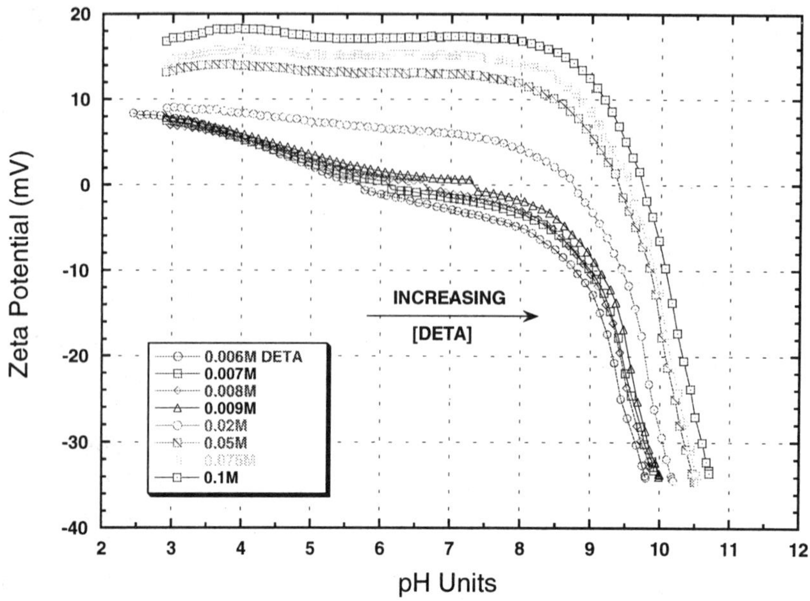

Figure 3. Concentration dependence of the zeta potential-pH curves for diethylenetriamine (DETA) and silica. The i.e.p. shifts with increasing concentration of DETA.

A similar kind of behavior, yet even now more pronounced, is observed for the case of the QMVA agent STAR. In Figure 4 are shown the apparent shifts of the i.e.p. with increasing concentrations of added STAR to a 1 % (v/v) colloidal dispersion of silica. We find that with an added concentration of as low as 5 mM STAR, the i.e.p. can shift to at least a pH value of 9. At slightly higher concentrations, the i.e.p. is shifted to even more alkaline values of the pH. It is interesting to note that in contrast to completely dissociated bases like TMAH, QMVA agents produce considerably more rapid sol-to-gel transitions as evidenced by the almost infinite slope of the zeta potential-pH curve when in proximity to the i.e.p. This characteristic feature of some QMVA gelling agents may permit the manufacture of gel materials with properties vastly different from those made by conventional methods.

The concentration dependence of the i.e.p. was investigated for a number of QMVA agents (including TMAH for comparison) and are shown in Figure 5. The experimental data shown are consistent with a Langmuir-type of adsorption

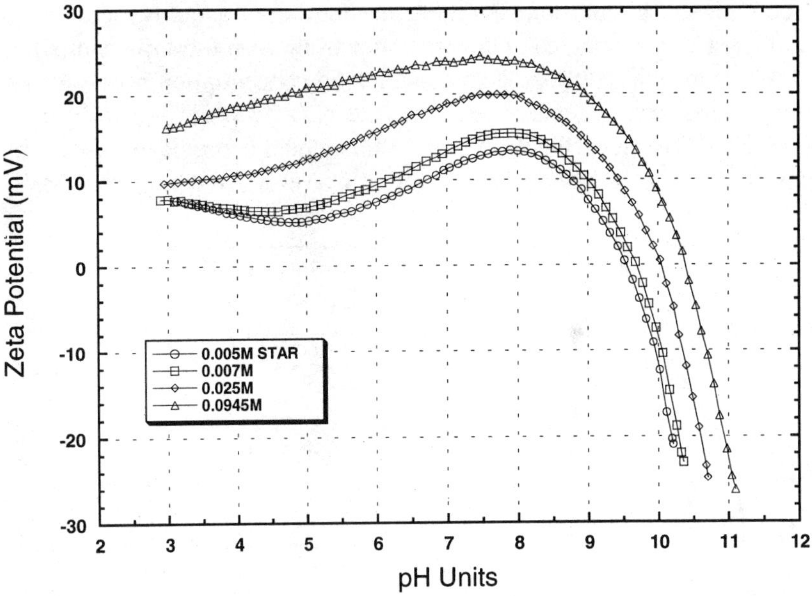

Figure 4. Concentration dependence of the zeta potential-pH curves for tris(2-aminoethyl) amine (STAR) and silica. The i.e.p. shifts with increasing concentration of STAR.

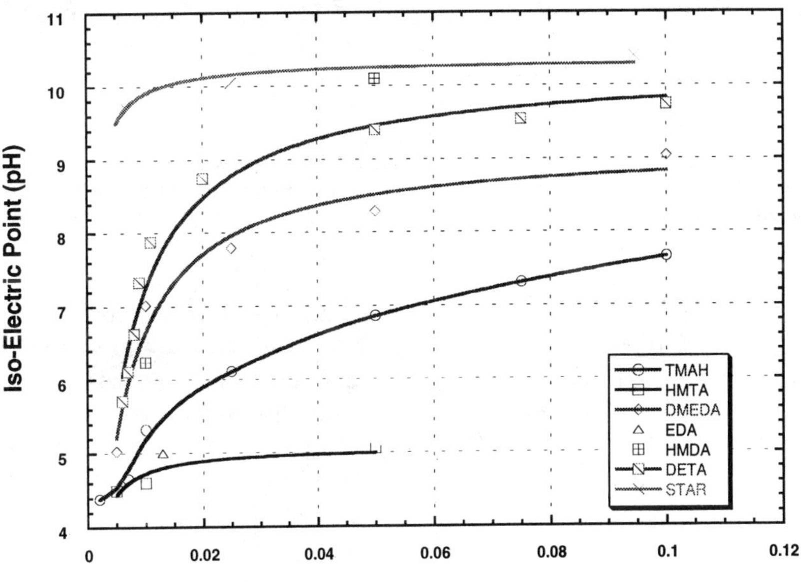

Figure 5. Concentration dependence of the i.e.p. for various QMVA agents (see text for description) and TMAH.

model for the ionized cationic state. As the concentration of agent is increased, a concomitant increase in the degree of charge compensation on the surface of silica occurs. This phenomenon reduces the equivalent acid concentration necessary for neutralization of the surface charge and thus necessarily leads to an i.e.p. at significantly higher pH values. Confirmation of this model is provided, in part, by careful analysis of the zeta potential-pH titration curves for the various QMVA agents.

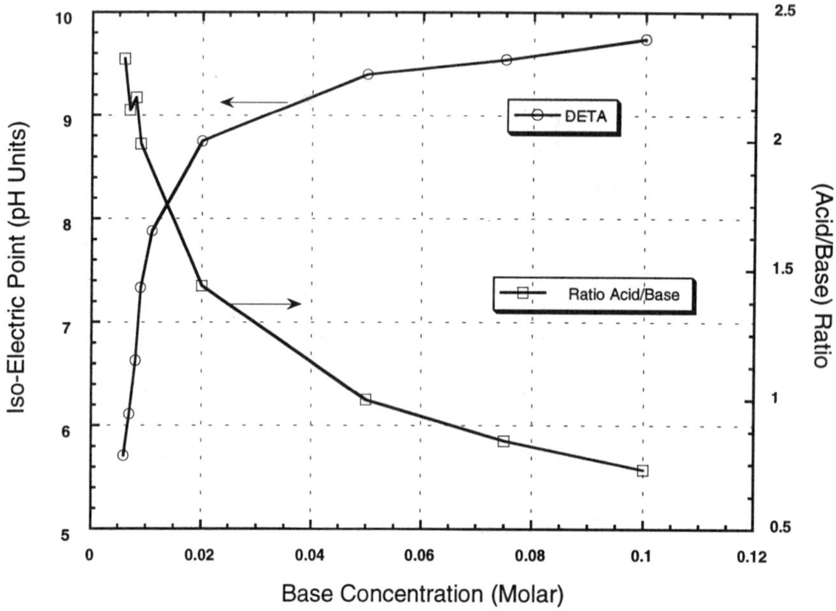

Figure 6. Correspondence between shifts of the i.e.p. with increasing base concentration (DETA) and the requisite acid/base molar ratio to fully compensate the surface charge on silica.

Figures 6 and 7 show a comparison of the i.e.p. and the acid/base ratio as a function of the base concentration for DETA and STAR, respectively. The acid/base ratio refers to the quantity of equivalents of acid necessary to titrate the zeta potential of colloidal silica down to its i.e.p. The graphical data clearly illustrates that as the concentration of base increases (shifting the i.e.p. to higher pH values), less acid is required to reach the i.e.p. For the case of STAR, considerably less than a 1:1 acid/base ratio is needed to reach the i.e.p. For TMAH (shown in Fig. 8), the acid/base ratio remains constant (~1.0) and is essentially independent of the base concentration. These results suggest that for monovalent systems such as TMA^+, an almost stoichiometric (1:1) correspondence exists in the neutralization of charge on the surface of silica. For divalent DETA, a 2:1 acid/base ratio is required at low base concentrations but decreases to less than stoichiometric values at higher concentrations. For trivalent STAR, considerably less than the stoichiometric value (3:1) drops the zeta potential of the suspension to its i.e.p. An estimate of the efficacy of QMVA agents for producing surface charge compensation can be made by comparing the experimentally determined acid/base ratios to the corresponding stoichiometric value. This will, of course, depend on the specific QMVA moiety and its spatial distribution of cationic charge. Whereas TMAH is close to stoichiometric, DETA is stoichiometric at low concentrations yet is more than twice as effective at charge compensation at higher concentrations (~0.1M). STAR is thrice effective at low concentrations and almost an order of magnitude more effective at higher concentrations than can be predicted solely on the basis of stoichiometry. Such results confirm, at least in part, the enhanced charge compensation on the surface of silica by the use of QMVA agents.

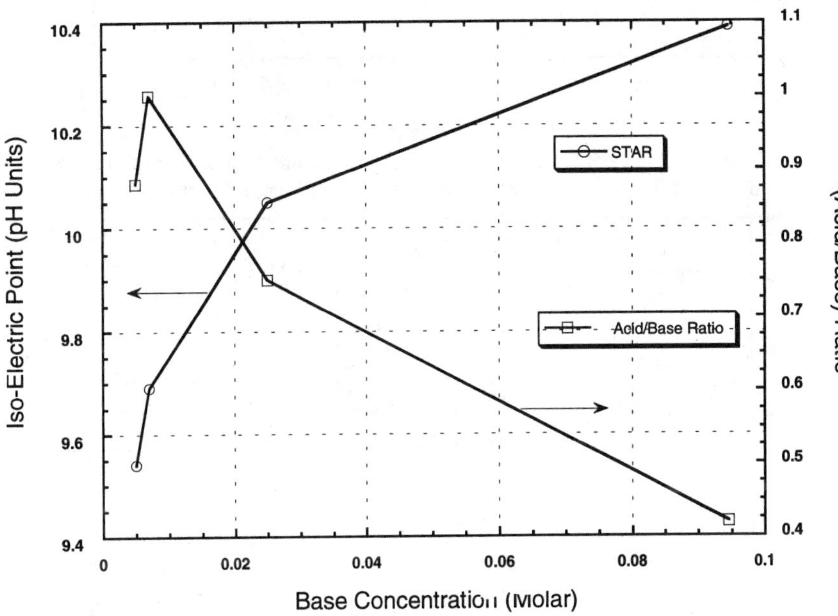

Figure 7. Correspondence between shifts of the i.e.p. with increasing base concentration (STAR) and the requisite acid/base molar ratio to fully compensate the surface charge on silica.

A process mechanism based upon the multiple ionization of QMVA agents, whereby effective surface charge compensation leads to gelation, suggests the existence of a relationship between the i.e.p. and the pK_a of the QMVA. The i.e.p. as a function of the mean pK_a value is plotted in Figure 9 for various QMVA agents having proton ionizations ranging from 1 to 4. Mean pK_a values were

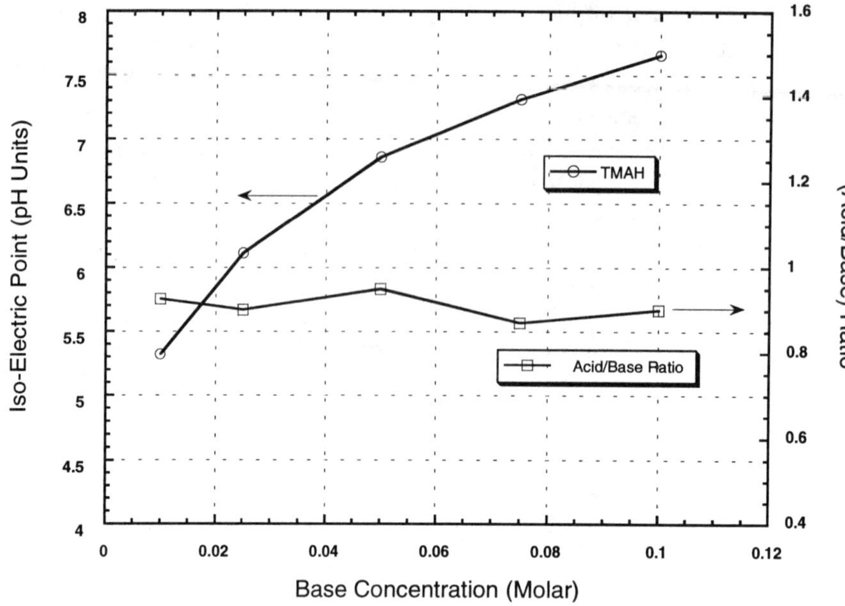

Figure 8. Correspondence between shifts of the i.e.p. with increasing base concentration (TMAH) and the requisite acid/base molar ratio to fully compensate the surface charge on silica.

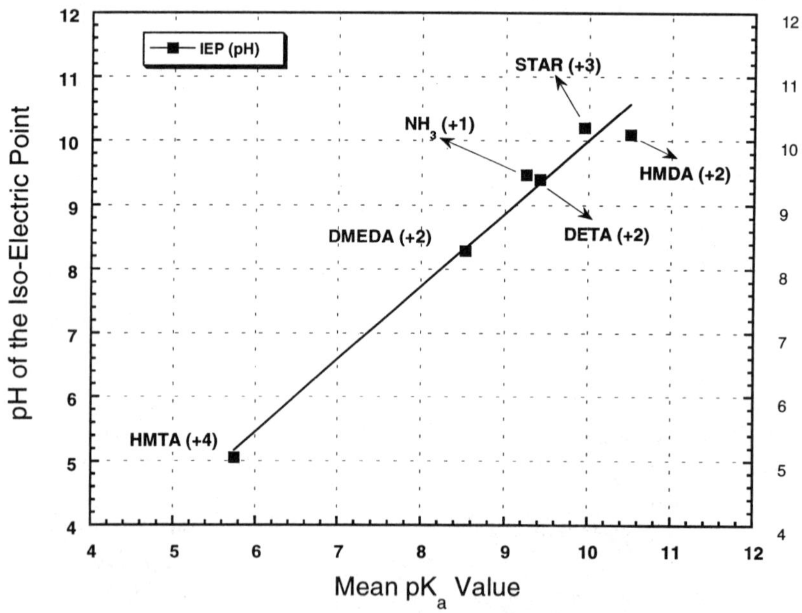

Figure 9. Universal correlation for partly ionized amine-type sol-gel agents depicting the relationship between the mean ionization pK_a value of the agent and the i.e.p. of silica.

calculated from thermodynamic values found in the literature[7] and the i.e.p. was determined experimentally from the zeta potential-pH curves measured for each QMVA and at a common concentration of 0.1M (the value for NH_4OH was measured at 1M). It is evident from Figure 9 that a universal correlation does indeed exist between the i.e.p. and pK_a This correlation provides a very useful and powerful tool that permits the prediction of conditions necessary for gelation to occur simply on the basis of a knowledge of the thermodynamic ionization properties of the QMVA base agent. It also lends additional evidence supporting the mechanism of agent ionization and charge compensation for this new method of processing sol-gel materials. Monovalent amine agents, (such as NH_3), which are not as efficient at charge compensation as QMVA, can fall on this universal correlation only at significantly higher concentrations. Zeta potential-pH curves for NH_4OH (Fig. 10) indicate that at least an order of magnitude higher concentrations are required to coincide the i.e.p. with the ionization pK_a value of 9.25 for ammonia. The use of QMVA agents hence provides a new and effective means for substantially reducing the amount of additives used in sol-gel processing.

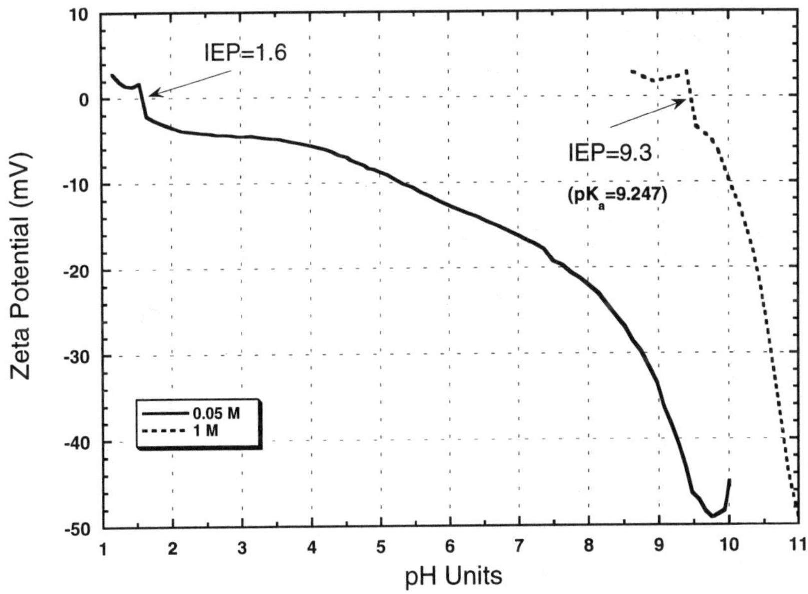

Figure 10. Stabilized silica at different ammonia/water concentrations. Monovalent systems require a significantly higher concentration of base to fully compensate the surface charge so as to coincide the i.e.p. with the thermodynamic pK_a value.

5. Conclusions

The present paper involving alternative reaction chemistries for sol-gel processing provides a number of key advantages for the production of sol-gel bodies in manufacture. With this approach, it is now possible to prescribe both the concentration and specific QMVA agent to produce silica-based sol-gel bodies under a broad range of processing conditions. This translates into enhanced process latitude for sol-gel. The sol-gel process chemistry is now engineered in such a way such that a single chemical component provides electrostatic stabilization of the sol (in the same manner as TMAH), while also serving to generate the agent *in situ* that is actually responsible for gelation. The pH-lowering additive, which may be chemically identical to the prior art "gelling agent", now serves only to catalyze further ionization of the pH-increasing QMVA ("stabilizing agent"). The attainment of a multiplicity of ionizations on a single QMVA induces gelation at considerably lower QMVA concentrations and thus in turn facilitates their subsequent removal during the later stages of the process. The ability to pick and choose from a wide-range of QMVA agents and the potential for the tuning of process parameters in a systematic way so as to achieve the production of gel bodies with tailored physico-chemical properties are evidence to the opportunities that exist for implementing alternative reaction chemistries in the processing of sol-gel materials.

References

1. T. Oja, G.L. Petersen, and D.W. Cannon, US Patent No. 4,497,208 (1985).

2. R.W. O'Brien, B.R. Midmore, A. Lamb, R.J. Hunter, "Electroacoustic Studies of Moderately Concentrated Colloidal Suspensions", *Faraday Discuss. Chem. Soc.*, **90** , 301-312 (1990).

3. R.W. O'Brien, "Electro-Acoustic Effects in a Dilute Suspension of Spherical Particles", *J. Fluid Mech.*, **190** , 71-86 (1988).

4. R.W. O'Brien, "The Electroacoustic Equations for a Colloidal Suspensions", *J. Fluid Mech.*, **212** , 81-93 (1990).

5. R.J. Hunter, *Zeta Potential in Colloid Science*, p. 229, Academic Press, New York, 1981.

6. E. Matijevic, pp. 328-369 in *Principles and Applications of Water Chemistry*. Edited by S.D. Faust and J.V. Hunter. Wiley, New York, 1967.

7. D.D. Perrin, B. Dempsey, and E.J. Serjeant, p. 27 in *pK_a Prediction for Organic Acids and Bases*, Chapman & Hall, New York, 1981.

Surface Characterization of Inorganic Colloids by the Colloid Vibration Potential

Frank Hinze, Michael Stintz and Siegfried Ripperger
Technische Universität Dresden, Institut für Verfahrenstechnik und Umwelttechnik, D01062 Dresden, Germany

Abstract

To describe the granulometric state of a dispersion in the submicron range, not only the particle size, but also the surface potential must be considered. This paper discusses the detection of the zeta potential in highly concentrated suspensions with a new measurement technique based on the colloid vibration potential. The obtained zeta potentials of different material systems are compared to results received by other measurement techniques. The major part of this paper deals with the influence of concentration on the measured potential. The potential decreases with increasing concentration. It is shown, how important knowledge about the zeta potential is for practical applications. Finally as an example the change of viscosity of a highly concentrated suspension as a function of the zeta potential is presented.

1. Introduction

If the particle size of a dispersion becomes very small, the surface related phenomena gain more significance. Such phenomena are e.g. chemical reactions, dissolution, and adsorption processes. Especially, the electrostatic properties have a great influence on the stability of dispersions. Therefore agglomeration is a major problem when working with submicron particles in liquids. It can happen that the degree of particle agglomeration becomes the dominating effect in a production process and thus more meaningful than the particle size.

The rising number of products with particles smaller than 1 μm makes it necessary to characterize the electrostatic surface behaviour. The knowledge of this behaviour, often a domain of physical chemists only, is also becoming important for chemical engineers.

The zeta potential is used to describe the surface charge. Normally the zeta potential can be measured at very small particle concentrations (0,01 % v/v) only, but for a characterization which is relevant in practice it is necessary to measure it in the original state (often at high concentrations up to 50 % v/v). The electroacoustic measurement technique presented here is only one of a few possibilities to measure the zeta potential in highly concentrated systems.

2. Colloid Vibration Potential

In recent years some progress has been made in electroacoustic-based measurement techniques for monitoring zeta potential and particle size distribution. Now instruments are available to measure the zeta potential by the electrokinetic sonic amplitude (ESA)[1,2] or the colloid vibration potential (CVP)[3]. This paper is concerned with a new instrument which measures the colloid vibration current (CVI)[4].

As a sound wave crosses a suspension, the particles start to oscillate because of the visco-inertial effect between particle and fluid. The resulting relative motion between the fluid and particles leads to a disturbance of the diffuse ion layer around the particle. The particles behave like dipoles analogous to the sedimentation potential effect. These electroacoustic formed dipoles induce a macroscopic colloid vibration potential (CVP). In CVP, the complex conductivity must be known for the calculation of an electroacoustic mobility or zeta potential from the CVP signal. In the device used here (AcoustoPhor 8000, Pen Kem Inc.) the colloid vibration current (CVI) is measured. The theory is similar to that for the CVP measurement. The advantage of the CVI is that it is not necessary to determine the complex conductivity. The measurement probe works at a single frequency of 2 MHz. The sample is pumped continuously through the measurement cell. In a second probe the device can detect the particle size distribution by acoustic spectrometry in a size range from 10 nm to 10 μm. This is important for the calculation of the zeta potential from the colloid vibration current, because the particle size must be known. Measurements up to 50 % by volume are possible (manufacturer's information[5]).

3. Experimental

It was necessary to evaluate this new electroacoustic CVI measurement probe as this was the first time this CVI probe was applied. Questions of the reproducibility of measurement results, the comparison of results, and the limits of the measurement technique are very important for the user. Unfortunately, there are no standards or reference materials for the zeta potential measurement in highly concentrated suspensions.

A reference material should possess a number of characteristics. Firstly, a high density contrast to the fluid (if possible $\rho_s / \rho_f > 2$) is required to obtain a measurable colloid vibration potential (CVP) based on the visco-inertial effect.

Secondly, a spherical or spheroidal shape for the particles and a narrow particle size distribution are also useful as the theories work best with monodisperse spherical particles. Besides, the particle size should be smaller than 3 μm to exclude sedimentation during the measurement. Finally, for comparative measurements with a microelectrophoresis cell, the size should be larger than 300 nm.

4. Results

4.1 Influence of the Concentration

Several metal oxides (silica, titania, alumina) were measured at different concentrations. Figure 1 shows the result of measurements from Gel Tech 1500 silica particles. This material possesses a narrow (monodisperse) particle size distribution, a density of 2.2 g/cm^3 and a particle size ($x_{50,3}$) of 1520 nm.[6] The concentration series was made at pH 9. This silica suspension has in the basic environment a high zeta potential and is quite stable. The concentration was varied from 3 to 43 % (v/v). The differently concentrated samples were prepared by dilution of a highly concentrated "master" suspension with supernatant. The supernatant was obtained by filtration. The measurement results (see Figure 1) show a strong change in the absolute potential values over the concentration range.

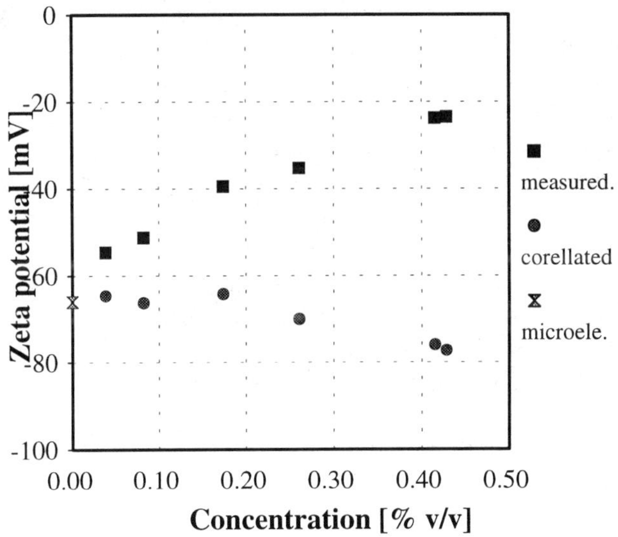

Figure 1. Gel Tech 1500 silica suspension. Zeta potential measured at different concentrations and multiplied with the correlation factor for the concentration dependence. Microele. = potential obtained by microelectrophoresis measurement.

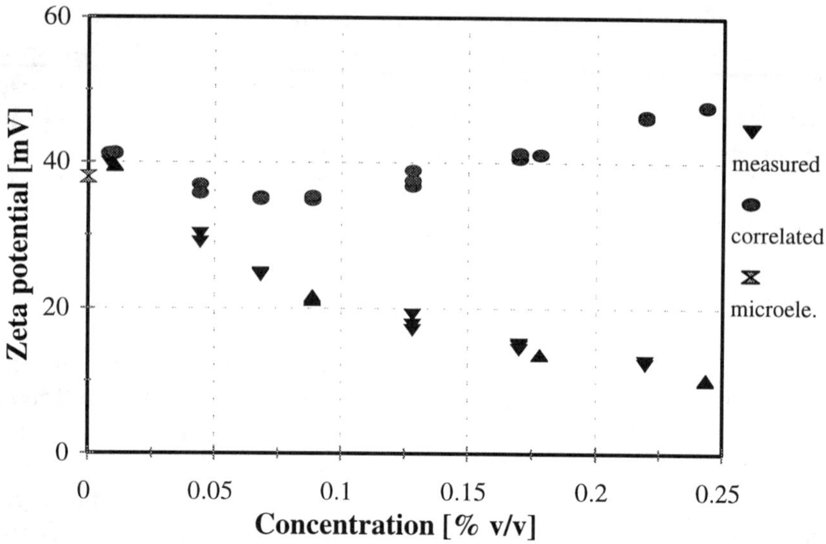

Figure 2. Titania R 900 suspension. Zeta potential measured at different concentrations and multiplied with the correlation factor for the concentration dependence. Sample prepared with ▲ supernatant or ▼ with ultra clean water, microele. = potential obtained by microelectrophoresis measurement.

Titania R 900 (Du Pont) was also used. It has a density above 4 g/cm^3 and the particle size ($x_{50,3}$) is 300 nm.[6] This material has a broader size distribution and a more ellipsoid shape. The particle surface is coated with a thin alumina layer. The isoelectric point is around pH 9. The samples were also prepared by dilution of a highly concentrated "master" suspension with supernatant or ultra clean water (18 MΩ), with the same conductivity and pH value for all samples. Concentrations from 0.8 to 26 % (v/v) were measured. Also, for this material there is a strong dependence of the zeta potential on the concentration (see Figure 2), similar to the results for Gel Tech in Figure 1. The different sample preparation methods seem to have no influence on the zeta potential. This rules out the possibility that the concentration dependence of the measured potentials is an artefact of the sample preparation.

4.2 Interpretation

Particle interactions become more significant as the concentration rises. Compression of the diffuse ion layer can occur because of the short interparticle distances and the higher possibility of particle collisions. Usually it is assumed that the double layer disturbance, and the consequent reduction of the measurable potential, starts at concentrations about 15 % v/v. For the present results it is surprising that the decrease of the zeta potential begins already after the first measured point (~ 1 % v/v). The zeta potential as a function of the concentration seems quite linear for the Gel Tech silica material. For the titania R 900 (Figure 2) it is more complex. For this product it seems that there are two concentration regions, with a quite linear relationship connected by a nonlinear transitional region. The first region is from 1 to above 7 % v/v with a strong decrease of the measured potential by an increase of the particle concentration. The second is the region over 7 % v/v up to the highest concentrations. The decrease of the potential in the second concentration region is less strong. It is just the opposite of the usually assumed behaviour of zeta potentials in higher concentrations. For rigid solid particles with thin diffuse layers the zeta potential should be stable until the double layer overlap starts. Then the zeta potential should decrease strongly. The concentration dependence of the potential of the CVI measurements seems to be related to the measurement technique.

In a paper by James[7] the concentration influence of the ESA signal is described. James explains this fact as one of various gain factors of the measurement apparatus. His measurement series with different materials (Silica-Ludox™, 3 different Latex materials) give a decrease of the electrodynamic mobility over the concentration. He finds that up to 5 % v/v the relationship is linear. With a further increase of the concentration the dependence gets less powerful and non linear. James suggests an empirical calibration factor for the ESA measurement technique.

To correlate the measured zeta potential with results at low concentrations one can suggest a function with two empirical factors:

$$\zeta_c = \zeta_0 + \phi \cdot C_2 \; ; \qquad \zeta_0 = S_m \cdot C_1$$

with the corrected potential ζ_c, the uncorrected potential ζ_0, which is the measured raw potential S_m multiplied with a calibration factor C_1, the concentration ϕ and the correlation factor for the influence of concentration C_2.

The linear regression analysis gives different values depending on the selected concentration range. For the silica sample the correlation factor is around $C_2 = 1.2$ ($\phi < 5$ % v/v). This factor may be inaccurate because there are only two measured potentials. The correlation factor for the concentration higher than 5 % v/v is above 0.63. This value corresponds very well to the concentration range of the titania R 900 results. But for concentrations below 5 % the value was found to be about 2.5. Again, there were only a few measurement points to define the correlation factor. In the paper from James a value for C_2 around 1.54 is discussed.

This was applied to the measurement series (Figures 1 and 2). The potential correlated in this way is approximately constant over the concentration range for the silica suspension. The titania sample agrees also very well to this factor at the lower concentrations. There is only one value at 0.8 % (v/v) which is slightly higher than the other potentials.

The whole concentration range can not be described by one correlation factor correctly. Whether a second linear correlation for the higher concentrations can be used, has to be found out in further work. It seems that this effect of the concentration influence on the potential is the same for the measurement devices which work on the ESA and the CVP measurement (less 5 % v/v ~ 1,5). It may be a general phenomena for the electroacoustic measurement technique. Not discussed here is an influence of the different densities of the samples like James[7] has done. The presented measurement data are insufficient to assess this effect. Also, a better understanding of the zeta potential at high concentrations is necessary.

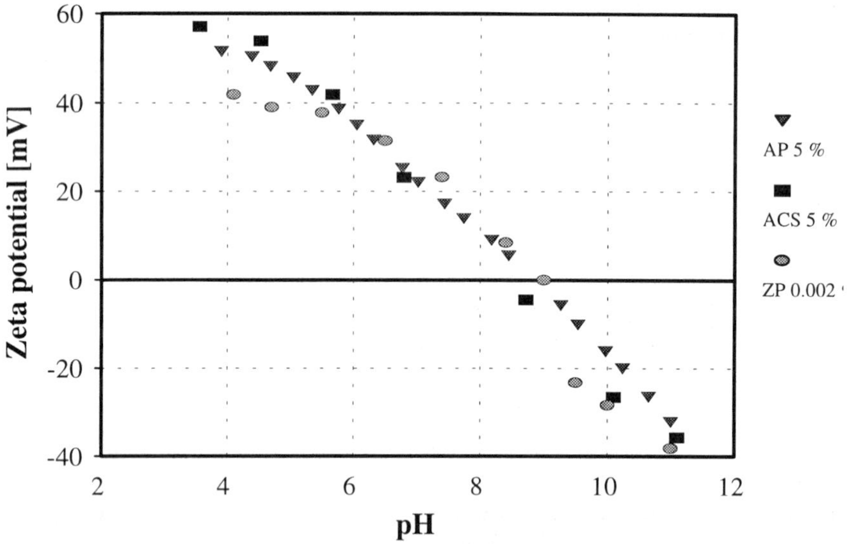

Figure 3. Comparison of measured zeta potential as function of the pH value for titania R 900 suspensions with the AP-AcoustoPhor (5 % v/v), the ACS-AcoustoSizer (5 % v/v) and the ZP-ZetaPlus (0.002 % v/v).

4.3 Comparison to Other Measurement Techniques

The advantage of applying the correlation function is the possibility to compare measured potential values from different concentrations and measurement techniques, but one should be aware of the limits of this correlation. For the following results, the correlated potential values (C_2 = 1.5) were used. The concentration was less than 5 % (v/v). Figure 3 shows results of a pH titration by a constant concentration of the titania R 900. The thin layer of alumina covering the particle surface determines the electrokinetic behaviour, thus the isoelectric point was found at pH 9, instead of the acid range as it is common for titania. Figure 3 shows good agreement between the results obtained by the AcoustoPhor (5 % v/v) and with an electrophoretic measurement device, the ZetaPlus (Brookhaven Inc., 0.01 % v/v).

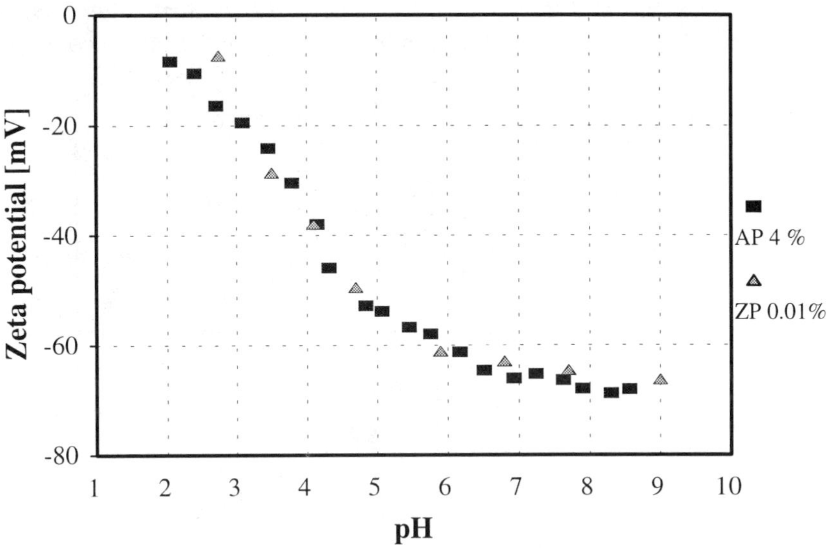

Figure 4. Comparison of measured zeta potential as a function of the pH value for Gel Tech 1500 silica suspensions with the AP-AcoustoPhor (4 % v/v), and the ZP-ZetaPlus (0.01 % v/v).

For the Gel Tech 1500 silica (Figure 4) the dependence of the zeta potential on the pH value was also measured. For this product, no isoelectric point could be detected. When it was tried to decrease the pH value below pH 2.2, with a large amount of acid, only small changes in the pH value could be obtained. The corresponding conductivity of the sample was increased enormously, making the comparison of the zeta potential values in that range impossible. The two curves obtained by the electroacoustic and the electrophoretic measurement technique are very similar. Repeating the CVI measurements shows good reproducibility. Differences appeared when samples were not well prepared (different conductivities, temperatures, and pH values or the sample was not well dispersed).

5. Application of Zeta Potential Knowledge

The presented results were measured under ideal conditions. The samples were prepared with ultra pure water and with a defined ion background. The suspensions were dispersed completely before the measurements were started. In practice there are hardly ever such conditions. Much more possibilities to influence the surface behaviour can appear. If the particles are very small (< 300 nm) it is not possible to obtain the supernatant of the suspension. Often impurities are in the sample, and the ion background becomes a complex mixture, often containing unknown ions. A change of the dispersion behaviour upon dilution becomes likely.

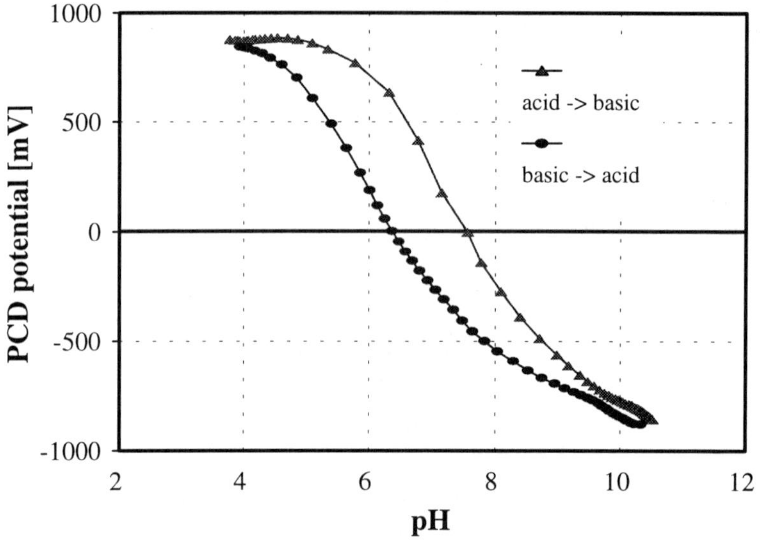

Figure 5. Surface potential versus pH value for a zirconia suspension (0.1% v/v) measured by PCD pH 03 (Mütek).

In Figure 5 a measurement result from a zirconia sample (Degussa) is shown. The sample was prepared without adding surfactants. For this material the relationship between surface charge and pH value depends on the titration direction. Hence a hysteresis in the potential curves can be observed. Figure 6 shows the zeta potential versus pH value at different concentrations. Clearly visible is the change of the isoelectric point from pH 6.5 at 0.01 % v/v to above pH 7.3 for a sample of 0.1 % v/v to above pH 9 at 2 % v/v. This means that the surface chemistry of this zirconia product will be modified by changing the concentration. Figure 6 further shows that the curve given by the producer is not comparable to the presented results. Without an exact description of the measurement conditions, the zeta potential values for this material can not be used.

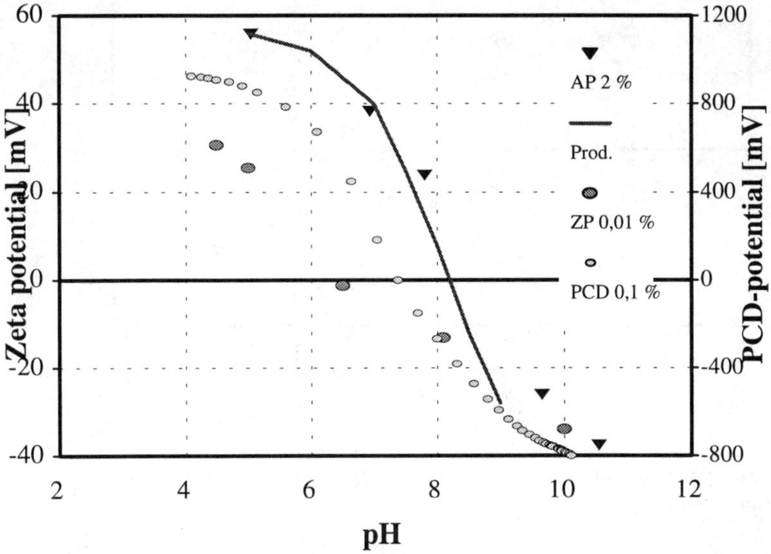

Figure 6. Influence of the concentration on the function zeta potential versus pH value measured with the AP AcoustoPhor at 2 % (v/v), the PCD pH 03 at 0.1 %, the ZP-ZetaPlus at 0.01 %, and the curve given by the producer of this zirconia material.

5.1 Change in the Viscosity

In practice the viscosity of a suspension is often of great interest. In pumping or mixing processes the viscosity influences the whole process. As an example of the uses of zeta potential knowledge, the change in the viscosity as a function of the zeta potential will be presented. Figure 7 shows the zeta potential of an alumina sample (Degussa) as a function of the pH value. The viscosity (see Figure 8) was measured with a rotational viscometer at different speeds. At the isoelectric point the viscosity of the suspension increases to a value 90 times higher than the viscosity of a stable suspension. The viscosity is strongly dependent on the energy input for the non-stable suspension. If the velocity increases, the viscosity decreases. The suspension has pseudoplastic behaviour. The difference of the viscosity for stable/nonstable suspensions becomes less for a high energy input, but the difference itself remains. The zeta potential explains the effect of the electrostatic destabilisation. The degreeof destabilization follows the change in surface charge.

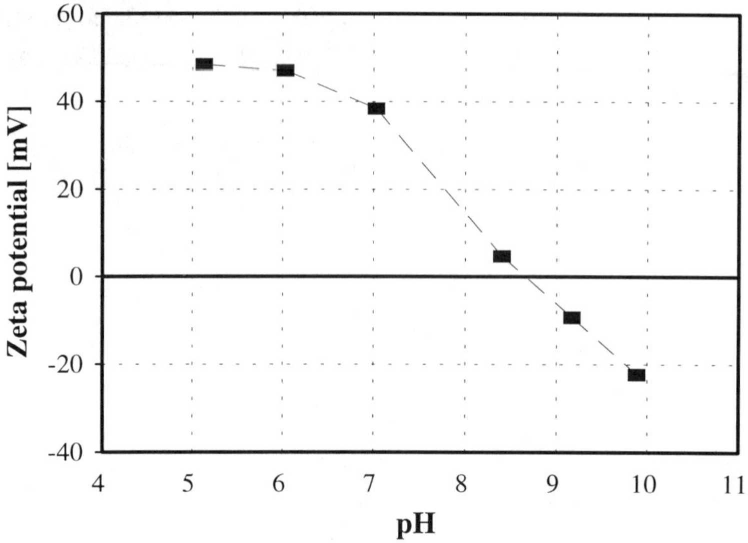

Figure 7. Zeta potential as function of the pH value for an alumina suspension of 4 % (v/v), measured with the AcoustoPhor.

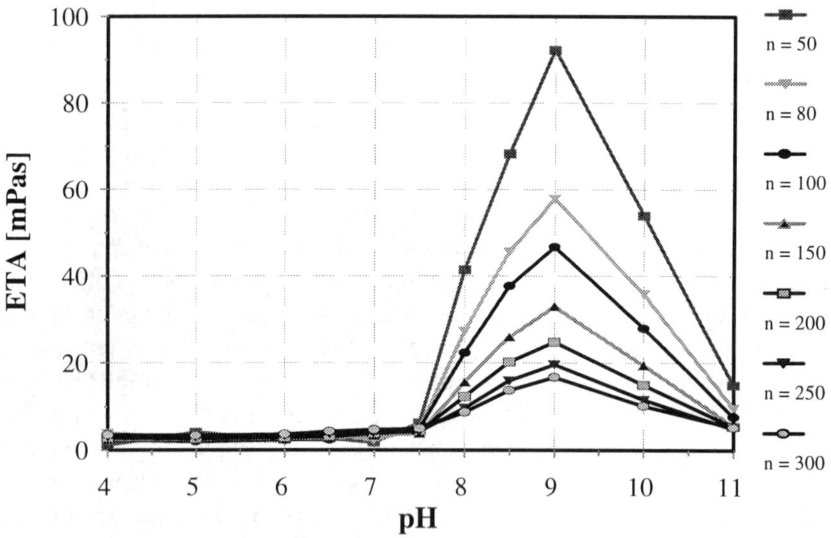

Figure 8. pH influence on the viscosity for an alumina suspension of 4 % (v/v), measured with a viscometer Rheotest RN, the rotational speed was varied between 50 and 300 min^{-1}.

6. Conclusions

For the exact surface characterization of a highly concentrated suspension it is necessary to measure it in its original state. With the electroacoustic CVI measurement technique it is possible to measure zeta potentials in high concentrations up to 50 % v/v. The measured CVI signal shows a significant dependence on the concentration, which seems to be similar to that of ESA measurement results. A linear empirical correlation factor can be applied up to 5 % v/v. Whether a second linear correlation factor for the higher concentrations can be used will require further work. A proper analysis of the interactions between concentration and measured potential should be given in the near future.

There are interpretation problems regarding zeta potential at higher particle concentrations. A big step towards better understanding could be accomplished with an appropriate reference material.

The electroacoustic measurement technique is a young, but promising technique, which is likely to be used for a large number of industrial applications in the near future.

Acknowledgement

The authors would like to thank the German Research Association (DFG) for financial support.

References

1. R.W. O'Brien, B.R. Midmore, A. Lamb, and R.J. Hunter, “Electroacoustic Studies of Moderately Concentrated Colloidal Suspensions”, *Faraday Discuss. Chem. Soc.*, **90**, 301-312 (1990).
2. R.W. O'Brien, W.N. Rowlands, and R. J. Hunter , "Determining Charge and Size with the AcoustoSizer", pp. 1-22 in *Electroacoustics for Characterization of Particulates and Suspensions*, NIST Special Publication 856. Edited by S.G. Malghan, U.S. Department of Commerce, Technology Administration, National Institute of Standards and Technology, Gaithersburg, MD, 1993.
3. B.J. Marlow, D. Fairhust, and H. P. Pendse, “Colloid Vibration Potential and the Electrokinetic Characterization of Concentrated Colloids”, *Langmuir*, **4**, 611-626 (1988).
4. A. Dukhin and P. Goetz "Acoustic and Electroacoustic Spectroscopy", *Langmuir*, **12**, 4985-4986 (1996).
5. User's Manual, AcoustoPhor 8000, PenKem Inc., 1995.
6. F. Hinze, M. Stintz, and S. Ripperger, "Particle Characterization in Highly Concentrated Submicron Suspensions by Ultrasonic Spectrometry", First European Congress of Chemical Engineering, Florence, 1997.
7. R.O. James, “Calibration of Electroacoustic (ESA) Apparatus Using Lattices and Oxide Colloids”, pp. 67-91 in *Electroacoustics for Characterization of Particulates and Suspensions*, NIST Special Publication 856. Edited by S.G. Malghan, U.S. Department of Commerce, Technology Administration, National Institute of Standards and Technology, Gaithersburg, MD, 1993.

Applications in Emulsions and Organic Particulate Suspensions

Ultrasonic Characterization of Food Emulsions

David Julian McClements
Department of Food Science, University of Massachusetts, Amherst, MA 01003 USA

Abstract

Ultrasonics is ideally suited for characterizing the concentrated and optically opaque emulsions that are common in the food industry. The use of ultrasound to determine the concentration, size distribution, crystallization and flocculation of emulsion droplets in foods is briefly reviewed. The major focus of this chapter is the recent development of a novel ultrasonic imaging technique for monitoring creaming and sedimentation profiles in food emulsions. This technique is capable of measuring both the size and concentration of droplets at any vertical or horizontal position within an emulsion and can therefore be used to provide a full profile of gravitational separation. In addition, critical technical issues that need to be addressed concerning the application of ultrasonic techniques in the food industry are identified.

1. Introduction

Many foods consist either partly or wholly as emulsions, or have been in an emulsified state sometime during their production, including milk, cream, butter, margarine, ice-cream, salad dressings, mayonnaise, sauces, coffee whiteners and soups.[1-3] The appearance, shelf-life and texture of these products depends on their composition and microstructure, *e.g.,* disperse phase volume fraction, droplet size distribution, droplet crystallinity and droplet-droplet interactions. Most food emulsions are concentrated and optically opaque and are therefore not amenable to characterization using traditional analytical techniques, such as electrical pulse counting or light scattering.[2,4] Developments in the application of ultrasound to emulsions have provided food scientists with a powerful new means of characterizing food products.[5,6]

Instruments based on ultrasound can be fully automated, are capable of rapid and precise measurements, are non-destructive and non-invasive and can be used to characterize systems which are concentrated and optically opaque.[5,7] Ultrasonics has been used to measure disperse phase volume fractions, droplet size distributions, droplet crystallinity, droplet-droplet interactions and gravitational separation in emulsions.[7-19] The most important applications of ultrasound in the food industry are described in this article.

2. Ultrasonic Characterization of Emulsions

The physicochemical properties and microstructure of an emulsion can be related to its ultrasonic velocity, c, and attenuation coefficient, α, using ultrasonic scattering theory. In the long wavelength limit:

$$\left(\frac{K}{k}\right)^2 = \left(1 - \frac{3i\phi A_0}{k^3 r^3}\right)\left(1 - \frac{9i\phi A_1}{k^3 r^3}\right) \tag{1}$$

where K and k are the complex wave numbers of the emulsion and continuous phase (= $\omega/c + i\alpha$), ω is the angular frequency ($=2\pi f$), f is the frequency of the applied ultrasonic wave, r is the droplet radius, ϕ is the disperse phase volume fraction, and A_0 and A_1 are the monopole and dipole scattering coefficients of an individual droplet. The magnitude of the scattering coefficients depends on the thermophysical properties of the component phases, including the adiabatic compressibility, density, viscosity, thermal conductivity, specific heat capacity and thermal expansivity.[20,21]

2.1 Determination of Droplet Concentration and Size

The bulk physicochemical and sensory properties of food emulsions are largely determined by the concentration and size of the droplets they contain and therefore it is important to have techniques to measure these properties. The disperse phase volume fraction and particle size distribution (PSD) of food emulsions can be determined using ultrasonic spectrometry.[10,14,22] The ultrasonic velocity and attenuation coefficient of an emulsion are measured over a range of frequencies (typically 0.1 to 100 MHz). The resulting ultrasonic spectra are then analyzed using ultrasonic scattering theory to determine the values of ϕ and PSD which give the best fit between experiment and theory (Equation 1). Ultrasound has already been used to determine disperse phase volume fractions and PSDs of a number of food emulsions, including sunflower oil-in-water emulsions, salad cream and milk.[6] The major advantage of this technique over traditional technologies, such as light scattering and electrical pulse counting, is that concentrated emulsions can be analyzed without the need of any dilution. Consequently, it is possible to develop ultrasonic sensors which are capable of analyzing the properties of emulsions on-line, which is extremely valuable in the food industry. It should be pointed out that Equation 1 gives excellent agreement with experimental measurements up to fairly high droplet concentrations (30-40%) under certain conditions.[23] Nevertheless, it needs to be modified to account for the effects of multiple scattering and droplet interactions before it can be used to analyze ultrasonic measurements in concentrated emulsions.[10,24] As this application of ultrasound is covered in many of the other chapters in this book, we will not discuss it further here.

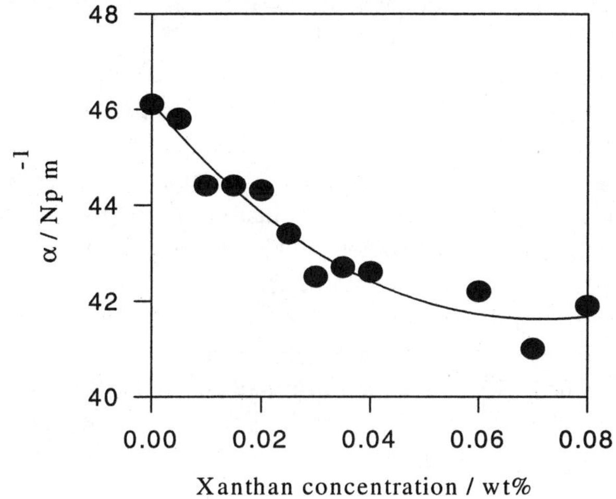

Figure 1. Dependence of the attenuation coefficient of a series of 20 wt % corn oil-in-water emulsions on xanthan concentration. The attenuation decreases because droplet flocculation reduces the energy loss.

2.2 Monitoring of Droplet Flocculation

Emulsions become flocculated when the attractive forces between the droplets dominate the repulsive forces.[2,4] The extent of flocculation, and the structure of any flocs formed, depends on the nature of the colloidal and hydrodynamic interactions between the droplets.[4] Flocculation causes an increase in the viscosity of an emulsion and alters the rate at which droplets cream or sediment.[1] It is therefore important for food scientists to have experimental techniques to monitor droplet flocculation.

Equation 1 assumes that the droplets in an emulsion are isolated and randomly distributed throughout the volume of the sample, which is obviously invalid in flocculated emulsions. In a flocculated emulsion, the amount of energy lost due to the thermal and visco-inertial loss mechanisms associated with the droplets is reduced, which causes a decrease in attenuation.[10,24] However, it is also possible to have an increase in attenuation because of thermal, visco-inertial and scattering loss mechanisms associated with the flocs themselves. As a consequence the attenuation of an emulsion may have a fairly complex dependence on the extent of droplet flocculation. Even so, the dependence of the ultrasonic properties of an emulsion on the flocculation of the droplets indicates that ultrasound could be used to probe droplet interactions. Nevertheless, work is still required to develop a theory which can be used to interpret measurements from flocculated emulsions.

The attenuation coefficient of a series of 20 wt% corn oil-in-water emulsions containing different concentrations of xanthan in the aqueous phase was measured at 30 °C and 5 MHz (Figure 1). Xanthan causes emulsion droplets to flocculate above a critical concentration because of an osmotic attraction between the droplets associated with the exclusion of biopolymer molecules from a narrow region surrounding the droplets.[2,4] The attenuation decreased appreciably as the xanthan concentration was increased from 0 to 0.04% and then leveled off at higher concentrations. These results suggest that the droplets become increasingly close together as the xanthan concentration is increased, until a critical level is reached where they are fully flocculated and cannot get any closer. Work is needed to develop a more quantitative theory for interpreting ultrasonic spectra in emulsions that have undergone substantial flocculation.

2.3 Measurement of Droplet Crystallization

Many oil-in-water emulsions used in the food industry contain droplets which are capable of undergoing phase transitions during their manufacture, storage or consumption. The appearance, texture and stability of these emulsions is largely determined by the amount of crystalline oil within the droplets.[2] Droplets which are partly crystalline are susceptible to *partial coalescence,* which is the process whereby a crystal from one droplet penetrates into another droplet causing an aggregate to form.[2] Extensive partial coalescence causes a pronounced decrease in the stability of an emulsion and an increase in its viscosity. In some foods partial coalescence is undesirable because it causes a decrease in shelf-life (*e.g.* cream, cream liqueurs), whereas in other foods it is an essential step in their manufacture (*e.g.,* ice-cream, butter, margarine and whipped cream). Consequently, it is important for food scientists to be able to monitor droplet crystallisation in emulsions.

The ultrasonic velocity of solids is usually greater than that of liquids, and so the amount of crystalline material in a sample can be ascertained using ultrasound. The fraction of crystalline droplets (ϕ) in an emulsion can be determined from the following equation:[12]

$$\phi = \left(\frac{1}{c^2} - \frac{1}{c_L^2}\right)\left(\frac{1}{c_S^2} - \frac{1}{c_L^2}\right)^{-1} \tag{2}$$

Here c is the ultrasonic velocity of the emulsion, and c_L and c_S are the ultrasonic velocities in the emulsion if all the droplets were completely liquid or completely solid, respectively. These values are determined by extrapolating measurements from the higher and lower temperatures into the region where the oil is partially crystalline or by using ultrasonic scattering theory to calculate their values. The solid contents determined using ultrasound are in good agreement with those determined using traditional techniques such as dilatometry and NMR.[2] In addition, ultrasound is more sensitive to low concentrations of solid and is therefore more suitable for application to low fat products. This technique has been used to monitor phase transitions in non-food oil-in-water emulsions, triglyceride oil-in-water and water-in-oil emulsions, margarine and butter, shortening, meat and various triglyceride/oil mixtures.[2]

In some systems large increases in the attenuation coefficient and appreciable velocity dispersion have been observed during melting and crystallisation because of a relaxation mechanism associated with the solid-liquid phase equilibrium.[2] The ultrasonic wave causes periodic fluctuations in the temperature and pressure of the material which perturb the phase equilibrium. When a significant proportion of the material is at equilibrium a large amount of ultrasonic energy is absorbed due to this process. This effect depends on the ultrasonic frequency, the relaxation time for the phase equilibrium and the amount of material undergoing phase equilibrium, and can cause large deviations in both the velocity and attenuation coefficient.[6] In systems where this phenomenon is important it is not possible to use Equation 2 to interpret the data. Nevertheless, it may be possible to use ultrasound to obtain valuable information about the dynamics of the phase equilibrium.

2.4 Monitoring of Gravitational Separation

Gravitational separation is one of the major forms of emulsion stability in the food industry. The droplets in an emulsion move upwards if they have a lower density than the surrounding liquid ("creaming"), and downwards if they have a higher density ("sedimentation"). Food scientists need analytical techniques which can be used to monitor this process for two reasons: (i) so that they can elucidate the major factors which determine the rate at which this process occurs, and (ii) so that they can predict the long-term stability of food emulsions from measurements made in the initial stages of the process. In the past, it was difficult to monitor this process because of the lack of analytical techniques to non-destructively analyze food emulsions that are concentrated and optically opaque.

Over a decade ago, a number of researchers developed non-destructive techniques based on ultrasound which could be used to monitor gravitational separation in concentrated emulsions.[15-18] Gravitational separation is monitored by measuring the ultrasonic velocity or attenuation coefficient of an emulsion as a function of height, and then using a suitable equation to convert the measurements into a droplet concentration profile.[19] Recently, we have extended this principle by developing an ultrasonic imaging technique which is capable of generating two-dimensional images of the spatial distribution of droplets within an emulsion.

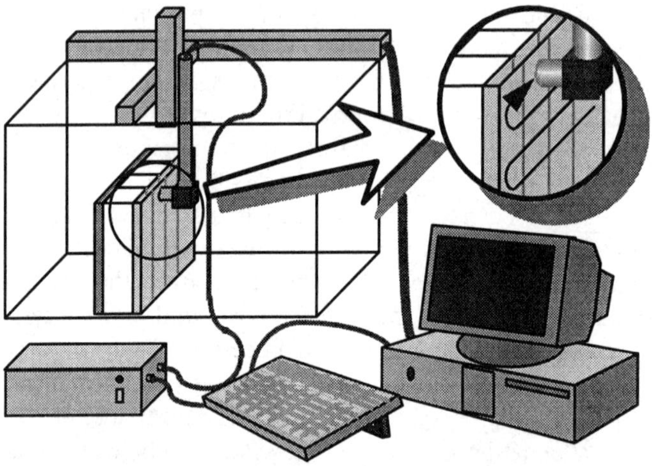

Figure 2. Schematic diagram of the ultrasonic imaging equipment. The ultrasonic transducer is scanned vertically and horizontally over the thermostated cell which contains the emulsions and an A-scan is acquired at each location.

An ultrasonic imaging device that was originally developed to determine the presence of imperfections in polymers and metals was modified so that it could be used to monitor gravitational separation. We custom-made a thermostated measurement cell to hold the emulsions, and modified the computer software so that it could be used to convert the ultrasonic signals into velocity and attenuation measurements. An ultrasonic transducer (5 MHz) is scanned vertically and horizontally over the surface of the measurement cell containing the sample to be analyzed (Figure 2). At each location within the cell an A-scan is acquired and stored by the computer.

The A-scans are analyzed in order to provide an image of the spatial distribution of the droplets within an emulsion. It is possible to analyze each individual A-scan so that ultrasonic velocity and attenuation spectra are obtained at every location within an emulsion. These spectra can then be converted into a disperse phase volume fraction and droplet size distribution using ultrasonic scattering theory [Equation 1]. In reality, this procedure is time consuming and requires large amounts of computer memory. For this reason, it is often more convenient to simplify the signal analysis in order to reduce the amount of time and memory required.

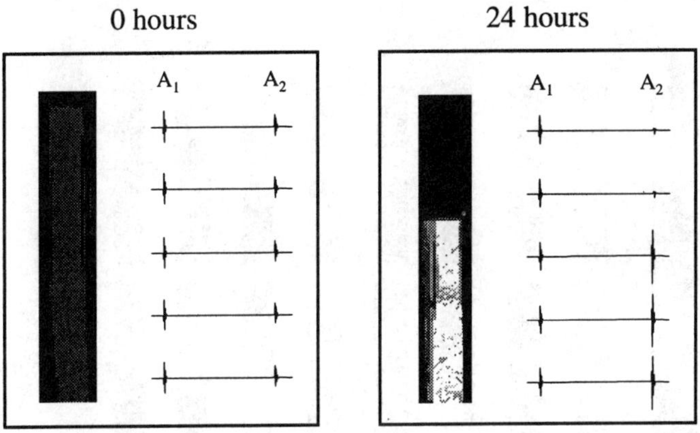

Figure 3. An ultrasonic image can simply be constructed by representing the amplitude of the second echo (A_2) at each position as a gray-scale. The darker the image, the greater the attenuation, and the higher the oil concentration.

We used two different methods to simplify the analysis of the data:

(i) *Qualitative two-dimensional imaging of creaming.* The amplitude of the ultrasonic pulse which travels through the emulsion (A_2) decreases as the droplet concentration increases because of adsorption and scattering of the ultrasonic wave caused by the presence of the droplets (Fig. 3). It is therefore possible to use a single parameter (A_2) to characterize the properties of an emulsion, rather than the whole of the A-scan. A two-dimensional image representing the droplet concentration profile is generated by using a gray-scale to represent A_2: where black corresponds to complete attenuation of the pulse (high droplet concentration) and white corresponds to no attenuation of the pulse (zero droplet concentration). Typical two-dimensional images of an emulsion before and after creaming are shown in Figure 3.

(ii) *Quantitative one-dimensional determination of creaming profile.* A more quantitative assessment of the extent of creaming is obtained by analyzing the A-scans in order to determine the ultrasonic velocity and attenuation spectra. Rather than carrying out this procedure at every location within an emulsion it is possible to use a single vertical slice through the center of the emulsion. The ultrasonic velocity and attenuation are then converted into droplet concentrations using a suitable mathematical theory or empirical equation.

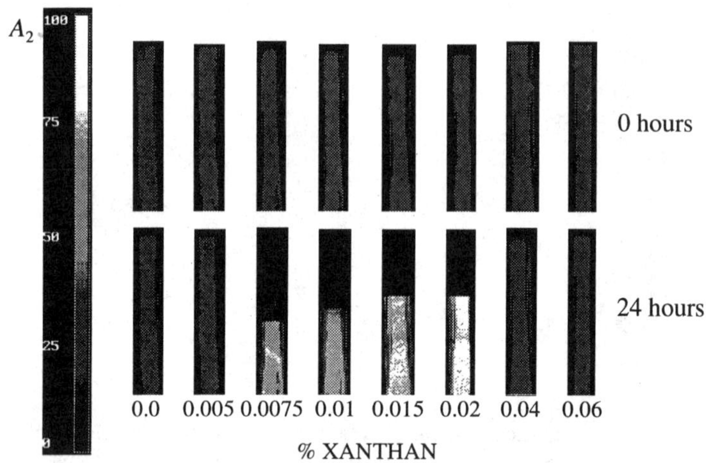

Figure 4. Two dimensional images of the extent of creaming in 20 wt% oil-in-water emulsions containing different concentrations of xanthan.

A series of 20 wt% corn oil-in-water emulsions containing different concentrations of xanthan in the aqueous phase were prepared. Ultrasonic images of the emulsions were acquired after 0 and 24 hours (Figure 4). At the beginning of the experiment, all the images had a uniform gray appearance, which corresponded to homogeneous emulsions with a droplet concentration of 20 wt%. After 24 hours, a significant amount of creaming was observed in the emulsions containing between 0.0075 and 0.02 wt% xanthan (there being an increase in the darkness of the image at the top of the emulsions, and a decrease in the darkness of the image at the bottom). The dependence of the creaming on xanthan concentration can be explained in terms of the influence of xanthan on the colloidal interactions between emulsion droplets and on the rheology of the continuous phase. At low xanthan concentrations the droplets exist as small individual entities and therefore creaming is slow. At intermediate xanthan concentrations the droplets become flocculated because of an osmotic attraction between them which leads to accelerated creaming. At high xanthan concentrations, the increase in viscosity caused by the biopolymer prevents the droplets from moving and therefore no creaming is observed.

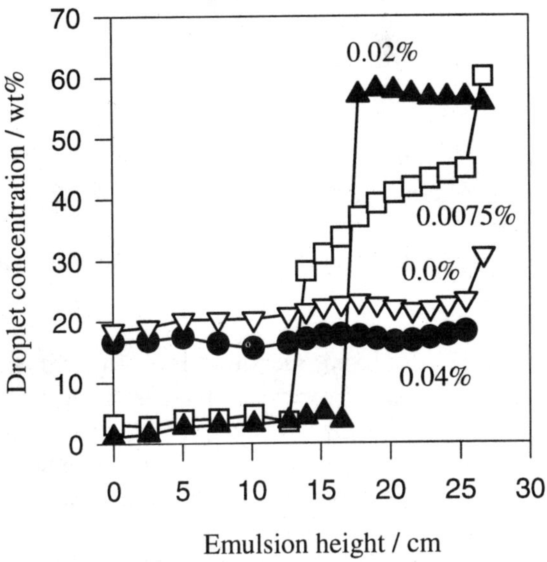

Figure 5. Creaming profiles of the variation of the droplet concentration with vertical height in 20 wt % oil-in-water emulsions containing different concentrations of xanthan.

The ultrasonic images provide a clear qualitative indication of the extent of creaming in the emulsions. Nevertheless, it is often important to have a more quantitative measure of gravitational separation, *i.e.* information about the variation in droplet concentration with emulsion height. For this reason, we also determined the creaming profiles by measuring the ultrasonic velocity as a function of sample height in a vertical plane through the center of the emulsions. The ultrasonic velocity measurements were then converted into disperse phase volume fractions using a previously determined calibration curve. These creaming profiles give a much clearer insight into the details of the creaming process (Figure 5). There was a sharp boundary between the droplet-depleted lower layer and the droplet-rich upper layer in the emulsion containing 0.02 % xanthan. At a xanthan concentration of 0.0075%, the boundary between the upper and lower layers is much less distinct, and there is evidence of compaction of the droplets at the very top of the emulsion. These results indicate that the ultrasonic imaging technique is capable of providing detailed information about the spatial distribution of droplets in concentrated and optically opaque emulsions.

3. Critical Technical Issues

Ultrasound has enormous potential as an analytical tool for characterizing emulsions in the food industry. Nevertheless, there are a number of technical issues that still need to be addressed before the technique finds more widespread use:

1. The theory used to interpret ultrasonic measurements gives excellent agreement with experimental data for semi-dilute emulsions containing non-interacting droplets. Nevertheless, it needs to be extended so that it can be applied to more concentrated emulsions and to take into account the effects of droplet flocculation.

2. The interpretation of ultrasonic measurements requires knowledge of a number of the thermophysical properties of oil and water phases, which are often not readily available in the literature or are difficult to measure. It is therefore important to develop a data base of the relevant thermophysical properties for the oil and aqueous phases commonly found in foods.

3. The presence of small gas bubbles can obscure the signal from emulsion droplets because of their ability to strongly scattering ultrasound. Strategies of overcoming this problem need to be developed, *e.g.* degassing the sample or making measurements at frequencies where the gas bubbles do not scatter significantly.

4. On-line versions of the ultrasonic techniques developed to characterize emulsions should be made commercially available to the food industry.

5. The advantages of the ultrasonic technique over other technologies must be demonstrated for a wide range of food products, so that the food industry gains confidence in the technique.

References

1. E. Dickinson and G. Stainsby, *Colloids in Foods*, Applied Sciences, London, 1982.

2. E. Dickinson and D.J. McClements, *Advances in Food Colloids*, Blackie Academic, Glasgow, 1995.

3. D.J. McClements, "Food Emulsions: Principles and Practice," CRC Press, NY, In Preparation.

4. R.J. Hunter, *Foundations of Colloid Science*, Oxford University Press, Oxford, 1986.

5. D.J. McClements, "Ultrasonic Characterization of Emulsions and Suspensions," *Adv. Colloid Interface Sci.*, **37**, 33 (1991).

6. D.J. McClements, "Ultrasonic Characterization of Foods and Drinks: Principles, Methods and Applications," *Critical Reviews in Food Science and Nutrition*, **37**, 1-46 (1997).

7. K.-E. Forysa and O. Nesse, pp. 437-468 in *Emulsions and Emulsion Stability*. Edited by J. Sjoblom. Marcel Dekker, NY, 1996.

8. D.J. McClements and M.J.W. Povey, "Scattering of Ultrasound by Emulsions," *J. Phys. D*, **22** 38 (1989).

9. D.J. McClements, M.J.W. Povey, M. Jury and E. Betsanis, "Ultrasonic Characterization of a Food Emulsion," *Ultrasonics*, **28**, 266 (1990).

10. N. Herrmann, "Application of Ultrasonic Techniques to the Study of Dispersions," Ph.D. thesis, University of Louis Pasteur, Strasbourg, France, 1997.

11. Y. Hemar, N. Herrmann, P. Lemarechal, R. Hocquart and F. Lequeux, "Effective Medium Model for Ultrasonic Attenuation due to the Thermo-Elastic Effect in Concentrated Emulsions," *J. Phys. II France,* **7**, 637 (1997).

12. D.J. McClements, E. Dickinson and M.J.W. Povey, "Crystallization of Hydrocarbon-in-Water Emulsions Containing a Mixture of Solid and Liquid Droplets," *Chem. Phys Letts.*, **172**, 449 (1990).

13. D.J. McClements, E. Dickinson, S.R. Dungan, J.E. Kinsella, J.G. Ma and M.J.W. Povey, "Effect of Emulsifiers Type on the Crystallization Kinetics of Oil-in-Water Emulsions Containing a Mixture of Solid and Liquid Droplets," *J. Colloid Interface Sci.*, **160**, 293 (1993).

14. U. Riebel and F. Loffler, "The Fundamentals of Particle Size Analysis by Means of Ultrasonic Spectrometry," *Part. Part. Syst. Charact.*, **6**, 135 (1989).

15. D.J. Hibberd, A.M. Howe and M.M. Robins, "Use of Concentration Profiles in Emulsions to Determine Phase Coexistence," *Colloids Surf.*, **31**, 347 (1988).

16. A.M. Howe, A.R. Mackie and M.M. Robins, "Technique to Measure Emulsion Creaming by Velocity of Ultrasound," *J. Disp. Sci. Tech.*, **7**, 231 (1986).

17. A.M. Howe and M.M. Robins, "Determination of Gravitational Separation in Dispersions from Concentration Profiles," *Colloid Surf.*, **43**, 83 (1990).

18. D.J. Hibberd, A.M. Howe, A.R. Mackie, P.W. Purdy and M.M. Robins, "Measurement of Creaming Profiles in Oil-in-Water Emulsions," in *Food Emulsions and Foams*. Edited by E. Dickinson. Royal Society of Chemistry, London, 1987.

19. V.J. Pinfield, E. Dickinson and M.J.W. Povey, "Modeling of Concentration Profiles and Ultrasound Velocity Profiles in Creaming Emulsion: Importance of Scattering Effects," *J. Colloid. Int. Sci.*, **166**, 363 (1994).

20. P.S. Epstein and R.R. Carhart, "The Absorption of Sound in Suspensions and Emulsions: I. Water Fog in Air," *J. Acoust. Soc. Am.,* **25**, 533 (1953).

21. J.R. Allegra and S.A. Hawley, "Attenuation of Sound in Suspensions and Emulsions: Theory and Experiment," *J. Acoust. Soc. Am.,* **51**, 1545 (1972).

22. D.J. McClements, "Principles of Ultrasonic Droplet Size Determination," *Langmuir*, **12**, 3454 (1996).

23. D.J. McClements, "Comparison of Multiple Scattering Theories with Experimental Measurements in Emulsions," *J. Acoust. Soc. Am.*, **91**, 849 (1992).

24. D.J. McClements, "Ultrasonic Determination of Depletion Flocculation in Oil-in-Water emulsions," *Colloids Surf. A.*, **90**, 25 (1994).

Time Domain Dielectric Spectroscopy: Electrode Polarization Correction for Applications to Aqueous Organic Particulate Suspensions

Yuri Feldman
Department of Applied Physics, The Hebrew University of Jerusalem, 91904, Jerusalem, Israel

1. Introduction

The recent appearance of Time Domain Dielectric Spectroscopy (usually called TDS) precision measuring systems[1-5] may help resolve the problems of accuracy and sensitivity associated with this method when investigating different kinds of substances. However, even in the case of two channel bridge measuring systems it is necessary to make special calibration measurements and considerable calculations in order to overcome many of the disadvantages inherent in TDS, such as shortcomings in the construction of coaxial lines, samplers, step generators, etc. These distortions and instabilities appear in the form of an altered shape of the incident pulse that cannot be compensated for either by the ordinary difference method[6-8] or by apparatus correction of the pulse shape. In particular, these unwanted distortions have been found to be crucial in the case of systems with parallel time nonuniform sampling.[9] Corrections of the incident pulse shape by apparatus adjustments in the case of nonuniform sampling of the signals do not lead to noticeable improvement because of procedure limitations such as time borders of the "time window". As a rule, distortions increase outside time interval bounds. In addition, TDS systems inherently have various general hardware problems due to instability in the time axis including jitter, time drift, and variation of the sweep speed in the oscilloscope. These problems are all associated with the high frequency limitations of the measuring system. On the other hand, the slow drifts of the base line and effects of truncation at long finite time limits are primarily connected with low-frequency limitations.[1,2,10]

Recent developments in high speed sampling techniques and digital processing, together with advances in the basic theory of TDS, have brought about a new generation of TDS measurement systems. We describe here such an instrument based on Time Domain Measuring Set (TDMS) especially designed for TDS hardware support. The system utilizes the difference method of measurement,[8] and the registration of primary signals with parallel time nonuniform sampling.[2,9]

2. Basic Principles of the Method

TDS is derived from transmission line theory in the time domain that aids in the study of heterogeneities in the coaxial lines according to the change of the test signal shape.[1,2] Until the line is homogeneous the shape of this pulse is not changed. But, in the case of heterogeneity in the line (the dielectric put into it, for example) the signal is partly reflected from the air-dielectric interface and partly passes through it. Variations include mainly the construction of the measuring cell and its position in the coaxial line. These variations lead to the different kinds of expressions for the values that are registered during the measurement and dielectric characteristics of the samples under study.

The technique used in this work is called the *lumped capacitance method.*[2,11-14] In this method the sample cell may be treated like a capacitor with capacitance $\varepsilon^*(\omega)C_0$ where $\varepsilon^*(\omega)$ is the sample permittivity and C_0 is the capacitance of the empty capacitor, which can be calculated from its geometric parameters. The voltage V(t) applied to the sample terminated to a coaxial line can be written in the following way:[11]

$$V(t)=V_0(t)+R(t) \tag{1}$$

where $V_0(t)$ and R(t) are the incident and the reflected signals respectively. The expression for the current flow through the sample is:

$$I(t)=\frac{1}{Z_0}\left[V_0(t)-R(t)\right] \tag{2}$$

where Z_0 is the characteristic impedance of the empty line.

The total current through the conducting dielectric is composed of the capacitor charging current $I_Q(t)$, and the low-frequency current between the capacitor electrodes $I_R(t)$. Since the active resistance at zero frequency of the cell containing sample is:[11-13]

$$r = \lim_{t\to\infty} \frac{V(t)}{I(t)} = Z_0 \lim_{t\to\infty} \frac{V_0(t)+R(t)}{V_0(t)-R(t)} \tag{3}$$

the low-frequency current can be expressed as:

$$I_R(t) = \frac{V(t)}{r} = \frac{V_0(t)+R(t)}{Z_0} \lim_{t\to\infty} \frac{V_0(t)-R(t)}{V_0(t)+R(t)} \tag{4}$$

Thus relation (2) can be written as:

$$I_Q(t) = \frac{1}{Z_0}\left\{\left[V_0(t)-R(t)\right]-\left[V_0(t)+R(t)\right] \lim_{t\to\infty} \frac{V_0(t)-R(t)}{V_0(t)+R(t)}\right\} \tag{5}$$

Relations (1) and (5) present the basic equations, related to the experimental signals recorded. The electric charge of the capacitor filled with the sample $Q(t)$ can be related to the dielectric response function Φ(t) and the applied voltage $V(t)$ by the relation:

$$Q(t) = C_0\left[\varepsilon_\infty V(t) + \int_0^t \dot{\Phi}(t-t')\, V(t')\, dt'\right] \tag{6}$$

Where C_0 is the empty cell capacitance. Taking into account that

$$Q(t) = \int_0^t I_Q(t')\, dt' \tag{6'}$$

and performing Fourier transform on Eq. (6), using the general equation[12]

$$\varepsilon^*(\omega) = \varepsilon_\infty + \int_0^\infty \Phi(t)\, e^{-i\omega t}\, dt \tag{6''}$$

one arrives at an expression for the complex dielectric permittivity of the sample:

$$\varepsilon^*(\omega) = \frac{1}{i\omega C_0 Z_0} \frac{L[I_Q]}{L[V_0(t)+R(t)]} \tag{7}$$

where L is the Laplace transform operator:

$$F(s) = L[f(t)] = 0\int_0^{\infty} f(t)\, e^{-st}\, dt \qquad (7')$$

where $s=\gamma+i\omega$ is complex frequency .

Usually the various modifications of differential methods are used.[2] The essence of the differential method is in recording standard sample signals with known dielectric characteristics and signals of the object being tested. This approach allows the compensation for undesirable reflections within the coaxial line and the increase in sensitivity while measuring weak signals (for example the measurement of dilute solutions of some samples with respect to a solvent)[8]. However, the ordinary difference method does not lead to the desired results because in addition to the two signals from the reference and unknown samples, a signal from the short circuit line is necessary as well.[7,8] Therefore, the distortions can only be partially compensated.

3. Experimental Tools

3.1 Hardware

This Time Domain Measuring Set (TDMS) has been especially designed for TDS hardware support. TDMS offers a comprehensive, high precision, and automatic measuring system. It has a small jitter-factor (< 1.5 ps), important for rise-time; small flatness of incident pulse (<0.5 % for all amplitudes), and a unique option for parallel time nonuniform sampling of the signal. It consists of a signal register, 2-channel sampler, and a pulse generator. The generator produces 200 mV pulses with 10μs duration and 28 ps rise time. Two sampler channels are characterized by an 18 GHz bandwidth and 1.5 mV noise (RMS). Both channels are triggered by one common sampling generator that provides their time correspondence during operation. The form of the voltage pulse thus measured is digitized and averaged by the digitizing block of TDMS. Time base is responsible for major metrology TDMS parameters.[2] Two program-controlled devices provide the TDMS time scale: a digital delay unit (DDU), and an analog delay unit (ADU).

A sequential DDU-ADU connection provides full time scale up to 10240 ns with unlimited step and initial delay within this time. TDMS can form variable step discretization sweeps. It gives us the unique opportunity of special sweeps for realization nonuniform sampling. Thus, it is possible to perform oscillation spectrum analysis, containing different time constants, corresponding to relaxation times of tested substances (from 10 ps to 10 μs). The minimum discrete step is 0.1 ps. The maximum equivalent sampling rate is 1000 GHz. Due to refined design methods, time jitter of TDMS is less than 1.5 ps (RMS). ADU non-linearity and drift are compensated for by the addition of special devices to the basic circuitry. ADU drift measurement is performed by checking two reference points of a 2 GHz sine wave. The 2 GHz generator is synchronized by quartz and a special calibration programs. The first reference point must be chosen at the beginning of the ADU time band; the second point at the end. The difference between these values is proportional to ADU drift. The sweep is given by means of 16-bit DAC and ADU on diode with the charge accumulation. A microprocessor program calculates this difference, inverts it and writes it to the TDAC buffer for drift compensation. The resulting error of the sweep within the range up to 40 nsec does not exceed 5 ps. A microprocessor program calculates this difference, inverts it and writes it to the TDAC buffer for drift compensation.

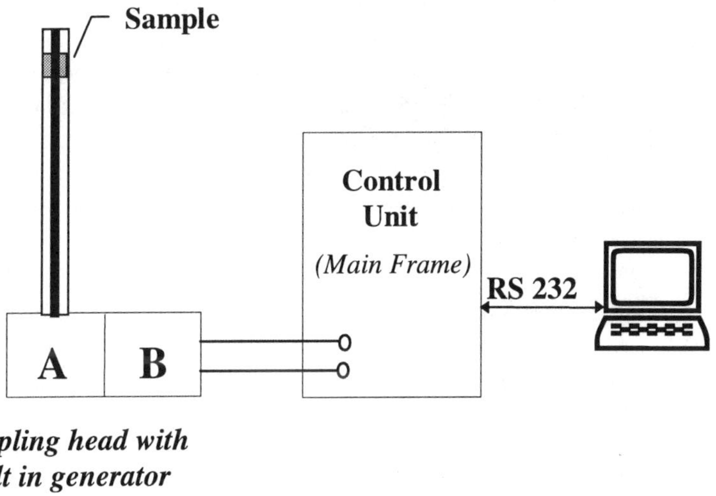

Figure 1. Circuit diagram of a TDS set-up.

Pulse generator and sampler have their own independent drifts. Such drifts are eliminated by an autostabilization program called "auto center" through autostabilization ADC control. For ADC control two DAC are used. One is controlling the vertical position of the signal, another one the time delay position. This program checks the position of the reference signal that is formed from the generator pulse in the microwave measuring circuit, and is received by any sampler channel. Two reference points (one at the stable part of the reference signal, and the other one at the pulse edge) provide data for the program about vertical and time drifts. ADC control codes are calculated on the basis of this information analysis. Drift compensation and autostabilization programs run in "hidden" mode, after each sweep cycle. It allows the reduction of overall signal drifts up to 1 ps and significantly increases precision and performances of TDMS. A block diagram of this set-up TDS-2 (Dipole TDS, Ltd., Jerusalem) is presented in Figure 1.

3.2 Software

Measurement procedures, registration, storage, time referencing, and data analysis are carried out automatically. The process of operation is done in an on-line mode and the results can be presented both in the frequency domain and in the time domain. There are several features of the modern software that control the process of measurement and calibration. One can define the time windows of interest that may be overlapped by one measurement. During the calibration procedure the precise determination of front edge position is carried out and the setting of internal auto center on this position applies to all of the following measurements. The precise determination and settings of horizontal and vertical positions of calibration signals are carried out also. All parameters are saved in a configuration file. This allows one to do a series of measurements with the same parameters without additional calibration. Some details of this software will be described in the following section.

i) The principle of building of the multi-window time scale: The implemented time scale is the piecewise approximation of the logarithmic scale. It includes $nw \subseteq 16$ sites with the uniform discretization step determined by the following formula:

$$\delta_{nw} = \delta_1 \acute{I} 2^{nw}, \tag{8}$$

where δ_1=5 ps is the discretization step at the first site, and the number of points in each step except for the first one , are set equal to npw=32. At the first site, the number of points $npw_1 = 2\acute{I}npw$. Doubling the number of points at the first site arises from the necessity of having a formal zero time position, which is impossible in the case of a strictly logarithmic structure of the scale. In addition, a certain number of points located in front of the zero time position are added. They serve exclusively for the visual estimation of the stability of the time position of a

signal and are not used for the data processing. For a given version of the program this number is equal to 25. This is the number of the point where unprocessed signals join if the difference procedure is applied.

This structure of the time scale allows the overlapping of the time range from 5 ps to 10 μs during one measurement, which results in a limited number of registered readings. At will, the overlapped range can be shortened, resulting in a decreasing number of registered points and reducing the amount of time required for data recording and processing.

The major advantage of the multi-window time scale is the ability to get more comprehensive information. The signals received by using such a scale contain information within a very wide time range and the researcher merely decides which portion of this information to use for data processing. Also, this scale provides for the filtration of registered signals close to the optimal one already at the stage of recording.

ii) Calibration: For calibration, the short circuit terminator must be used in a configuration where the terminator is connected to the measuring line. Then, by using the appropriate choice, the calibration program bootstraps. The calibration procedure runs automatically if the base block works normally and the operator makes no error. Otherwise, the diagnostic part of the program determines a possible reason of failure and a message on how to operate in this case is generatied. The result of calibration is the formation of the time scale with the corresponding parameters and the recording of the detected parameters into a storage file. The calibration also determines the adjusted coefficients of the channel sensitivity, which later can be defined and used for normalization of the measured signals.

iii) Measurements: The sample cell must be terminated at the measuring line, the appropriate menu choice must be made, and the measurement program must be run. With every access to the program the state of the base block is checked, and the adjusted coefficients of the channel sensitivity, baseline position, and the amplitude of the probing signal are defined. These parameters are all used for the signal scaling at the registration stage. If errors or failures in the base block operation are found and if they can be eliminated by the operator, a help message appears.

iv) Data Processing: The data processing software includes the options of signal corrections, correction of electrode polarization, and dc conductivity, and different fitting procedures in the time and frequency domains.[2] For the case of nonuniform sampling the Samoloon algorithm[13] of numerical Fourier transform is used, and can be presented in the following way:

$$F(\omega_j)=\frac{\delta_1}{1-e^{-i\omega_j\delta_1}}\sum_{n=2}^{N_{int}}\{f(n\delta_1)-f[(n-1)\delta_1]\}e^{-i\omega_j n\delta_1}$$

$$+\sum_{k=2}^{N_{int}}e^{-i\omega_j\sum_{m=1}^{k-1}T_m}\frac{\delta_k}{1-e^{-i\omega_j\delta_k}}\sum_{n=1}^{N_{int}}\{f(n\delta_k)-f[(n-1)\delta_k]\}e^{-i\omega_j n\delta_k} \qquad (9)$$

where: N_{int} is the number of uniform measurement intervals, δ_k is the sampling rate in a uniform time interval (k=1,..., N_{int}), N_k is the number of points in uniform interval k (k=1,..., N_{int}), ω_j is a frequency point, in which Fourier Transform is calculated, and T_m is the total current time interval. The fitting programs allow one to determine spectroscopic parameters and correlation analysis in time domain.

3.3 Sample Holders

A universal sample holder that can be used for both liquid and solid samples in the low and high frequency borders of the TDS method is unfortunately as yet unavailable. The choice of its configuration depends on the measurement method and data treatment procedure. In the framework of the lumped capacitor approximation one can consider three general types of sample holders:[2,5] (a) A cylindrical capacitor filled with sample. Such a cell can also be regarded as a coaxial line segment with the sample having an effective γd (γ is the ratio of the vacuum capacitance per unit length of the cell to that of the matched coaxial cable) length characterized in this case by the corresponding spread parameters. This makes it possible to use practically identical cells for various TDS method modifications. For the total reflection method it is the most frequently used configuration;[1,3,4,6,10,14,15] (b) a plate capacitor terminated to the central electrode on the end of the coaxial line;[4,12-13] (c) an open-ended coaxial line sensor.[5] In the case of the lumped capacitance approximation, (a) and (b) configurations have high frequency limitations and for highly polar systems one must take into account the finite propagation velocity of the incident pulse or, in other words, the spread parameters of the cell.[1] The choice of cell shape is determined to a great extent by the aggregate condition of the sample studied. While cell (a) is convenient for measuring liquids, configuration (b) is more suitable for the study of solid disks and films. Both cell types can be used to measure powder samples. While studying anisotropic systems (liquid crystals, for instance) the investigator may replace a coaxial line by a strip line or construct the cell with the configuration providing the measurements under various directions of the applied electric field.[3,5] A third, (c) type, cell is used only when it is impossible to put the sample into an (a) or (b) type cell.[5,16] The fringing capacity of the coaxial line end is the working capacity for such a cell. The sample holders described above can be thermostabilized.

4. Errors

The error sources for various TDS modifications are practically identical.[2,17] All of the sources are independent, so the total error is equal to the sum of all the components. We will classify two kinds of errors of different nature: hardware errors and computational errors. The level of hardware errors can be estimated by their influence on the measuring time signals. The evaluation of computional errors is based on the calculation of the permittivity spectrum of model signals in time domain produced by a hypothetical Debye-type dielectric with a known permittivity. A unique aspect of the technique proposed is the differentiation of hardware errors relative to certain sources, while computed errors are considered on the whole. The accuracy of this system depends on the frequency range, dielectric permittivity, and working capacitance of the sample cell.[20-22]

5. Methodical Aspects of TDS Measurements

Here we present some of the TDS measurement difficulties that can be encountered in the study of different dielectric samples. It must be noted that the accuracy and resolution of the set-up are strictly dependent on the polarity and dc conductivity of the sample. The most convenient and simple samples for TDS measurements are the pure polar nonconductive liquids that can be used as reference samples for TDS set-up testing. Figure 2 presents the Cole-Cole plots of pure alcohols. The range of dielectric permittivities allows one to be in the framework of lumped capacitance up to about 8 GHz.

The automatic fitting parameters are presented in Table I, and are in good agreement with those found in the literature.[23,24] The TDS-2 system shows reasonable accuracy for such a system when tested in both time and frequency domains.

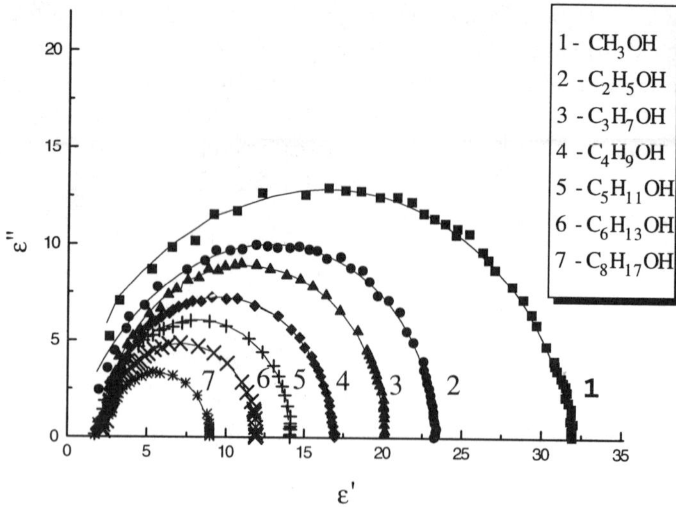

Figure 2. Cole-Cole plot for normal alcohols at 30°C.

In contrast to simple samples, complex samples such as aqueous suspensions of organic nanoparticulates need a wide range of frequencies to be overlapped by one measurement. For example, biological systems, solutions, mixtures, microemulsions, emulsions, and nanoparticulate dispersions, etc., have several relaxation processes in different time scales. The multiwindow system is very useful for the study of such systems.

Table I. Dielectric fitting parameters for normal Alcohols at 30°C.

Alcohol	ε_S	ε_∞	τ (ps)	Standard deviation
methanol	31.9	2.8	54	0.048
ethanol	23.6	2.5	132	0.044
propanol	19.6	2.4	330	0.022
butanol	16.7	2.4	425	0.012
pentanol	14.2	2.2	665	0.0064
hexanol	12.1	2.4	810	0.0028
octanol	8.9	2.3	1190	0.0012

In the case of low polarity systems (polymers, polar molecules in non-polar solvent, etc.) the main difficulty of TDS set-up resolution is the weak signal. There are two options for increasing the sensitivity of the spectrometer. One is to increase the working capacitance of the sample cell and another is to increase the accumulation number. Thus, for the evaluation of the reliable dielectric spectrum for the samples of polymer films or rubber one has to apply $\sim 10^5$ accumulation, which takes almost 40 min. for single-signal registration.[2]

Many dielectric materials, such as aqueous suspensions of nanoparticulates, are conductive. The TDS study of conductive samples is very complex because of electrode polarization and the need to correct for the effect of low frequency conductivity. Usually for a low conductive system the value of dc conductivity can be evaluated as described above.

One of the greatest obstacles in TDS measurements of conductive systems is the parasitic effect of electrode polarization. The accumulation of charge on the surfaces of the electrodes leads to the formation of electrical double layers. The magnitude of the capacitance caused by electrode polarization is so great that its correction is absolutely necessary for obtaining accurate measurements with conductive samples, particularly for aqueous biological systems and for aqueous suspensions in general. Many approaches[7,25-27] have been tried in order to correct for these effects of electrode polarization. However, the nature of this phenomenon remains unclear, such that no simple technique for its correction is available. Here we will briefly describe one way to correct for this phenomenon.[29,30] For more than fifty years the highest frequency limit of electrode polarization was considered to be of the order of 200-500 kHz. The Time Domain Dielectric Spectroscopy (TDDS) measurements of aqueous solutions of proteins and cell suspensions in the frequency range 200 kHz to 5 GHz [27] showed that this parasitic phenomenon cannot be neglected at much higher frequencies. Moreover, this effect strongly depends on the conductivity of the system being examined and on the material of the electrodes. Schwan was the first to mention the porous or fractal nature of electrode polarization phenomena[27] and called it a "mystery". He also mentioned the increase of this influence with increasing frequency.

Electrical properties of fractal electrodes usually are displayed in complex impedance relationship. In the frequency region this fact is reflected in the appearance of specific dependencies of complex impedance $Z(j\omega)$ from frequency[29,30]

$$Z_{\nu}(j\omega) = A(\nu)(j\omega)^{-\nu}, \tag{9}$$

where the fractional exponent ν is a value lying in the interval $0<\nu<1$, $A(\nu)$ is a constant, and the frequency ω is located in a limited frequency range $\omega_{min} \leq \omega \leq \omega_{max}$, due to the finite scaling factors of electrode surface. The exponent ν is connected with the fractal dimension of an electrode surface. This impedance behavior which occurs for a wide class of electrodes allows us to introduce a new element with impedance characteristics with negative exponent ν.[29,30] We call this element a <u>recap</u> element which is formed from two words re(sistance) and cap(acitance). The electric and fractal properties for the electrical branch circuits in terms of recap element can be presented as follows:

$$\Lambda_{\nu_i}(p) = R(RC)^{-\nu}\, p^{-\nu} \quad \text{for } 0 < \nu < 1. \tag{10}$$

This element for various ν occupies an intermediate position between R (ν=0) and C (ν=1) and realize its impedance only in *finite interval of frequencies.* Using this definition, one can rearrange the equivalent circuit of the measuring sample cell with electrolyte solution as shown on Figure 3.

$$\Lambda_{\nu_i}(p) \equiv \Lambda_{\nu_i} p^{-\nu_i} = R_i\,(R_i C_i)^{-\nu_i}\, p^{-\nu_i} \tag{11}$$

are two *recap* elements (*i*=1,2) taking into account the fractal properties of two electrodes, C and R are capacitance and resistance of the measuring cell respectively. Also, it is interesting to note that in the time domain the multiplication operation is equivalent to the linear operation of fractional integration[29]

$$p^{-\nu}\, I(p) = \frac{1}{\Gamma(\nu)} \int_0^t (t-t')^{\nu-1}\, I(t')\, dt \tag{12}$$

where I(t') is an a.c. current, Γ(ν) is a gamma-function, and p is the complex frequency of the Laplace transform:

$$I(p) = \frac{1}{\Gamma(\nu)} \int_0^t I(t)\, e^{-pt}\, dt \tag{13}$$

This current allows us to consider the electrical properties of the cell and establish a relationship between the voltage and the current.

$$V(t) = \frac{\Lambda_1}{\Gamma(\nu_1)} \int_0^t (t-t')^{\nu 1-1}\, I(t')\, dt' + \frac{\Lambda_2}{\Gamma(\nu_2)} \int_0^t (t-t')^{\nu 2-1}\, I(t')\, dt'$$

$$+ \frac{1}{c} \int_0^t e^{-(t-t')/\tau_c}\, I(t')\, dt' \qquad (14)$$

where τ_c=RC. Equation (14) allows us to consider relationships between V(t) and I(t) for different kinds of measuring cells with electrolyte solution that can be described by the equivalent circuit presented in Figure 3. In the case of TDS measurements this approach can be successfully applied for most sample cells. Let us consider the case where I(t) = constant. For this case one can write:

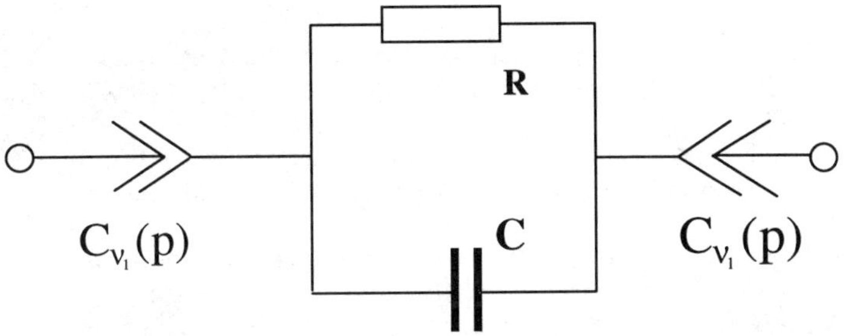

Figure 3. The equivalent circuit for a conductive dielectric sample with electrode polarization impedance described by recap, $C_\nu(p)$.

$$V(t) = \left(\frac{t}{\tau_{\nu 1}}\right)^{\nu 1} + \left(\frac{t}{\tau_{\nu 2}}\right)^{\nu 2} + R\,(1 - e^{-t/\tau_c}) \qquad (15)$$

where τ_c will define the time limits of the real contribution of *recap* elements (electrode polarization) to V(t) applied to the sample holder containing a conductive aqueous suspension. Assuming that both electrodes of sample cell have the same fractal surface (Λ_1=Λ_2) and that there is no dispersion of the conductive suspensions in the asymptotically long time window, Equation (15) can be rewritten in the following way:

$$V(t) = A_1\, t^{\nu 1} + R\, t/\tau_c \quad \text{for } t/\tau_c << 1; \qquad (16)$$

$$V(t) = A_1 \, t^{\nu_1} \quad \text{for} \;\; t/\tau_c >> 1. \tag{17}$$

In thc case of TDS measurements, the relation (17) can be very useful in order to subtract the contribution of electrode polarization ($A_l t^{\nu_l}$) from V(t). In the case of conductive liquids both the voltage and the current flow are influenced by electrode polarization. The applied voltage V(t) has an asymptotically increasing component in the TDS observed time window and the current I(t) has an asymptotically decreasing component. The subtraction of the function $A_1 t^{\nu_l}$ from V(t) and adding this function to I(t) will correct the original signals for electrode polarization parasitic effect in all time interval of TDS system.

References

1. R.H. Cole, J.G. Berberian, S. Mashimo, G. Chryssik, A. Burns, and E. Tombari., "Time Domain Reflection Methods for Dielectric Measurements to 10 GHz," *J. Appl. Phys.*, **66** [24] 793-802 (1989).

2. Yu. Feldman, A. Andrianov, E. Polygalov, G. Romanychev, I. Ermolina, Yu. Zuev, and B. Milgotin "Time Domain Dielectric Spectroscopy: An Advanced Measuring System," *Rev. Sci. Instrum.,* **67** [9] 3208-3216 (1996).

3. R. Nozaki and T.K. Bose., "Broadband Complex Permittivity Measurements by Time-Domain Spectroscopy," *IEEE Trans. Instrum. Meas.* **39** [6] 945-951 (1990).

4. N.E. Hager III, "Broadband Time-Domain-Reflectometry Dielectric Spectroscopy using Variable-Time-Scale Sampling," *Rev. Sci. Instrum.,* **65** [3] 887-891 (1994).

5.Yu.D. Feldman, Yu.F. Zuev, E.A. Polygalov, and V.D. Fedotov, "Time Domain Spectroscopy. A New Effective Tool for Physical Chemistry Investigation," *Colloid Polym. Sci.,* **270** [8] 768-780 (1992) .

6. R.H. Cole, S. Mashimo, and P. Winsor, IV, "Evaluation of Dielectric Behavior by Time Domain Spectroscopy 3. Precision Difference Methods," *J. Phys. Chem.*, **84** [3] 786-780 (1980).

7. S. Bone, "Time Domain Reflectometry: The Difference Method Applied to Conductive Aqueous Solutions," *Biochim. Biophys. Acta.*, **967** [4] 401-407 (1988).

8. Yu.D. Feldman, Yu.F. Zuev, I.V. Ermolina, and V.A. Goncharov, "Time Domain Dielectric Spectroscopy with Nonuniform Signal Discretization", *Russian J. Phys. Chem.* **62** [2] 269-271 (1988).

9. I.V. Ermolina, E.A. Polygalov, G.D. Romanychev, Yu.F. Zuev, and Yu.D. Feldman, "Time Domain Dielectric Spectroscopy with Nonuniform Signal Discretization", *Rev. Sci. Instrum.,* **62** [8] 2262-2265 (1991).

10. D. Bertolini, M. Cassettari, G. Salvetti, E. Tombari, and S. Veronesi, "Time Domain Reflectometry to Study the Dielectric Properties of Lliquids: Some Problems and Solutions," *Rev. Sci. Instrum.*, **61** [12] 450-456 (1990).

11. Yu.D. Feldman, V.A. Goncharov, Yu.F. Zuev, and V.M. Valitov, "Treatment of TDS Data for the Lumped Capacitance Method; Time Domain Treatment," *Chem. Phys. Lett.,* **58** [2]

304-308 (1978).

12. Yu.D. Feldman, V.A. Goncharov, Yu.F. Zuev, and V.M. Valitov, "Treatment of TDS Data for the Lumped Capacitance Method; Frequency - Domain Treatment," *Chem. Phys. Lett.*, **65** [1] 69-70 (1979).

13. V.A. Goncharov and Yu.D. Feldman, "Treatment of TDS Data for the Lumped Capacitance Method. Further Consideration," *Chem. Phys. Lett.*, **71** [3] 513-518 (1980).

14. J.G. Berberian and R.H. Cole, "Cell Design for Low-Temperature Time-Domain Reflectance Measurements," *Rev. Sci. Instrum.*, **63** [2] 99-103 (1992).

15. J.G. Berberian, "Time Domain Reflectometry: Bilinear Correction and Extending the Range of Analysis beyond the Quarter and Half Wavelength Conditions", *J. Molec. Liquids.*, **56** [1] 1-18 (1993).

16. C. Gabriel, E.H. Grant, and I.R. Young, "Use of Time Domain Spectroscopy for Dielectric Properties with a Coaxial Probe," *J. Phys. E: Sci. Instrum.*, **19** [7] 843-846 (1986).

17. A.W. Dawkins, R.J. Sheppard, and E.H Grant, "An On-Line Computer-Based System for Performing Time Domain Spectroscopy II. Analysis of the Errors in Total Reflection TDS", *J. Phys. E: Sci. Instrum.*, **14** [8] 1260-1265 (1981).

20. M.A. Rzepecka and S.S. Stuchly, "A Lumped Capacitance Method for the Measurement of the Permittivity and Conductivity in the Frequency and Time Domain - A Further Analysis," *IEEE Trans. Instrum. Meas.*, **IM-24** [1] 27-32 (1975).

21. K. Baba and T. Fujimura, "Time Domain Analyses for Various Types of Dielectric Relaxations by a Lumped Capacitor Method," *Jpn. J. Appl. Phys.*, **25** [3] 487-490 (1986).

22. G. Romanychev, I. Ermolina, E. Polygalov, Yu. Zuev, D. Upshinsky, and Yu. Feldman, "Analysis of Error Sources in Time Domain Dielectric Spectroscopy Method," *Measurement Technique R.*, **35** [2] 61-63 (1992).

23. S.K. Garg and C.P.Smyth, "Analysis of Error Sources in Time Domain Dielectric Spectroscopy Method," *J. Phys. Chem.*, **69** [9] 1294-1301 (1965).

24. V.V.Levin and Y.D.Feldman, "Dipole Relaxation in Normal Aliphatic Alcohols," *Chem. Phys. Lett.*, **87** [1] 162-164 (1982).

25. H.P. Schwan, "Electrode Polarization Impedance and Measurements in Biological Materials," *Ann. New York Acad. Sci.*, **148,** 191-209 (1968).

26. S. Takashima, *Electrical Properties of Biopolymers and Membranes*, IOP Publishing Ltd., 1989.

27. H.P. Schwan, "Linear and Nonlinear Electrode Polarization and Biological Materials," *Annals of Biomedical Engineering*, **20**, 269-288 (1992).

28. R. Lisin, B. Ginzburg, M. Schlesinger, and Yu. Feldman, "Time Domain Dielectric Spectroscopy Study of Human Cells. I. Erythrocytes and Ghosts," *BBA*, **1280,** 34-40 (1996).

29. R. Nigmatullin, Ya. Ryabov, E. Polygalov, and Yu. Feldman "The Fractal Nature of Electrode Polarization," pp. 164-167 in *Proceedings of the 7-th International Conference on Dielectric Materials Measurements & Applications,* University of Bath, UK, 23-26 September 1996.

30. Yu. Feldman, R. Nigmatullin, E. Polygalov, and J. Texter, "The Electrode Polarization Correction in Time Domain Dielectric Spectroscopy," submitted to *Phys. Rev. E*, (1998).

Dielectric Relaxation Spectroscopy of Biological Cell Suspensions

Koji Asami
Institute for Chemical Research, Kyoto University, Uji, Kyoto 611, Japan

Abstract

This paper is concerned with dielectric behavior of biological cell suspensions in the radio frequency range. Attention is focused on the dielectric relaxation arising from interfacial polarization (Maxwell-Wagner mechanism). The following topics related to electrical cell models and measurement techniques are described: (1) analyses of dielectric relaxation spectra of various biological cells using electrical cell models in which intracellular structure is taken into account; (2) effects of cell shape on the dielectric relaxation spectra, which are examined using an ellipsoidal cell model; (3) dielectric monitoring of yeast growth in culture by the electrode-less method with an inductive probe, which is free from an artifact due to electrode polarization that is a serious problem in monitoring; (4) attempt at dielectric imaging of individual biological cells using the "scanning dielectric microscope", which can image local capacitance and conductance measured with a fine probe electrode that is scanned over a sample in an aqueous medium.

1. Introduction

Dielectric spectroscopy is of growing importance in characterization of colloidal dispersions, including biological cell suspensions, because of its capability of non-invasive and real-time measurement. Its application is no longer limited to laboratory research, but also includes industrial process control. This is mainly due to recent developments in the dielectric theories of dispersed systems and the measurement techniques.

The first applications of dielectric spectroscopy to biological cell suspensions date back to the early 1900's. In the early works, dielectric relaxation was found in the radio frequency range, which indicated that cells can be regarded as a conducting core covered with an insulating thin membrane. Fricke, a pioneer in this field, first estimated the membrane capacitance of red blood cells from their dielectric relaxation and calculated the membrane thickness to be 3nm.[1] Following his studies, many biological cell suspensions were measured and their dielectric spectra were analyzed by theories developed with a simple electrical cell model, the so-called "single-shell" model, that is a sphere covered with a concentric shell.

In the early 1980's dielectric spectroscopy was drastically changed by the advent of computer-controlled instruments capable of precise and rapid measurements. This enabled us not only to investigate unstable biological cells, but also to follow their dynamic behavior.[2-7] In addition to the progress of instrumentation, the use of personal computers permitted precise analysis of the dielectric relaxation spectra by applying complicated, realistic cell models to a huge amount of data collected by automated instruments.

The purpose of this chapter is to summarize some facets of recent theoretical and technical developments, with which we have been concerned in the past two decades. For a comprehensive account of this subject, the reader should consult excellent reviews and books.[8-15]

2. Summary of Dielectric Relaxation of Biological Cell Suspensions

The dielectric relaxation spectra (DRS) of biological tissues and cell suspensions show three distinct dispersions (or relaxations), termed α-, β- and γ-dispersions, over a frequency range of 1 Hz to 10 GHz (Fig. 1). This well-cited classification was introduced by Schwan.[8] Although the detailed features and underlying mechanisms of these relaxation processes have been described in his review articles (e.g., see reference 8), a brief account of them would be helpful.

The α-dispersion, which appears at frequencies below a few kHz, is not easy to measure because of interference from an electrode polarization artifact. Thus, a few studies have been reported for cell suspensions,[16-18] and the relaxation process has not been well understood although several theories were proposed for it. According to the theories, the α-dispersion could originate from the displacement of counterions surrounding charged membranes.[19-21]

The β-dispersion, which is usually observed in the frequency range of 1 kHz to 1 GHz, has been extensively and intensively studied for various cells, i.e., bacteria and bacteria-like organisms,[18,22-26] yeast,[27-29] animal cells,[30-35] and plant cells.[36] The β-dispersion is caused by interfacial polarization, or the Maxwell-Wagner mechanism, and is mainly attributed to the existence of the insulating plasma membrane surrounding a cell, since, if the membrane is disrupted or permealized by detergents, the dispersion almost disappears. Likewise, dielectric dispersion also arises from the intracellular organelles covered with membranes, usually appearing at higher frequencies than does dispersion due to the plasma membrane. Studies of the β-dispersion, therefore, can give important information on the electrical and structural properties of the plasma membranes and intracellular organelles.

The γ-dispersion that results from reorientation of water molecules lies above 1 GHz.[37] Between the β- and γ-dispersions there may be small dispersions that could be accounted for by relaxations of biopolymers and bound water.

In the following sections, we will focus on the β-dispersion and describe the related theories and measurement techniques.

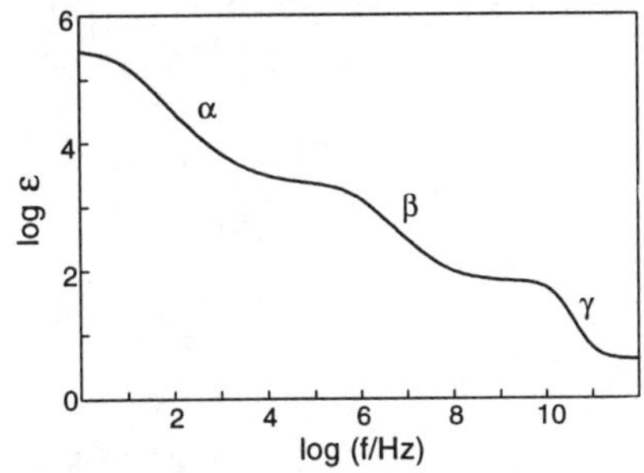

Figure 1. An idealized dielectric relaxation spectrum of a biological cell suspension. The relative permittivity ε shows three major dispersions: α-, β- and γ-dispersions.

3. Electrical Cell Models and Related Dielectric Theories

A dielectric theory for a dilute cell suspension has been developed for the single-shell model.[38] The theory has been extended to a concentrated cell suspension[39] and to more realistic cell models, the so called "composite cell models", in which the intracellular organelles are taken into account[31,40] Garcia et al.[41] modified the theory of the single-shell model by taking into account the diffusive effect of ions in the internal and external phases, although this contribution is small under physiological conditions. For non-spherical cell suspensions, theories have been developed with ellipsoidal cell models,[6,26,42-44] and a numerical calculation method for arbitrarily shaped cells has been reported.[45] The dielectric theories with complicated, realistic cell models can deal with the dielectric behavior of various types of biological cells through numerical calculation using a personal computer.

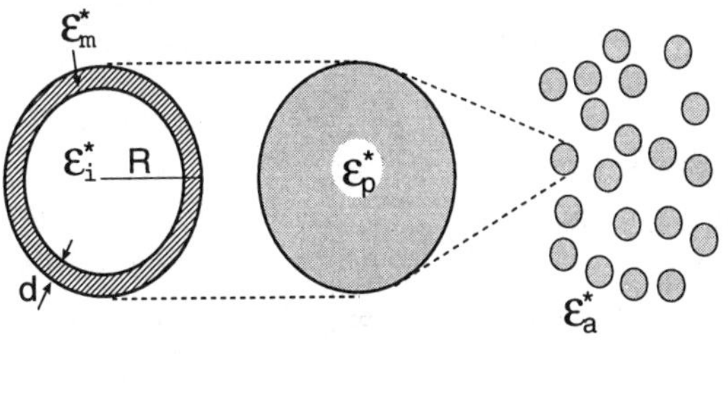

Figure 2. A simple electrical model of a spherical cell suspension. Spherical cells that are represented by the single-shell model are dispersed in a continuous phase.

3.1 Single-Shell Model

The dielectric theory for the single-shell model (Fig. 2) is based on the two equations first quoted in the book of Maxwell,[46] which provide the resistance of a suspension of homogeneous spheres and the equivalent resistance of the shell-sphere. Instead of the original equations, it is usual to use alternative equivalent equations that are expressed in complex permittivity defined as $\varepsilon^* = \varepsilon - j\kappa/\omega\varepsilon_0$, where ε is relative permittivity, κ conductivity, ω angular frequency, ε_0 the permittivity of vacuum and $j=(-1)^{1/2}$. The equivalent, homogeneous complex permittivity of the shell-sphere ε_p^* composed of a core of $\varepsilon_i^*(\varepsilon_I^* = \varepsilon_i - j\kappa_i/\omega\varepsilon_0)$ and a shell of $\varepsilon_m^*(\varepsilon_m^* = \varepsilon_m - j\kappa_m/\omega\varepsilon_0)$ is given by

$$\frac{\varepsilon_p^* - \varepsilon_m^*}{\varepsilon_p^* + 2\varepsilon_m^*} = \frac{\varepsilon_i^* - \varepsilon_m^*}{\varepsilon_i^* + 2\varepsilon_m^*} v \tag{1}$$

$$v = \left(1 - \frac{d}{R}\right)^3 \tag{2}$$

where R is the outer radius of the shell-sphere and d is the thickness of the shell. The complex permittivity ε^* of a system in which spheres of ε_p^* are dispersed in a continuous medium of $\varepsilon_a^*(\varepsilon_a^* = \varepsilon_a - j\kappa_a/\omega\varepsilon_0)$ at volume fraction P is given by

$$\frac{\varepsilon^* - \varepsilon_a^*}{\varepsilon^* + 2\varepsilon_a^*} = \frac{\varepsilon_p^* - \varepsilon_a^*}{\varepsilon_p^* + 2\varepsilon_a^*} P \tag{3}$$

This equation is known as Wagner's mixture equation.[47] Starting from Eqs. 1 and 3, Pauly and Schwan[38] derived an equation that can describe the frequency dependence of complex permittivity of a shell-sphere suspension as:

$$\varepsilon^* = \varepsilon_h + \frac{\Delta\varepsilon_p}{1 + j\omega\tau_p} + \frac{\Delta\varepsilon_q}{1 + j\omega\tau_q} + \frac{\kappa_l}{j\omega\varepsilon_0} \tag{4}$$

This equation predicts two relaxation processes of Debye-type, designated P-and Q-dispersions for the low and high frequency dispersions, respectively. The quantities in Eq. 4 are: $\Delta\varepsilon$ is the relaxation magnitude, τ the relaxation time, ε_h the high frequency limit of the relative permittivity and κ_l the low frequency limit of the conductivity. The relaxation parameters ($\Delta\varepsilon_p$, τ_p, $\Delta\varepsilon_q$, τ_q, ε_h, κ_l) can be expressed by the electrical and morphological parameters involved in the single-shell model (ε_a, κ_a, ε_m, κ_m, ε_i, κ_i, R, d, P). Pauly- Schwan's equation (Eq. 4) excellently explained the dielectric behavior of dilute erythrocyte ghost suspensions,[48] a single spherical bilayer membrane,[49] and a single microcapsule[50] as models for biological cells. In the case of biological cells, in which the membrane thickness is much smaller than the cell radius (d/R<<1) and the membrane conductivity is much less than the cytoplasmic and external conductivities (κ_m/κ_a<<1 and κ_m/κ_i<<1), the P-dispersion becomes much larger in magnitude than the Q-dispersion, and the relaxation magnitude $\Delta\varepsilon$ is approximately given by

$$\Delta\varepsilon = \frac{9P}{(2+P)^2}\frac{RC_m}{\varepsilon_0} \tag{5}$$

from which one can calculate the membrane capacitance C_m($C_m = \varepsilon_m\varepsilon_0/d$).

When the volume fraction of a cell suspension exceeds 0.2, Pauly-Schwan's equation may not be applicable because in Eq. 3 (Wagner's mixture equation) the electrical interaction between spheres is not considered. For concentrated particle suspensions, Hanai's mixture equation,[51] which was proposed as an extension of Bruggeman's equation[52] to the case of complex permittivity, provides a reasonable approximation.

$$\frac{\varepsilon^* - \varepsilon_p^*}{\varepsilon_a^* - \varepsilon_p^*}\left(\frac{\varepsilon_a^*}{\varepsilon^*}\right)^{\frac{1}{3}} = 1 - P \tag{6}$$

Hanai's mixture equation has been confirmed to be valid for colloidal dispersions of volume fractions up to 0.7.[9,53,54] The combination of Eqs. 1 and 6 may give an approximation for a concentrated suspension of cells, although strictly speaking Eq. 1 is not applicable to a shell-sphere placed in a non-uniform electric field. The validity of the combination was experimentally confirmed with suspensions of polystyrene microcapsules[55] and erythrocyte ghosts,[48] both having well defined structural and electrical properties. Accordingly, when one deals with relatively concentrated suspensions, an alternative equation to Eq. 5 may be given by

$$\Delta\varepsilon = \frac{3}{2}\left[1-(1-P)^{\frac{3}{2}}\right]\frac{RC_m}{\varepsilon_0} \tag{7}$$

3.2 Composite Cell Models

Since most biological cell types have organelles and membranous structure within the cytoplasm, the single-shell model may not be applicable to such cells. Indeed, we frequently experience some discrepancies between the observed dielectric spectra and the theoretical ones predicted from the single-shell model. In order to explain these discrepancies, various composite cell models including intracellular structure have been proposed. Irimajiri et al.[40] have developed the theory of the "multi-stratified shell model", which was first introduced by Fricke.[56] The "double-shell" model, a special case of the multi-stratified shell model, was applied to a lymphoma cell that possesses a sizable nucleus in it[31] and also to isolated mitochondria with double membranes.[57]

Here, we present DRS for three different types of biological cells, i.e., spherical erythrocytes, lymphocytes and plant protoplasts, in order to discuss the possible contribution of the intracellular structure to their DRS, using the composite cell models (Fig. 3). Figure 4 shows frequency dependence of the equivalent relative permittivity and loss factor of the cells, which were calculated from the data of the cell suspensions using Hanai's mixture equation (Eq. 6).

Spherical erythrocytes that have no intracellular organelles show one dielectric dispersion without any additional dispersion although the DRS of intact, biconcave erythrocytes includes two subdispersions because of its non-spherical shape, as will be discussed in the next subsection. The dispersion curve of spherical erythrocytes was fully simulated by the single-shell model.[35] Lymphocytes with a sizable nucleus show two dispersions: one main dispersion around 1 MHz and one additional dispersion centered about 10 MHz. Plant protoplasts isolated from leaves have a large vacuole and a thin cytoplasmic layer containing chloroplasts. The protoplasts show one large dispersion centered at about 1 MHz and one small dispersion around 6 MHz besides the dispersion around 10 kHz related to the plasma membrane.

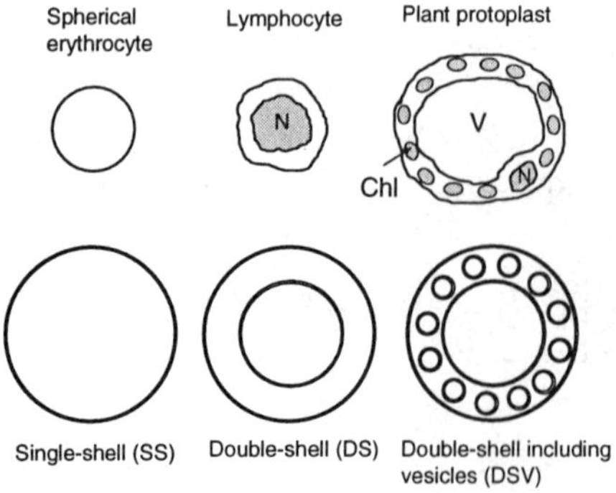

Figure 3. Idealized structures of spherical erythrocyte, lymphocyte and plant protoplast, and related electrical cell models. N, nucleus; V, vacuole; Chl, chloroplast.

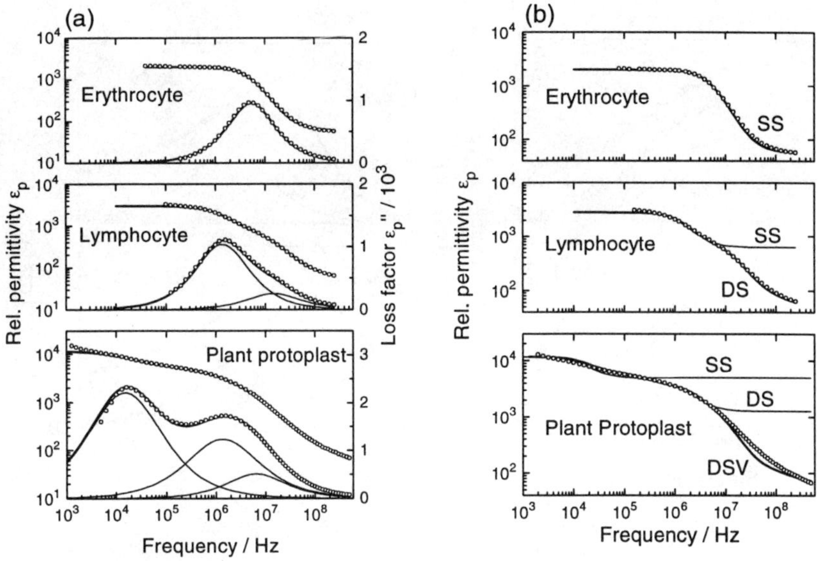

Figure 4. Dielectric relaxation spectra of three different types of biological cells: spherical mouse erythrocyte, mouse lymphocyte, and plant leaf protoplast. (a) The real part (relative permittivity, ε_p) and imaginary part (loss factor, ε_p'') of the equivalent complex permittivity of cells are plotted against frequency. The loss factor spectra for lymphocyte and plant protoplast were decomposed into two or three subdispersions of Cole-Cole type.[73] (b) The dielectric relaxation spectra were simulated by the three cell models shown in Fig.3: the single-shell model (SS), the double-shell model (DS) and the double-shell model including vesicles (DSV). (From Ref. 7. ©Elsevier Science)

Comparison of dielectric behavior between these three types of cells implies that intracellular organelles contribute to additional dispersions at high frequencies, which should be confirmed by simulations with the composite cell models shown in Fig. 3. The dispersion curve from lymphocytes was in good agreement with that calculated using the double-shell model, in which the inner shell-sphere corresponds to the nucleus.[35] The double-shell model including vesicles gave a better simulation for the dispersion curve of plant protoplasts than the single-shell and the double-shell model.[36] In this case, the inner-shell sphere corresponds to the vacuole and the vesicles to the chloroplasts. Hence, it is concluded that interfacial polarization due to intracellular organelles accounts for the additional high-frequency dispersions.

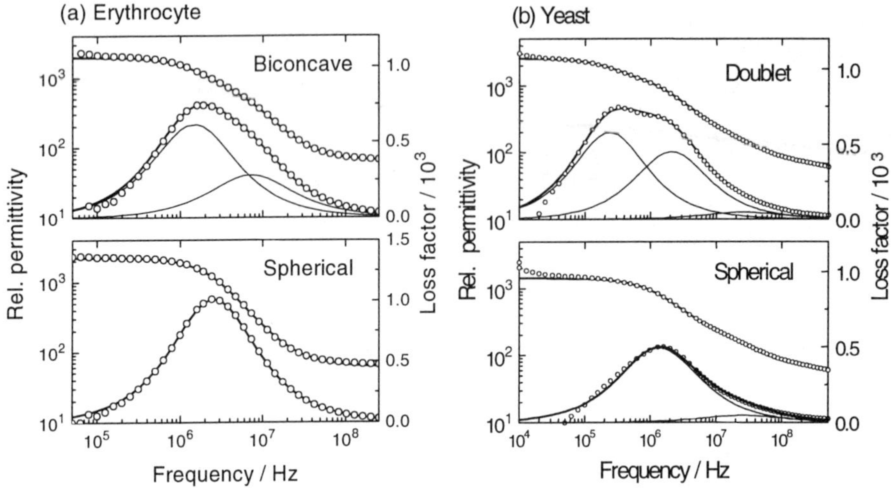

Figure 5. Effects of cell shape on dielectric relaxation spectra of (a) human erythrocyte and (b) yeast (*Saccharomyces cerevisiae*, cdc9 mutant) suspensions. The real part (relative permittivity) and imaginary part (loss factor) of the complex permittivity of the suspensions are plotted against frequency. (a) Intact erythrocytes have a biconcave shape in physiological saline, and spherical erythrocytes were prepared by swelling intact erythrocytes in a dilute saline. (b) Cells of cdc9 arrested at a stage of the cell cycle before cell separation (mitosis) and accumulated as doublets after shifting cells to restriction temperature (37°C). Cells that were normally grown and harvested at the stationary phase were nearly spherical. The additional high-frequency dispersions around a few tens of MHz found for both types of yeast cells are due to the contribution of the cell wall and vacuole.[29] The loss factor spectra of intact erythrocyte and yeast suspensions were decomposed into two or three subdispersions of Cole-Cole type.[73]

3.3 Non-Spherical Cell Models

The dielectric theory of an ellipsoidal cell model that is an ellipsoidal core covered with a confocal shell is more complicated than that of spherical cell models because the equivalent complex permittivity of a shell-ellipsoid ε_p^* has three different components along its three different axes.[26,42,43] In the case of a cell whose shape is an ellipsoid of revolution of two different axes, the suspension of randomly oriented cells would show approximately two dielectric subdispersions corresponding to the two components of ε_p^*.[6]

Figure 5 shows the DRS measured for two types of non-spherical cells: intact erythrocyte of biconcave shape and yeast as doublet. The DRS of an intact erythrocyte suspension was decomposed into two subdispersions, in contrast to that of a spherical erythrocyte suspension that showed only one dispersion. Using a temperature-sensitive cell division cycle mutant, cdc9, which could be arrested at the premitotic stage of cell cycle by incubation at 37°C, we can harvest the yeast cells as doublets. The DRS of the arrested yeasts has one more subdispersion around 200 kHz in addition to the DRS found for nearly spherical single cells that were harvested at the stationary state in normal culture at 25°C. The observed difference in DRS characteristic of these non-spherical cells was as expected in that simulations using oblate- and prolate-ellipsoidal models for the biconcave erythrocyte and the doublet yeast, respectively, were qualitatively successful. For a quantitative explanation, however, numerical analysis with models of more realistic shape is required.

4. Measurement Techniques and Related Problems

Dielectric spectra of biological cell suspensions can be easily obtained within a few minutes over a frequency range of 1 kHz to 1 GHz using rapid, automated, frequency-domain spectrometers (FDS) with high precision. Recently, time-domain spectrometers (TDS) have been improved in accuracy[58,59] to be applicable to biological cell suspensions.[60] With TDS the time required for measurements can be much shorter than with FDS, thereby enabling us to investigate faster phenomena.

However, measurements at frequencies below a few kHz still have a big problem due to electrode polarization.[61] Although various measurement and correction methods have been proposed for its solution, no established method is as yet available. If cells are suspended in a high-conductive medium at a low volume fraction ($P<0.01$), the electrode polarization effect becomes considerable even at several hundred kHz. This is especially a problem with measurements on culture broth in fermentation. In the next subsection we will introduce a new method, which can offer the dielectric spectra between 100 kHz and 10 MHz without any nuisance due to electrode polarization.

In the second subsection an attempt at dielectric imaging of individual biological cells is described. Although electrical impedance (dielectric) imaging techniques, called electrical impedance tomography, have been developed to investigate tissues and body for clinical application,[62,63] the techniques with multi-electrode systems are not appropriate for small biological cells. Dielectric imaging with a scanning probe technique would be a possible choice for biological cells.

4.1 Dielectric Monitoring of Cell Growth using an Inductive Probe

The application of dielectric spectroscopy to real-time estimation of microbial biomass in cell culture was first made by Harris et al.[4] by taking advantage of excellent linear relationships between the permittivity and the cell mass of cell suspensions. Since the volume fraction of cells in culture broth is a few percents at most, the artifact due to electrode polarization becomes relatively large compared with the magnitude of the dielectric relaxation due to the cells, which is usually less than one hundred in dielectric unit. Hence, a prerequisite for successful measurements is to eliminate the electrode polarization effect. To solve this problem, Harris et al.[4] adopted the four-electrode method, in which electrodes for current supply are separated from electrodes for voltage detection.[61,64] Recently, an alternative technique has been developed by Wakamatsu,[65] which uses an inductive probe instead of metal electrodes. The technique is regarded as an electrode-less method based on electromagnetic induction, being free not only from the electrode polarization effect but also from the interference of bubble formation on electrodes. This advantage allows real-time as well as *in situ* dielectric monitoring of cells in culture or fermentation.[7]

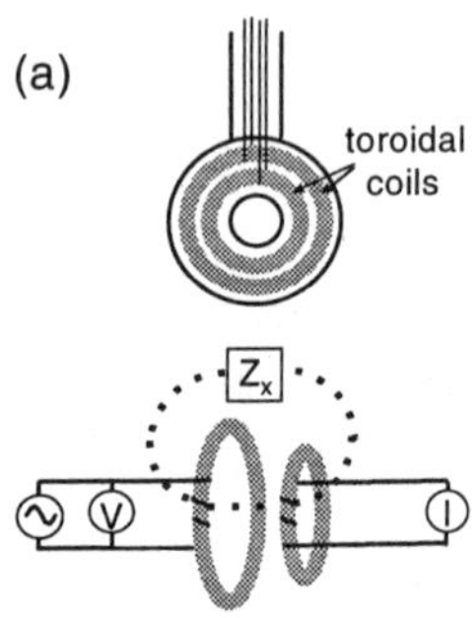

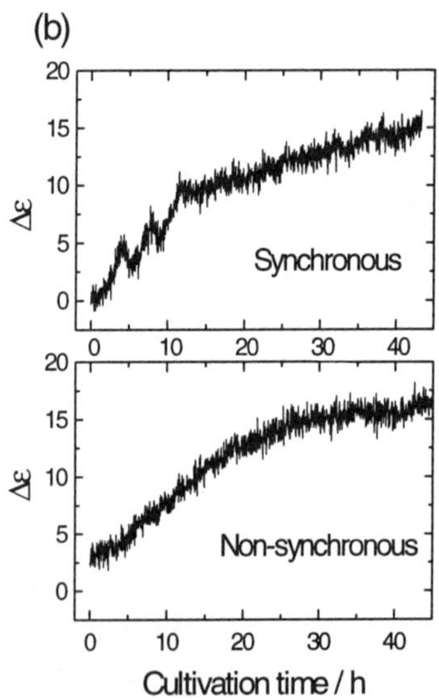

Figure 6. (a) An inductive probe for the electromagnetic induction method. The two toroidal coils are coupled through the impedance Z_x of a sample liquid surrounding them; (b) Dielectric monitoring of synchronous and non-synchronous growth of yeast using the inductive probe. The magnitude of dielectric relaxation $\Delta\varepsilon$ is plotted against cultivation time. Cells (*Saccharomyces cerevisiae*, cdc28-13 mutant) were cultured in a rich medium at 17°C.

The inductive probe consists of two concentric toroidal coils, which are molded in epoxy resin (Fig. 6a). For measurements, the probe is immersed in a sample liquid. The voltage applied to one toroidal coil is related to the current from the other coil through the impedance of the sample liquid around the probe. Thus, the relative permittivity and conductivity of the sample are determined from the input voltage and the output current.

As an example of its application, the results obtained for synchronous and non-synchronous yeast growth in culture are described. The yeast strain used is cdc28-13, which is a temperature-sensitive cell division cycle mutant that grows normally at the permissive temperature, but is arrested at the “start” point of its cell cycle after being shifted to the restriction temperature (37°C). The arrested cells start growing simultaneously when returned to the permissive temperature, resulting in a synchronous growth. Figure 6b shows the time course of the relaxation magnitude $\Delta\varepsilon$ in the non-synchronous and synchronous growth of cdc28-13 cells. There appeared periodical drops of $\Delta\varepsilon$ in the early synchronous growth in contrast to the smooth curve for the non-synchronous growth (Gheorghiu and Asami, in preparation). The drops of $\Delta\varepsilon$ correspond to the shape changes in cell separation (mitosis), which is confirmed by a theoretical simulation[6] and by the experimental evidence that cells as doublets before cell separation have larger $\Delta\varepsilon$ than single spherical cells when compared at the same volume fraction (see Fig. 5b).

Since dielectric spectra from biological cell suspensions provides integrated information on the cell properties, such as cell concentration, intracellular structure, cell shape and electrical properties, dielectric monitoring would provide a useful technique for investigating the dynamic behavior of biological cells.

4.2 Dielectric Imaging using a Scanning Fine Probe

The "suspension method" that estimates the average electrical properties of cells from the DRS of their suspension using an appropriate mixture equation cannot characterize individual cells. Electrical properties of individual biological cells can be investigated by methods that follow their motion in response to the force induced by ac fields, namely dielectrophoresis[66-68] and electrorotation,[69] and by a method using the micropipet technique.[70,71] An alternative approach may be the use of a "scanning dielectric microscope" (SDM) that can image local changes in capacitance and conductance sensed by a fine probe electrode that is scanned over cells at a constant height.[72] The SDM is designed for investigating dielectric behavior of particles in aqueous media over a wide frequency range.

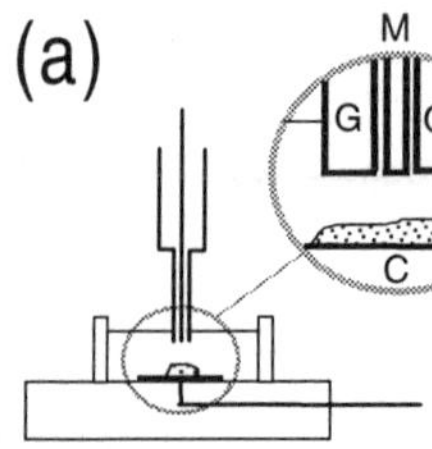

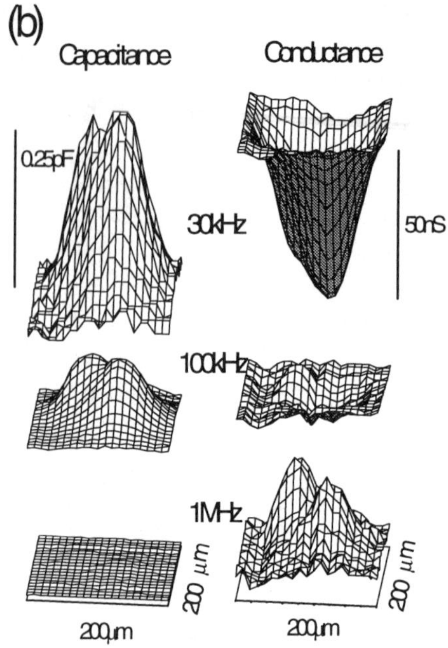

Figure 7. (a) A schematic diagram of the scanning dielectric microscope. A fine coaxial probe consists of a measuring electrode (M) and a guard (G) is scanned over a sample on a counter plate electrode (C); (b) Dielectric images of a MDCK-cell. The cells were grown on a stainless steel foil. The external medium was a 0.3M mannitol solution containing traces of the culture medium.

Figure 7a shows a schematic view of the SDM. The sample is placed on a plate electrode in a chamber filled with an aqueous solution. The coaxial probe consists of an inner probing electrode and an outer guard electrode. Dielectric measurements are made based on the three-terminal method, in which both the measuring and guard electrodes are held at the same potential against the counter electrode, thereby yielding a parallel field between the measuring and counter electrodes. Thus, the three-terminal configuration is effective to eliminate the fringing field and to restrict the measurement to a small area.

Figure 7b shows dielectric images of a single cell of MDCK, which originates from dog kidney, obtained by the SDM. The images depend on frequency. With increasing frequency, the capacitance of the cell decreases and the conductance increases; the cell shows dielectric dispersion. The SDM, therefore, would enable examination of the dielectric properties of individual biological cells, although there remain some problems: its resolution is not sufficient for close examination of cells and its application to date is limited to low-conductivity media only.

5. Concluding Remarks

The β-dispersion of biological cell suspensions can be easily measured with computer-controlled spectrometers and simulated well by dielectric theories based on interfacial polarization. By taking advantage of spectrometers capable of real-time measurement, dielectric monitoring of the dynamic behavior of biological cells becomes feasible. Dielectric imaging of biological cells would have the possibility of gaining an insight into the cell structure if those technical problems are solved.

To analyze dielectric behavior of cells with awkward shape and complicated internal structure, it is necessary to develop numerical calculation methods with the aide of the finite or boundary element method (K. Sekine, personal communication). The numerical methods would also provide a useful tool to examine the validity of the mixture equations and to analyze the dielectric images obtained by SDM.

Among the technical problems to be solved in the future is how to eliminate the electrode polarization effect at low frequencies. The electromagnetic induction method is available between 100 kHz and 10 MHz, but it should be possible, if difficult, to extend this approach down to frequencies below 100 kHz.

Acknowledgments

I wish to thank Dr. T. Hanai (Kyoto University), Dr. A. Irimajiri (Kochi Medical School) and Dr. S. Takashima (University of Pennsylvania) for their stimulating discussions and helpful advice. I am grateful to Mr. T. Yonezawa (Suntory), Mr. H. Wakamatsu, Mr. N. Koyanagi (Hewlett-Packard Japan), Dr. S. Siano (Hewlett-Packard) and Dr. E. Gheorghiu (National Institute of Biotechnology, Rumania) for helpful discussions on the measurements using the inductive probe. This research was supported by grants (03650791, 06680855, 09555252) from the Ministry of Education, Culture and Science, Japan.

References

1. H. Fricke, "The Electric Capacity of Suspensions with Special Reference to Blood," *J. Gen. Physiol.*, **9**, 137-152 (1925).

2. K. Asami and T. Hanai, "Dielectric Monitoring of Biological Cell Sedimentation," *Colloid Polym. Sci.*, **270**, 78-84 (1992).

3. A. Irimajiri, M. Ando, R. Matsuoka, T. Ichinowatari, and S. Takeuchi, "Dielectric Monitoring of Rouleaux Formation in Human Whole Blood. A Feasibility Study," *Biochim. Biophys. Acta*, **1290**, 207-209 (1996).

4. C.M. Harris, R.W. Todd, S.J. Bungard, R.W. Lovitt, G. Morris, and D.B. Kell, "Dielectric Permittivity of Microbial Suspensions at Radio Frequencies. A Novel Method for the Real-time Estimation of Microbial Biomass," *Enzyme Microb. Tchnol.*, **9**, 181-186 (1987).

5. K. Asami and T. Yonezawa, "Dielectric Analysis of Yeast Cell Growth," *Biochim. Biophys. Acta*, **1245**, 99-105 (1995).

6. K. Asami and T. Yonezawa, "Dielectric Behavior of Non-spherical Cells in Culture," *Biochim. Biophys. Acta*, **1245**, 317-324 (1995).

7. K. Asami, T. Yonezawa, H. Wakamatsu, and N. Koyanagi, "Dielectric Spectroscopy of Biological Cells," *Bioelectrochem. Bioenerg.*, **40**, 141-145 (1996).

8. H.P. Schwan, "Electrical Properties of Tissue and Cell Suspensions"; pp. 147-209 in *Advances in Biological and Medical Physics*, Vol. 5, Edited by J.H. Lawrence and C.A. Tobias, Academic Press, New York, 1957.

9. T. Hanai, " Electrical Properties of Emulsions"; pp. 353-478 in *Emulsion Science*, Edited by P. Sherman, Academic Press, London and New York, 1968.

10. K.S. Cole, *Membranes, Ions and Impulses*, University of California Press, Berkeley and Los Angeles, 1968.

11. E.H. Grant, R.J. Sheppard, and G.P. South, *Dielectric Behaviour of Biological Molecules in Solution*, Oxford University Press, Oxford, 1978.

12. R. Pethig, *Dielectric and Electronic Properties of Biological Materials*, John Wiley & Sons, New York, 1979.

13. R. Pethig and D.B. Kell, "The Passive Electrical Properties of Biological Systems: Their Significance in Physiology, Biophysics and Biotechnology," *Phys. Med. Biol.*, **32**, 933-970 (1987).

14. S. Takashima, *Electrical Properties of Biopolymers and Membranes*, Adam Hilger, Bristol, England, 1989.

15. K.R. Foster and H.P. Schwan, "Dielectric Properties of Tissues"; pp. 25-102 in *Handbook of Biological Effects of Electromagnetic Fields*, 2nd ed., Edited by C. Polk and E. Postow, CRC Press, Boca Raton, 1996.

16. C.W. Einolf and E.L. Carstensen, "Passive Electrical Properties of Microorganisms. IV. Studies of the Protoplasts of *Micrococcus lysodeikticu*," *Biophys. J.*, **9**, 634-643 (1969).

17. D.B. Kell, "Dielectric Properties of Bacterial Chromatophores," *Bioelectrochem. Bioenerg.*, **11**, 405-415 (1983).

18. C.M. Harris and D.B. Kell, "On the Dielectrically Observable Consequences of the Diffusional Motion of Lipids and Proteins in Membranes. 2. Experiments with Microbial Cells, Protoplasts and Membrane Vesicles," *Eur. Biophys. J.*, **13**, 11-24 (1985).

19. G. Schwarz, "A Theory of the Low-Frequency Dielectric Dispersion of Colloidal Particles in Electrolyte Solution," *J. Phys. Chem.*, **66**, 2636-2642 (1962).

20. S.S. Dukhin and V.N. Shilov, *Dielectric Phenomena and the Double Layer in Disperse Systems and Polyelectrolites*, John Wiley & Sons, New York and Toronto, 1974.

21. E. Gheorghiu, "The Dielectric Behavior of Suspensions of Spherical Cells: A Unitary Approach," *J. Phys. A: Math. Gen.*, **27**, 3883-3893 (1994).

22. H.P. Schwan and H.J. Morowitz, "Electrical Properties of the Membranes of the Pleuropneumonia-like Organism A 5969," *Biophys. J.*, **2**, 395-407 (1962).

23. E.L. Carstensen, H.A. Cox, Jr., W.B. Mercer, and L.A. Natale, "Passive Electrical Properties of Microorganisms. II. Resistance of the Bacterial Membrane," *Biophys. J.*, **7**, 493-503 (1967).

24. E.L. Carstensen, J. Maniloff and C.W. Einolf, Jr., "Electrical Properties and Ultrastructure of *Mycoplasma* Membranes," *Biophys. J.*, **11**, 572-581 (1971).

25. E.L. Carstensen, R.E. Marquis, S.Z. Child, and G.R. Bender, "Dielectric Properties of Native and Decoated Spores of *Bacillus megaterium*," *J. Bacteriology*, **140**, 917-928 (1979).

26. K. Asami, T. Hanai, and N. Koizumi, "Dielectric Analysis of *Escherichia coli* Suspensions in the Light of the Theory of Interfacial Polarization," *Biophys. J.*, **31**, 215-228 (1980).

27. C.M. Harris and D.B. Kell, "The Radio-frequency Dielectric Properties of Yeast Cells Measured with a Rapid, Automated, Frequency-Domain Dielectric Spectrometer," *Bioelectrochem. Bioenerg.*, **11**, 15-28 (1983).

28. V. Raicu, G. Raicu, and G. Turcu, "Dielectric Properties of Yeast Cells as Simulated by the Two-Shell Model," *Biochim. Biophys. Acta*, **1274**, 143-148 (1996).

29. K. Asami and T. Yonezawa, "Dielectric Behavior of Wild-Type Yeast and Vacuole-Deficient Mutant over a Frequency Range of 10 kHz to 10 GHz," *Biophys. J.*, **71**, 2192-2200 (1996).

30. H. Pauly and H.P. Schwan, "Dielectric Properties and Ion Mobility of Erythrocytes," *Biophys. J.*, **6**, 621-639 (1966).

31. A. Irimajiri, Y. Doida, T. Hanai, and A. Inouye, "Passive Electrical Properties of Cultured Murine Lymphoblast (L5178Y) with Reference to its Cytoplasmic Membrane, Nuclear Envelope, and Intracellular Phases," *J. Membr. Biol.*, **38**, 209-232 (1978).

32. A. Irimajiri, K. Asami, T. Ichinowatari, and Y. Kinoshita, "Passive Electrical Properties of the Membrane and Cytoplasm of Cultured Rat Basophil Leukemia Cells. I. Dielectric Behavior of Cell Suspensions in 0.01-500 MHz and its Simulation with a Single-Shell Model," *Biochim. Biophys. Acta*, **896**, 203-213 (1987).

33. A. Irimajiri, K. Asami, T. Ichinowatari, and Y. Kinoshita, "Passive Electrical Properties of the Membrane and Cytoplasm of Cultured Rat Basophil Leukemia Cells. II. Effects of Osmotic Perturbation," *Biochim. Biophys. Acta*, **896**, 214-223 (1987).

34. C.L. Davey, D.B. Kell, R.B. Kemp, and R.W.J. Meredith, "On the Audio- and Radio-Frequency Dielectric Behavior of Anchorage-Independent, Mouse L929-Derived LS Fibroblasts," *Bioelectrochem. Bioenerg.*, **20**, 83-98 (1988).

35. K. Asami, Y. Takahashi, and S. Takashima, "Dielectric Properties of Mouse Lymphocytes and Erythrocytes," *Biochim. Biophys. Acta*, **1010**, 49-55 (1989).

36. K. Asami and T. Yamaguchi, "Dielectric Spectroscopy of Plant Protoplasts," *Biophys. J.*, **63**, 1493-1499 (1992).

37. J.B. Hasted, *Aqueous Dielectrics*, Chapman and Hall, London, 1973.

38. H. Pauly and H.P. Schwan, "Uber die Impedanz einer Suspension von Kugelformigen Teilchen mit einer Schale (The Impedance of a Suspension of Spherical Particles Surrounded by a Shell)," Z. Naturforsch., **14b**, 125-131 (1959).

39. T. Hanai, K. Asami, and N. Koizumi, "Dielectric Theory of Concentrated Suspension of Shell-Spheres in Particular Reference to the Analysis of Biological Cell Suspensions," *Bull. Inst. Chem. Res., Kyoto Univ.*, **57**, 297-305 (1979).

40. A. Irimajiri, T. Hanai, and A. Inouye, "A Dielectric Theory of "Multi-Stratified Shell" Model with its Application to a Lymphoma Cell," *J. Theor. Biol.*, **78**, 251-269 (1979).

41. A. Garcia, R. Barchini, and C. Gross, "The Influence of Diffusion on the Permittivity of a Suspension of Spherical Particles with Insulating Shells in an Electrolyte," *J. Phys. D: Appl. Phys.*, **18**, 1891-1896 (1985).

42. H. Fricke, "The Electric Permittivity of a Dilute Suspension of Membrane-Covered Ellipsoids," *J. Appl. Phys.*, **24**, 644-646 (1953).

43. K. Asami, T. Hanai, and N. Koizumi, "Dielectric Approach to Suspensions of Ellipsoidal Particles Covered with a Shell in Particular Reference to Biological Cells," *Jpn. J. Appl. Phys.*, **19**, 359-365 (1980).

44. C. Boned and J. Peyrelasse, "Etude de la Permittivite Complex d'Ellipsoides Disperses dans un Milieu Continu. Analyses Theorique et Unmerique (Study of Complex Permittivity of Ellipsoids Dispersed in a Continuous Medium. Theoretical and Numerical Analyses)," *Colloid Polym. Sci.*, **261**, 600-612 (1983).

45. D. Vrinceanu and E. Gheorghiu, "Shape Effects on the Dielectric Behavior of Arbitrarily Shaped Particles with Particular Reference to Biological Cells," *Bioelectrochem. Bioenerg.*, **40**, 167-170 (1996).

46. J.C. Maxwell, *Treatise on Electricity and Magnetism*, Oxford University Press, London, 1873.

47. K.W. Wagner, "Erklarung der Dielectrishen Nachwirkungsvorgange auf Grund Maxwellscher Vorstellungen (Explanation of Dielectric Effects Based on Maxwell's Concept)," *Archiv fur Electrotechnik.*, **2**, 371-387 (1914).

48. H. Kaneko, K. Asami, and T. Hanai, "Dielectric Analysis of Sheep Erythrocyte Ghost. Examination of Applicability of Dielectric Mixture Equations," *Colloid Polym. Sci.*, **269**, 1039-1044 (1991).

49. K. Asami and A. Irimajiri, "Dielectric Dispersion of a Single Spherical Bilayer Membrane in Suspension," *Biochim. Biophys. Acta*, **769**, 370-376 (1984).

50. K. Asami and K.S. Zhao, "Dielectric Measurement of a Single Sub-Millimeter Size Microcapsule," *Colloid Polym. Sci.*, **272**, 64-71(1994).

51. T. Hanai, "Theory of the Dielectric Dispersion Due to the Interfacial Polarization and its Application to Emulsion," *Kolloid Z.*, **171**, 23-31(1960).

52. D.A.G. Bruggeman, "Berechnung Verschiedener Physikalischen Konstanten von Heterogenen Substanzen (Estimation of Different Physical Constants of Heterogeneous Substances)," *Ann. Phys., Lpz.*, **24**, 636-664 (1935).

53. T. Hanai, T. Imakita, and N. Koizumi, "Analysis of Dielectric Relaxations of W/O Emulsions in the Light of Theories of Interfacial Polarization," *Colloid Polym. Sci.*, **260**, 1029-1034 (1982).

54. M. Clausse, "Dielectric Properties of Emulsions and Related Systems"; pp. 480-715 in *Encyclopedia of Emulsion Technology*. Edited by P. Becher. Marcel Dekker, New York and Basel, 1983.

55. H. Z. Zhang, K. Sekine, T. Hanai, and N. Koizumi, "Dielectric Observations of Polystyrene Microcapsules and the Theoretical Analysis with Reference to Interfacial Polarization," *Colloid Polym. Sci.*, **261**, 381-389 (1983).

56. H. Fricke, "The Complex Conductivity of a Suspension of Stratified Particles of Spherical or Cylindrical Form," *J. Phys. Chem.*, **59**, 168-170 (1955).

57. K. Asami and A. Irimajiri, "Dielectric Analysis of Mitochondria Isolated from Rat Liver. II. Intact Mitochondria as Simulated by a Double-Shell Model," *Biochim. Biophys. Acta*, **778**, 570-578 (1984).

58. R.H. Cole, J.G. Berberian, S. Mashimo, G. Chryssik, A. Burns, and E. Tombari, "Time Domain Reflection Methods for Dielectric Measurements to 10 GHz," *J. Appl. Phys.*, **66**, 793-802 (1989).

59. Y.D. Feldman, Yu.F. Zuev, E.A. Polygalov, and V.D. Fedotov, "Time Domain Dielectric Spectroscopy. A New Effective Tool for Physical Chemistry Investigation," *Colloid Polym. Sci.*, **270**, 768-780 (1992).

60. R. Lisin, B.Z. Ginzburg, M. Schlesinger, and Y. Feldman,. "Time Domain Dielectric Spectroscopy Study of Human Cells. I. Erythrocytes and Ghosts," *Biochim. Biophys. Acta*, **1280**, 34-40 (1996).

61. H.P. Schwan, "Alternating Current Electrode Polarization," *Biophysik*, **3**, 181-201 (1966).

62. D.C. Barber, "Applied Potential Tomography," *J. Phys. E: Sci. Instrum.*, **17**, 723-733 (1984).

63. A.M Dijkstra, B.H. Brown, A.D. Leathard, N.D. Harris, D.C. Barber, and D.L. Edbrooke, "Clinical Applications of Electrical Impedance Tomography," *J. Med. Eng. Tech.*, **17**, 89-98 (1993).

64. H.P. Schwan and C.D. Ferris, "Four-Electrode Null Technique for Impedance Measurement with High Resolution," *Rev. Sci. Instrum.*, **39**, 481-485 (1968).

65. H. Wakamatsu, "A Dielectric Spectrometer for Liquid Using the Elelctromagnetic Induction Method," *Hewlett-Packard Journal*, **48**, 37-44 (1997).

66. H.A. Pohl, *Dielectrophoresis*, Cambridge University Press, London, 1978.

67. K.V.I.S. Kaler and T.B. Jones, "Dielectrophoretic Spectra of Single Cells Determined by Feedback-Controlled Levitation," *Biophys. J.*, **57**, 173-182 (1990).

68. K.V.I.S. Kaler, J.P. Xie, T.B. Jones, and R. Paul, "Dual-Frequency Dielectrophoretic Levitation of *Canola* Protoplasts," *Biophys. J.*, **63**, 58-69 (1992).

69. W.M. Arnold and U. Zimmermann, "Rotating-Field-Induced Rotation and Measurement of the Membrane Capacitance of Single Mesophyll Cells of *Avena sativa*," *Z. Naturforsch.*, **37c**, 908-915 (1982).

70. S. Takashima., K. Asami, and Y. Takahashi, "Frequency Domain Studies of Impedance Characteristics of Biological Cells Using Micropipet Technique," *Biophys. J.*, **54**, 995-1000 (1988).

71. K. Asami, Y. Takahashi, and S. Takashima, "Frequency Domain Analysis of Membrane Capacitance of Cultured Cells (HeLa and Myeloma) Using the Micropipette Technique," *Biophys. J.*, **58**, 143-148 (1990).

72. K. Asami, "The Scanning Dielectric Microscope," *Meas. Sci. Technol.*, **5**, 589-592 (1994).

73. K.S. Cole and R.H. Cole, "Dispersion and Absorption in Dielectrics. I. Alternating Current Characteristics," *J. Chem. Phys.*, **9**, 341-351 (1941).

Credits

Page 99 (Riebel, figure 1): Reprinted, by permission, from Particle and Particle Systems Characterization, Vol. 6, figure 4, 1989, published by Wiley-VCH.

Page 104 (Riebel, figure 5): Reprinted, by permission, from Particle and Particle Systems Characterization, Vol. 12, figure 5, 1995, published by Wiley-VCH.

Page 105 (Riebel, figure 6): Reprinted, by permission, from Particle and Particle Systems Characterization, Vol. 11, figure 5, 1994, published by Wiley-VCH.

Page 106 (Riebel, figure 7): Reprinted, by permission, from Particle and Particle Systems Characterization, Vol. 12, figure 8, 1995, published by Wiley-VCH.

Page 106 (Riebel, figure 8): Reprinted, by permission, from Particle and Particle Systems Characterization, Vol. 12, figures 9, 1995, published by Wiley-VCH.

Page 167 (Pednse, figure 1): Reprinted, by permission, from The Proceedings of the First International Particle Technology Forum, Part I, P. 146, figs. 1 and 2. © 1994 American Institute of Chemical Engineers.

Page 168 (Pednse, figure 2): Reprinted, by permission, from The Proceedings of the First International Particle Technology Forum, Part I, P. 146, figs. 3 and 4. © 1994 American Institute of Chemical Engineers.

Page 235 (Bergström, figure 4): Reprinted from J. European Ceramic Society, 17, Richard Greenwood and Lennart Bergström, "Electroacoustic and Reological Properties of Aqueous Ce-ZrO_2 (Ce-TZP) Suspensions,: pages 537–548, Copyright 1997, with kind permission from Elsevier Science Ltd., The Boulevard, Langford Lane, Kidlington 0X5 1GB, UK.

Page 236 (Bergström, figure 5): Reprinted from J. European Ceramic Society, 17, Richard Greenwood and Lennart Bergström, "Electroacoustic and Reological Properties of Aqueous Ce-ZrO_2 (Ce-TZP) Suspensions,: pages 537–548, Copyright 1997, with kind permission from Elsevier Science Ltd., The Boulevard, Langford Lane, Kidlington 0X5 1GB, UK.

Page 339 (Asami, figure 4): Reprinted from Bioelectrochemistry and Bioenergetics Volume 40, K. Asami, T. Yonesawa, H. Wakamatsu, and N. Koyanage, "Dielectric Sprectroscopy of Biological Cells,: figures 1–3, 1996, with kind permission from Elsevier Science S.A., PI.O. Box 564, 1001 Lausanne, Switzerland.

KEYWORD AND AUTHOR INDEX